D1625723

0 552 923 9

Chemical Applications of
Graph Theory

CHEMICAL APPLICATIONS OF GRAPH THEORY

Edited by

A. T. Balaban

*Institute of Atomic Physics
and Dept. of Organic Chemistry, Polytechnic
Bucharest, Roumania*

1976

ACADEMIC PRESS
LONDON   NEW YORK   SAN FRANCISCO
A Subsidiary of Harcourt Brace Jovanovich, Publishers

ACADEMIC PRESS INC. (LONDON) LTD.
24/28 Oval Road,
London NW1

United States Edition published by
ACADEMIC PRESS INC.
111 Fifth Avenue
New York, New York 10003

Library of Congress Catalog Card Number 75–19615
ISBN: 0 12 076050 9

Typeset in IBM Press Roman by
PREFACE LTD.
Salisbury, Wiltshire
Printed in Great Britain by
Whitstable Litho Ltd., Whitstable, Kent

Contributors

BALABAN, ALEXANDRU, T. *Institute of Atomic Physics, and Dept. of Organic Chemistry, Polytechnic, Bucharest, Roumania.*

BROCAS, JEAN, *Faculty of Sciences, University of Brussels, Brussels, Belgium*

DUBOIS, JACQUES-EMILE, *Laboratory of Physical Organic Chemistry, University of Paris, Paris, France.*

DUGUNDJI, JAMES, *Department of Mathematics, University of Southern California, Los Angeles, California, U.S.A.*

GIELEN, MARCEL, *Faculty of Sciences, University of Brussels, Brussels, Belgium.*

GILLESPIE, PAUL, *Department of Chemistry, University of Southern California, Los Angeles, California, U.S.A.*

GORDON, MANFRED, *Institute of Polymer Science and Department of Chemistry, University of Essex, Colchester, England.*

HARARY, FRANK, *Department of Mathematics and Institute for Social Research, University of Michigan, Ann Arbor, Michigan, U.S.A.*

MARQUARDING, DIETER, *Wissenschaftliches Hauptlaboratorium der Farbenfabriken Bayer A. G., Leverkusen, West Germany.*

PALMER, EDGAR M., *Department of Mathematics, Michigan State University, East Lansing, Michigan, U.S.A.*

RAMIREZ, FAUSTO, *Department of Chemistry, State University of New York, Stony Brook, New York, U.S.A.*

READ, RONALD C., *University of Waterloo, Department of Combinatorics and Optimization, Waterloo, Ontario, Canada.*

ROBINSON, ROBERT W., *Department of Mathematics, University of Newcastle, New South Wales, Australia*

ROUVRAY, DENNIS H., *Institut für Strahlenchemie, Max-Planck-Institut für Kohlenforschung, Mülheim an der Ruhr, German Federal Republic.*

TEMPLE, WILLIAM B., *Institute of Polymer Science and Department of Chemistry, University of Essex, Colchester, England.*

UGI, IVAR, *Organisch-Chemisches Institut der Technischen Universität, München, German Federal Republic*

Preface

For a mathematician, a graph is the application of a set on itself (i.e. a collection of elements of the set, and of binary relations between these elements). For a chemist, however, the geometrical realization of a graph is more appealing, namely a collection of *points* (i.e. elements of the set) and of *lines* joining some of these points either to other points or to themselves. The name *graph* originates from this geometrical realization and was introduced by Sylvester.

Since no specification is usually made as to the shape or length of lines, or to angles between lines, graphs are topological rather than geometrical objects, having as the most important feature the vicinity relationships between points.

Two kinds of correspondence between graphs and chemical categories have found numerous applications in chemistry: (i) a graph corresponds to a molecule, i.e. points symbolize atoms and lines symbolize chemical covalent bonds (these may be called *structural* or *constitutional graphs*), and (ii) a graph corresponds to a reaction mixture, i.e. points symbolize chemical species and lines symbolize conversions between these species (these may be called *reaction graphs*). The former type of graph gave Cayley the incentive to develop a procedure for counting the constitutional isomers of alkanes; later it led Pólya towards the discovery of his powerful counting theorem, which can be applied even to stereochemical problems, as discussed in this volume. Thus chemistry is an acknowledged origin for the beginning and development of graph theory, and mathematicians like Cayley and Pólya (especially the latter) published papers in chemical journals.

Having chemistry as one of the breeding grounds, graph theory is well adapted for solving chemical problems, both by the high degree of abstraction evidenced by the generality of such concepts as points, lines, and neighbours, as well as by the combinatorial derivation of many graph-theoretical concepts which correspond to the essence of chemistry viewed as the study of combinations between atoms.

The time is now ripe for an 'inverse osmosis', namely for chemistry to profit from the progress of graph theory. Indeed, this has happened for the past few decades, especially the last one, in all main areas of chemistry: the Hückel matrix in quantum chemistry has a graph-theoretical significance; reaction graphs play an ever increasing role in explaining and rationalizing rearrangements, especially in inorganic, and organometallic chemistry; structural graphs lead to applications in organic chemistry which include such topics as coding and nomenclature, correlation of properties, automated structural formula search in mass spectrometry, strategy of synthetical approaches, definitions and enumerations of various kinds of isomers such as valence isomers.

The present monograph includes eleven chapters by authors who all contributed themselves to progress or breakthrough in applying graph theory to chemical problems. It is, of course, impossible to exhaust all applications, especially the most recent ones. Some areas which are not reflected in the contents of this volume concern topics which were reviewed recently, e.g. applications in chemical kinetics by Yatsimirsky, K. B., (1973). *Z. Chem.,* **13**, 201; or some other areas mentioned in the first issue of *MatCh* (1975). Topological indices, the uses of graphs in chemical documentation or in the strategy of synthesis design, are barely touched upon. Interesting controversies occurred on the problem whether the characteristic polynomial of a graph determines uniquely the topology of the molecule: Balaban, A. T., and Harary, F., (1971). *J. Chem. Docum.,* **11**, 258; Kudo, Y., Yamasaki, T., and Sasaki, S.-I., *Ibid.,* (1973). **13**, 225; Herndon, W. C., *Ibid.,* (1974). **14**, 150.

The editor expresses his sincere thanks to the authors who contributed chapters to this book, and hopes that chemists of all specialities (inorganic, organic, theoretical chemists) and ages (students, teachers, professors and research workers) will become acquainted with the interdisciplinary area belonging to chemical graphs. If this happens, and if some of these will realize the fascination of this field, when for instance one sees that chemical doodles with pencil on paper can have interesting mathematical significance, and if they will be enticed to think in terms of graphs and apply this relatively recent branch of mathematics to new chemical problems, the purpose of this book will be attained.

Foreword

For more than a century most chemists have used constitutional formulae without realizing that by representing 'connectedness' of atoms such formulae are graphs or multigraphs. In contrast, graph theory has received early important impulses from organic chemistry: Cayley's and Polya's contributions, discussed in the first chapters of this book, are examples. Chemists became aware of the general usefulness of graph theory only when they started to use graphs in molecular orbital theory, for representing complex sequences of reaction steps, and formal interrelationships between isomers. Pictorial representations of graphs are so easily intelligible that chemists are often satisfied with inspecting and discussing them without paying too much attention to their algebraic aspects, but it is evident that some familiarity with the theory of graphs is necessary for deeper understanding of their properties.

This book should help to bridge the gap (or even abyss) between chemical and mathematical literature. It should thereby provide chemists with the mathematical background, help them to become acquainted with many recent applications, and prepare them for future developments, which can be expected to be rapid.

The editor of this first comprehensive monograph on chemical applications of graph theory is himself a pioneer and promotor of reaction graphs. He has been fortunate to find for all chapters authors who have made original contributions to the subject and who are still actively involved in it. This gives the book a special timeliness.

May it serve its purpose and stimulate the development of this important interdisciplinary area.

Zürich, den 18. Juli 1975   V. PRELOG

Contents

CHAPTER 1

Early History of the Interplay between Graph Theory and Chemistry

Alexandru T. Balaban

Institute of Atomic Physics, and Department of Organic Chemistry, Polytechnic,
Bucharest, Roumania

Frank Harary

Department of Mathematics and
Institute for Social Research,
University of Michigan, Ann Arbor,
Michigan, U.S.A.

Although graph theory is now an integral branch of combinatorial analysis, it began as a part of topology and even today, most electrical engineers and many chemists, working in network theory still consider 'topology' to be entirely synonymous with 'graph theory'.

Graph theory was independently discovered[1] on several occasions and three names deserve special mention — Euler, Kirchhoff, and Cayley.

Euler published the first known paper[2] on graph theory, in which he resolved the Königsberg bridge problem. The Pregel river in Königsberg (today Kaliningrad) has two islands, and at that time there were seven bridges connecting these islands and the banks of the river (Fig. 1).

The problem was to stroll across each bridge exactly once and return to the starting point. (The inhabitants tried this by taking random walks, of course, but without success.) Euler replaced each land area by a point and each bridge by a line joining two points, obtaining the graph (strictly, 'multigraph') of Fig. 2.

He showed that the problem is insoluble and gave a necessary and sufficient condition for any graph to be so traversable, i.e. that there is a solution only for connected graphs in which every point has an even degree, that is an even number of lines meeting at each point, a property now called 'eulerian'. Thus, if the line (bridge) joining points A and D were replaced by one joining points B and C, the problem could be solved in a variety of ways, as the reader may see for himself (Fig. 3).

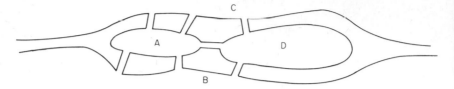

FIG. 1   The Königsberg bridges in 1736.

Kirchhoff[3] discovered graphs while solving problems involving the calculation of currents in electrical networks. He showed that the solution could be accomplished efficiently by using the 'spanning trees' of the graph represented by the given network.

Organic chemistry became the third breeding ground for graph theory. The best known early organic chemists, who founded the structure theory, were Couper, Butlerov, and Kekulé. They found it convenient to represent a covalent bond between two atoms as a line joining two points; thus every structural formula is a graph (or, if multiple bonds are involved, a multigraph). In order to illustrate the impact of this theory, a quotation from Sylvester[4] is appropriate:

> 'By the *new* atomic theory, I mean that sublime invention of Kekulé which stands to the *old* in a somewhat similar relation as the astronomy of Kepler to Ptolemy's, or the System of Nature of Darwin to that of Linnaeus. Like the latter, it lies outside of the immediate sphere of energetics, basing its laws on pure relation of form, and like the former as perfected by Newton, these laws admit of exact arithmetical definitions.'

Chemists were now able to predict the number of isomers of small alkanes and small alkanols on the basis of simple graphical constructions. Even today, students of organic chemistry are taught that C_3H_8, C_4H_{10}, and C_5H_{12} have 1, 2, and 3 isomers, respectively, because 'one can only construct that many structural formulas'. The enumeration of the chemical isomers, in particular the constitutional isomers of alkanes C_nH_{n+2}, was a challenging mathematical problem to which, in 1874–5, Cayley applied the graphical concept of a tree put forward by him[5] in 1857.

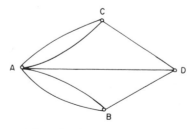

FIG. 2   The multigraph of the Königsberg bridge problem.

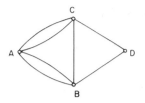

FIG. 3   Modified Eulerian multigraph of the Königsberg bridge problem.

If only carbon atoms of alkanes are depicted, there is a one-to-one correspondence between these isomers and trees (Fig. 4) whose points have at most degree 4. To solve the problem of finding the number of such trees, Cayley altered the problem, enumerating first the rooted trees (in which one point is distinguished from the others) and then the unrooted trees. He developed a solution for the initial problem[6-8] which enabled him to count constitutional isomers of alkanes. However, his figures are only correct for alkanes with 1−11 carbon atoms.

Other chemists also began to investigate this problem,[9-14] and in 1931, Henze and Blair produced a satisfactory method for counting the isomers of alkanes[15] and alkanols.[16] However, their calculations of these numbers contained errors which were corrected by Perry[17], whose figures were later verified by computers.[18]

Between 1932 and 1934, Blair and Henze succeeded in enumerating all the structural isomers of alkanes[19] and alkyls,[20] considering not only constitutional isomers, as Cayley had done, but also stereoisomers.* They also provided methods for obtaining isomer numbers for unsaturated hydrocarbons from the ethylene[21] and acetylene series,[22] and for the constitutional isomers of the main types of aliphatic compounds (organic acids, aldehydes, alkyl halides, amides, amines, esters, ethers, ketones, nitriles, etc.).[23]

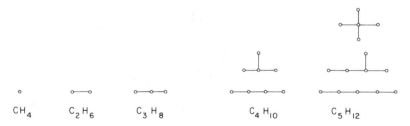

CH_4 C_2H_6 C_3H_8 C_4H_{10} C_5H_{12}

FIG. 4   The trees with 5 points at most.

*Constitutional (topological) isomerism includes chain isomerism as well as position isomerism, where distances and angles are unimportant; stereoisomerism includes diastereomerism and enantiomerism. Henze and Blair used 'structural isomerism' for 'constitutional isomerism'. The enumeration of achiral and chiral alkanes was accomplished only recently.[25]

The name of J. K. Senior[24] should be mentioned for his papers during 1927–1951. By 1930 the early, empirical, approach to chemical isomer enumeration had become obsolete and had been superseded by more rigorous methods. During 1935–1937, Pólya[27] developed his powerful theorem, enabling him to solve counting problems such as those examined above in a more direct and elegant manner. His contribution to chemical enumeration, which is examined in chapter 3, marks the beginning of a new era in graphical enumeration.

For further details concerning the pioneers of isomer enumeration, ref. 26 should be consulted.

REFERENCES

1. Harary, F. (1969). "Graph Theory", chapter 1, Addison–Wesley, Reading Mass. 1969.
2. Euler, L. (1736). *Comment. Acad. Sci. I. Petropolitanae*, 8, 128; translated in *Sci. Amer.* 189, 66 (1953).
3. Kirchhoff, G. (1842). *Ann. Phys. Chem.* 72, 497.
4. Sylvester, J. J. (1878). *Amer. J. Math.* 1, 64.
5. Cayley, A. (1857). *Philos. Mag.* 13, 19.
6. Cayley, A. (1874). *Philos. Mag.* 67, 444.
7. Cayley, A. (1875). *Ber. dtsch. chem. Ges.* 8, 1056.
8. Cayley, A. (1881). *Amer. J. Math.* 4, 266.
9. Schiff, H. (1875). *Ber, dtsch. chem. Ges.* 8, 1542.
10. Losanitsch, S. M. (1897). *Ber. dtsch. chem. Ges.* 30, 1917.
11. Losanitsch, S. M. (1897). *Ber. dtsch. chem. Ges.* 30, 3059.
12. Hermann, F. (1880). *Ber. dtsch. chem. Ges.* 13, 792.
13. Hermann, F. (1897). *Ber. dtsch. chem. Ges.* 30, 2423.
14. Hermann, F. (1898). *Ber. dtsch. chem. Ges.* 31, 91.
15. Henze, H. R. and Blair, C. M. (1931). *J. Amer. Chem. Soc.* 53, 3077.
16. Henze, H. R. and Blair, C. M. (1931). *J. Amer. Chem. Soc.* 53, 3042.
17. Perry, D. (1932). *J. Amer. Chem. Soc.* 54, 2918.
18. Lederberg, J. (1969). *In* "The Mathematical Sciences. A Collection of Essays" p. 37. (G. A. W. Boehm, editor). MIT Press, Cambridge, Mass.
19. Blair, C. M. and Henze, H. R. (1932). *J. Amer. Chem. Soc.* 54, 1538.
20. Blair, C. M. and Henze, H. R. (1932). *J. Amer. Chem. Soc.* 54, 1098.
21. Henze, H. R. and Blair, C. M. (1933). *J. Amer. Chem. Soc.* 55, 680.
22. Coffman, D. D., Blair, C. M. and Henze, H. R. (1933). *J. Amer. Chem. Soc.* 55, 252.
23. Henze, H. R. and Blair, C. M. (1934). *J. Amer. Chem. Soc.* 56, 157.
24. Senior, J. K. (1951). *J. Chem. Phys.* 19, 865 and previous papers.
25. Robinson, R. W., Harary, F. and Balaban, A. T., *Tetrahedron* (in press).
26. Rouvray, D. H. (1975). *Endeavour, 34(121)* 28.
27. Pólya, G. (1937). *Acta Math.* 68, 145.

CHAPTER 2

An Exposition of Graph Theory

F. Harary

*Department of Mathematics
and Institute for Social Research
University of Michigan,
Ann Arbor, Michigan, U.S.A.*

This chapter follows the terminology of the book.[1] However, it uses as a primary source the author's work[2] which was prepared for a readership of theoretical physicists and, as a secondary source, the paper[3] in which basic definitions were given along with 27 open problems on the counting of graphs (many of which have since been solved[4]).

We define a *graph* G as a finite non-empty set V of 'points' together with a prescribed subset X of the collection of unordered pairs of distinct points; X is the set of 'lines' of G. Thus by definition, every graph is finite and has no loops or multiple lines. In various recent papers, the graphs shown in Fig. 1 are now called 'Michigan graphs' to distinguish them from other kinds.

Whenever two points u and v are joined by a line x, we say that $x = \{u,v\}$ is in X, that $x = uv$, and that points u and v are *adjacent* and that each is *incident* with line x. Two graphs are *isomorphic* (Fig. 2) if there exists a 1 to 1 correspondence between their point sets that preserves adjacency. If the definition of a graph is generalized to permit multiple lines, the result is called a *multigraph*. If, in addition to multiple lines, we further allow the presence of loops, then we have a

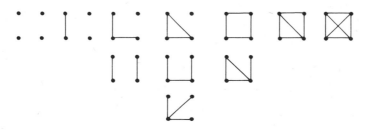

FIG. 1   All graphs with four points.

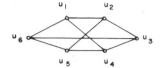
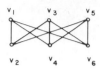

FIG. 2   Two isomorphic graphs.

general graph, also known as a *pseudograph*. If loops are allowed but not multiple lines, we have a *loop-graph* (Fig. 3).

A *walk* of a graph is a sequence, beginning and ending with points, in which points and lines alternate and each line is incident with the points immediately preceding and following it. A walk of the form $v_0, x_1, v_1, x_2, \ldots, v_n$ is said to *join* v_0 and v_n. The *length* of a walk is the number of occurrences of lines in it. A *trail* is a walk in which all lines are distinct; a *path* is a walk in which all points are distinct. An *open walk* has distinct first and last points; a *closed walk* is not open. A *cycle* is a closed walk with more than two points all of which are distinct, except the first and last, which coincide. A *spanning walk* contains all the points of G. A graph is *connected* if each pair of its points are joined by a path. A *tree* is a connected graph with no cycles. The *distance* between two points is the length of a shortest path joining them. The *degree* of a point is the number of lines incident with it; the *partition* of a graph is the sequence of degrees of its points, often listed in non-increasing order. In a *regular graph*, all points have the same degree. A *cubic graph* is regular of degree 3.

We illustrate some of these concepts by the connected graph of Fig. 4. The walk $v_1 v_2 v_3 v_4$ is a path, $v_1 v_2 v_3 v_1$ is a cycle, and $v_5 v_2 v_1 v_4 v_3 v_2$ is a spanning trail which is not a path. The distance from v_1 to v_5 is 2, the degree of v_1 is 3, and the partition of G is $(3, 3, 3, 2, 1)$.

A *subgraph H* of a graph G with point set V and line set X consists of subsets of V and X which together form a graph (G is then a *supergraph* of H). In a *spanning subgraph* of G, all the points of V occur. The most important kinds of such subgraphs are spanning paths and cycles (also called *Hamiltonian*), and spanning trees. The subgraph of the graph G of Fig. 4 shown in Fig. 5 is a spanning tree of G. The cycle $u_1 u_2 u_3 u_4 u_5 u_6$ of the graph shown in Fig. 2 is a Hamiltonian cycle.

The *complete graph* K_p has p points and every pair of points are adjacent. The graphs K_p for p = 1 to 5 are shown in Fig. 6.

FIG. 3   A multigraph, a loop-graph, and a general graph.

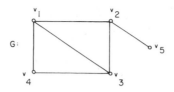

FIG. 4   A connected graph G.

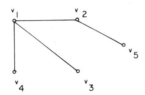

FIG. 5   A subgraph (spanning tree) of the graph of Fig. 4.

A *directed graph* (or more briefly a *digraph*) consists of a finite set V of points together with a prescribed collection of *ordered pairs* of distinct points of V. Each such ordered pair (u,v) is called an *arc* or a *directed line* and is also denoted uv when the directedness is clear by context. The definition of isomorphism of digraphs is analogous to that of graphs.[5]

The number of points in any type of graph is called its *order* and is denoted by p. The number of lines is denoted by q.

The degree* of a point v_i in graph G is denoted d_i or deg v_i, and is the number of lines incident with v_i. Since every line is incident with two points, it contributes the number two to the sum of the degrees of the points. We thus arrive at the first theorem of graph theory, due to Euler:[6] *The sum of the*

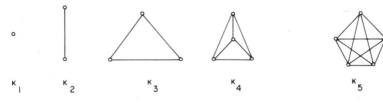

FIG. 6   The first five complete graphs.

*In some papers on graphs 'valency' is used for 'degree' and 'trivalent' is used for 'cubic'. This would not be appropriate here because of the potential confusion with chemical concepts.

FIG. 7   A forest consisting of 3 trees.

degrees of the points of a graph G is twice the number of lines,

$$\sum_i d_i = 2q.$$

As corollaries, we see at once that in any graph the number of odd-degree points is even, and that the order of a cubic graph is even.

It is convenient to have names for points of small degree. The point v is *isolated* if deg $v = 0$; it is an *endpoint* if deg $v = 1$.

A *component* of a graph is a maximal connected subgraph. A *forest* (Fig. 7) is a graph with no cycles; thus a tree is a connected forest. A disconnected graph has therefore at least two components. For example, the forest of Fig. 7 has 3 components.

We denote by C_n the graph consisting of a cycle with n points (C_3 is a triangle) and by P_n the path with n points. The *girth* of a graph G which is not a forest is the length of a shortest cycle in G.

The *removal of a point* v from a graph G yields the subgraph $G - v$ of G, which consists of all points of G except v and all lines not incident with v (therefore $G - v$ is the maximal subgraph of G not containing v). The *removal of a line x* from G results in the spanning subgraph $G - x$ of G, which contains all lines of G except x (therefore $G - x$ is the maximal subgraph of G not containing x_j).

If u and v are not adjacent in G, the addition of line uv results in the smallest supergraph of G containing this line.

Two graphs are *homeomorphic* if both yield the same graph on ignoring points of degree 2, e.g. all cycles are homeomorphic.

The *complement G* of a graph G has the same point set $V(G)$, but two points are adjacent in G if and only if they are not adjacent in G.

A *bipartite* * graph G (or *bigraph*, or *bicolorable graph*) is a graph whose point set $V(G)$ can be partitioned into two subsets V_1 and V_2 with m and n points respectively ($p = m+n$) such that every line joins V_1 and V_2. For example, the graph of Figure 2 is a bigraph and every tree is one. If G contains every possible line joining V_1 and V_2, then G is the *complete bigraph* $G = K_{m,n} = K(m,n)$. König[7] proved that a graph is bipartite if and only if all its cycles are of even length.

A *planar graph* G can be drawn in the plane so that no two lines cross each other. Kuratowski proved the criterion: *G is planar if and only if G has no subgraph homeomorphic to the complete graph K_5 or to the complete bigraph $K_{3,3}$.*

It was demonstrated that trees have either a *centre* (one vertex) or a bicentre (two adjacent vertices) at the largest distance from the endpoints.

A tree, one point of which is distinct from the other ones, is called a *rooted tree*, and the distinguished point is called the *root.* If the root is an endpoint (its degree is 1), the tree is called a *planted tree*.

REFERENCES

1. Harary, F. (1969). "Graph Theory", Addison-Wesley, Reading.
2. Harary, F. (1967). Graphical enumeration problems, *In:* "Graph Theory and Theoretical Physics", (F. Harary, ed.), pp. 1–41 Academic Press, London.
3. Harary, F. (1960). Unsolved problems in the enumeration of graphs. *Publ. Math. Inst. Hungar. Acad. Sci.* 5, 63–95.
4. Harary, F. and Palmer, E. M. (1973). "Graphical Enumeration", Academic Press, New York.
5. Harary, F., Norman, R. Z. and Cartwright, D. (1965). "Structural Models: An introduction to the theory of directed graphs." Wiley, New York.
6. Euler, L. (1953). The Königsberg bridges, (translated into English) *Sci. Amer.* 189, 66–70.
7. König, D. (1936). Theorie der endlichen und unendlichen Graphen. Leipzig. Reprinted Chelsea, New York, (1950).

CHAPTER 3

Pólya's Contributions to Chemical Enumeration

Frank Harary, Edgar M. Palmer and Robert W. Robinson

Department of Mathematics,
University of Michigan,
Ann Arbor, Michigan, U.S.A.

Ronald C. Read

University of Waterloo,
Department of Combinatories and Optimization,
Waterloo, Ontario, Canada

1   PÓLYA'S HAUPTSATZ

Before developing the classical and powerful enumeration theorem of Pólya[1] we shall provide some simple examples involving generating functions so that the reader can gain some familiarity with the uses to which they can be put and some facility with the kinds of manipulations to be expected.

The solution of a chemical enumeration problem is often most conveniently expressed in terms of a polynomial or power series (generating function) whose coefficients display the solution. For example, if we are to enumerate the number of compounds that can be obtained by substituting chlorine atoms for the hydrogen atoms in a benzene ring, the solution is given by the generating function:

$$1 + x + 3x^2 + 3x^3 + 3x^4 + x^5 + x^6. \tag{1}$$

Here the coefficient of x^k is the number of substituted benzenes with exactly k

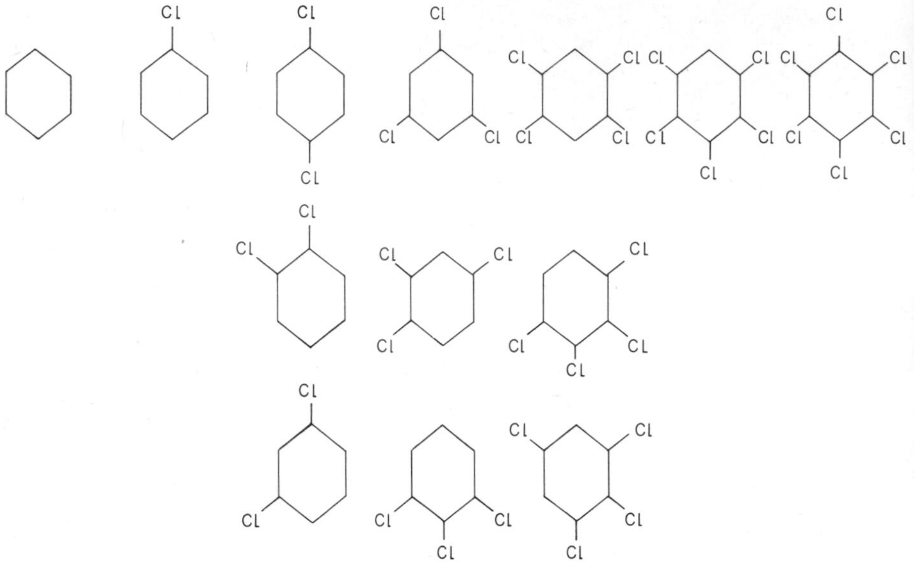

FIG. 1   All the substituted benzene rings.

chlorine atoms. The diagrams which correspond to these compounds are shown in Fig. 1.

We now turn to some combinatorial problems whose solutions can be expressed readily by using generating functions. Suppose we have a box and we wish to put into it one ball, colored either red, white, or blue, or no balls at all. Then a generating function which displays the 4 possible outcomes in a natural and obvious fashion is $1 + r + w + b$, with the obvious meaning of the variables. If there are two different boxes, say box 1 and box 2 without loss of generality, the generating function which shows all 16 ways of arranging at most one ball per box is the product $(1 + r + w + b)(1 + r + w + b)$. In general for n boxes, the solution is given by $(1 + r + w + b)^n$ and the coefficient of $r^i w^j b^k$ is the number of arrangements in which i boxes have a red ball, j have a white, k have a blue, and the rest are empty.

Next suppose we have an unlimited supply of red balls and just one box. Then the generating function

$$1 + r + r^2 + r^3 + r^4 + \ldots \tag{2}$$

has as its coefficients the number (just 1) of ways of having any number of red balls in the box. But if we wish to arrange the red balls in box 1 and box 2, we square this generating function, obtaining

$$1 + 2r + 3r^2 + 4r^3 + \ldots . \tag{3}$$

In this generating function, the coefficient $k + 1$ of r^k is the number of ways of dividing k red balls between two boxes. In general, for n boxes, the corresponding solution is obtained by raising the series (2) to the n'th power. Note that this latter result could be expressed in the form $(1 - r)^{-n}$ because by long division, or from the binomial theorem, (2) can be expressed as $(1 - r)^{-1}$. In fact, it follows from the binomial theorem that the coefficient of r^k in $(1 - r)^{-n}$ is

$$\binom{-n}{k} (-1)^k = \binom{n + k - 1}{k}$$

and hence this latter binomial coefficient is the number of ways of arrangements of k red balls among n different boxes.

Returning to our first example in which we discussed the substitution of chlorine atoms in the benzene ring, observe that the positions occupied by the hydrogen atoms can be regarded as six boxes. These boxes can be filled with a hydrogen atom or a chlorine atom. With just one box, we could use $1 + x$ as the counting series or generating function, with the 1 to indicate that the position is taken by hydrogen while x denotes chlorine. However $(1 + x)^6$ does not yield the correct result (1) because the boxes are not differently labeled boxes. They are arranged in a circular array and so the coefficients of $(1 + x)^6$ are too large and must be reduced by considerations that depend on the symmetries of the 6 positions in the benzene ring. We shall soon see how Pólya's Hauptsatz can be used to resolve this difficulty by coordinating the information carried by the series $1 + x$, standing for hydrogen or chlorine, with respect to the symmetries of the hexagon.

Pólya's classical Hauptsatz[1] can be viewed as a way of counting functions from a domain into a range with equivalence determined by a particular permutation group acting on the domain. It enables one to express the generating functions for a class of chemical compounds in terms of an appropriate permutation group and another generating function called the figure counting series. Its sweeping generality and ease of application make this result a most powerful tool in enumerative analysis. An extensive treatment of Pólya's Hauptsatz, with variations and numerous applications, can be found in the book by Harary and Palmer[2] on graphical enumeration. Earlier papers by Pólya on chemical enumeration[3-4] are steps towards his classical paper.[1]

We shall use the example of Fig. 1. in which chlorine can be substituted for hydrogen to illustrate the details of applying the Hauptsatz. First note that each substitution can be regarded as a function from the six positions (vertices of a hexagon) to the set $\{H, Cl\}$ consisting of a hydrogen atom and a chlorine atom. For example, consider the three functions f, g, and h from $\{1,2,3,4,5,6\}$ into

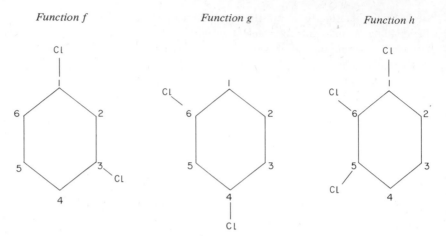

FIG. 2   The substituted benzenes determined by f, g, and h.

{H,Cl} defined by:

$$f(1) = f(3) = Cl, f(2) = f(4) = f(5) = f(6) = H$$
$$g(4) = g(6) = Cl, g(1) = g(2) = g(3) = g(5) = H$$
$$h(1) = h(5) = h(6) = Cl, h(2) = h(3) = h(4) = H$$

These functions are also represented by Fig. 2 in which the six positions are numbered 1 to 6 around the benzene ring. There are a total of 2^6 such functions but many of them represent the same chemical compound. For example, f and g are just two of the six functions which represent m-dichlorobenzene. These six functions, however, differ only by a rotation of the benzene ring. Furthermore, there is no rotation or reflection of the ring which can possibly cause the functions g and h to be regarded as the same. In general, the symmetries of the hexagon completely determine a partition of the 2^6 functions into classes and each class represents one of the compounds of Fig. 1. Specifically, here is a list of all 12 symmetries of the hexagon:

Rotations	Reflections
(1) (2) (3) (4) (5) (6)	(12) (36) (45)
(1 2 3 4 5 6)	(23) (56) (14)
(1 3 5) (2 4 6)	(34) (25) (16)
(6 5 4 3 2 1)	(1) (4) (26) (35)
(6 4 2) (5 3 1)	(2) (5) (13) (46)
(14) (25) (36)	(3) (6) (15) (24)

In our notation, the expression (1 2 3 4 5 6) denotes the clockwise rotation in which 1 is sent to 2, 2 to 3, . . . , and 6 to 1; while (1) (4) (26) (35) denotes the reflection in which 1 4 are fixed while 2 and 6 are interchanged, as are 3 and 5. This collection of symmetries are permutations which constitute a permutation group denoted by D_6, called the *dihedral group* of degree 6 and order 12.

It can be seen now that any two of the 2^6 different functions mentioned above, say f_1 and f_2, represent the same chemical compound if and only if there is a permutation α in D_6 such that

$$\alpha f_1 ((k)) = f_2(k) \tag{3.1}$$

for each $k = 1, \ldots, 6$. In our example, we have seen that f and g should be regarded as equivalent. Indeed, if we select permutation $\alpha = (34)(25)(16)$, then $g(\alpha k) = f(k)$ for all $k = 1, \ldots, 6$. Furthermore, there is no permutation in D_6 that causes f and h to be regarded as the same.

Part of the information required to enumerate these equivalence classes of functions is carried by a polynomial which incorporates the cycle structure of the permutation group under consideration. Let A be a permutation group with finite object set X and for each permutation α in A and each non-negative integer r let $j_r(\alpha)$ be the number of cycles of r objects in the disjoint cycle decomposition of q. In our example, the permutation $(34)(25)(16)$ has 3 cycles of order 2 so $j_2 = 3$, and $(1)(4)(26)(35)$ has $j_1' = j_2' = 2$. Then the *cycle index of A*, called the *Zyklenzeiger* by Pólya, is the polynomial in the variables $s_1, s_2, s_3, \ldots$ given by the following expression

$$Z(A) = \frac{1}{|A|} \sum_{\alpha \in A} \prod_r s_r^{j_r(\alpha)} . \tag{3.2}$$

In our example, it follows quickly from the list of permutations in D_6 that the cycle index is

$$A(D_6) = \frac{1}{12} \{s_1^6 + 3s_1^2 s_2^2 + 4s_2^3 + 2s_3^2 + 2s_6\} \tag{3.3}$$

Pólya's Hauptsatz enumerates classes of functions from X, the object set of the permutation group A, into another set Y of elements called *figures*. Two such functions f_1 and f_2 are in the same class or are *A-equivalent* if there is a permutation α in A such that $f_1(\alpha k) = f_2(k)$ for all objects k in X. Often the figures in the set Y have integral weights assigned to them and are also enumerated by a generating function called the *figure counting series*, denoted by $c(x) = c_0 + c_1 x + c_2 x^2 + \ldots$, which has as the coefficient of x^m the number of elements in Y of weight m. In our example, we have $X = \{1, \ldots, 6\}$ and $Y = \{H, Cl\}$. If we assign the weights of 0 and 1 to H and Cl respectively, then $c_0 = c_1 = 1$ and our figure counting series is just $1 + x$. In this case, we can define the weight of a function to be the sum of the weights of the images of all domain elements and as a consequence, the weight of a function in our example is just the number of chlorine atoms substituted for hydrogen. In general then, let w be any function from Y into the set of non-negative integers. Any function

f from X into Y has as its weight

$$\sum_{k \in X} w(f(k)).$$

With this definition, A-equivalent functions [in the sense of (3.1)] always have the same weight. Therefore we can also define the weight of each of these classes to be the weight of any function in the class.

The Hauptsatz relates the generating function $C(x)$, which enumerates these classes by weight to the cycle index $Z(A)$ and the figure counting series $c(x)$:

Pólya's Hauptsatz. *The generating function $C(x)$ which enumerates equivalence classes of function determined by the permutation group A is obtained by substituting the figure counting series $c(x)$ in the cycle index $Z(A)$ as follows. Each variable s_r in $Z(A)$ is replaced by $c(s^r)$. Symbolically we write:*

$$C(x) = Z(A, c(x)).$$

In the example, with $A = D_6$ and $c(x) = 1 + x$ and $Z(D_6)$ given by (3.3) we have

$$A(D_6, 1 + x) = \frac{1}{12} \{(1 + x)^6 + 3(1 + x)^2 (1 + x^2)^2 + 4(1 + x^2)^3$$

$$+ 2(1 + x^3)^2 + 2(1 + x^6)\}. \qquad (3.4)$$

On simplifying the right side of (3.4) we obtain the generating function $C(x)$ for the classes of these functions which correspond to the substituted benzenes shown in Fig. 1; c.f. (1):

$$C(x) = 1 + x + 3x^2 + 3x^3 + 3x^4 + x^5 + x^6.$$

Pólya[3] displayed this result in a chemically more familiar notation by $X = X/H$ and 'multiplying' by $C_6 H_6$ with the result:

$$C_6 H_6 + C_6 H_5 X + 3C_6 H_4 X_2 + 3C_6 H_3 X_3 + 3C_6 H_2 X_4 + C_6 HX_5 + C_6 X_6.$$

We conclude this section with an illustration of the theorem which allows us to enumerate the number of ways in which chlorine can be substituted for hydrogen in toluene. In Fig. 3 each of the positions number 1 through 8 can be occupied by H or Cl. As before we let $X = \{1, \ldots, 8\}$ and $Y = \{H, Cl\}$. From Fig. 3 we observe that positions 1, 2, 3 should be regarded as indistinguishable. Therefore we consider the symmetric group S_3 acting on $\{1,2,3\}$ whose permutations are:

$$(1) \,(23) \quad (123)$$
$$(2) \,(13) \quad (132).$$
$$(3) \,(12)$$

We also note that functions should be in the same class if they differ only by a

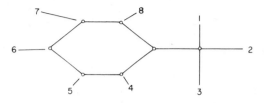

FIG. 3   Toluene

reflection of the hexagon around the horizontal axis. The group B associated with this reflection contains the identity permutation (4) (5) (6) (7) (8), and (6) (57) (48). For the group A we take the direct product or product of B and S_3. This group is denoted by BS_3 and consists of all products of permutations, the first from B and the second from S_3. Note further that the cycle index of such a product is always the product of the cycle indexes, $Z(BS_3) = Z(B)Z(S_3)$. Therefore, since we see at once from the permutations listed above,

$$Z(S_3) = \frac{1}{6}(s_1^3 + 3s_1 s_2 + 2s_3)$$

and since

$$Z(B) = \frac{1}{2}(s_1^5 + s_1 s_2^2),$$

it follows that

$$Z(BS_3) = \frac{1}{12}(s_1^8 + 3s_1^6 s_2 + 2s_1^5 s_3 + s_1^4 s_2^2 + 3s_1^2 s_2^3 + 2s_1 s_2^2 s_3).$$

From the Hauptsatz we can obtain the generating function for the substituted toluenes by setting $s_r = 1 + x^r$ in $Z(BS_3)$. In carrying out the details of this substitution we have the result:

$$C(x) = 1 + 4x + 10x^2 + 16x^3 + 18x^4 + 16x^5 + 10x^6 + 4x^7 + x^8.$$

Note that in setting $x = 1$ in this expression we obtain the total number 80 of toluenes substituted by one type of reagent, e.g. of all possible chlorotoluenes or of deuterated toluenes.

2   ROOTED TREES AND ALKYL RADICALS

At this stage it will be instructive to consider a few rather more elaborate applications of Pólya's Hauptsatz, and we shall look first at the enumeration of rooted trees. Let T_n be the number of rooted trees having n points (including

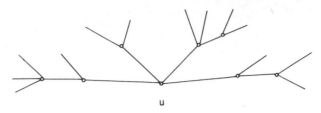

FIG. 4   A rooted tree.

the root); then the problem is to discover a method for computing the values of T_n and this we shall do by making these numbers the coefficients in a counting series

$$T(x) = T_1 x + T_2 x^2 + T_3 x^3 + \ldots \ldots \tag{A}$$

Consider a rooted tree, such as that in Fig. 4, rooted at the point u.

If we delete the root of this tree, together with the lines incident with it, we shall obtain a collection of rooted trees, as in Fig. 5.

Each of these trees is rooted, since the point that was adjacent to the original root u is thereby distinguished from the other points of the tree. The number of rooted trees obtained in this way will be k, the degree of the root.

Thus from a single rooted tree having n points, and for which the degree of the root is k, we obtain an unordered collection of k rooted trees, having $n - 1$ points in all. Conversely, from these k rooted trees we can uniquely reconstruct the original rooted tree by joining each of the roots to a new point u.

It is therefore a Pólya-type problem to enumerate rooted trees for which the root has degree k. We have k 'boxes', and in each box we may place a single rooted tree. The boxes have no special order, and can therefore be permuted by any element of S_k, the symmetric group of degree k. In the terminology of Pólya's Hauptsatz above, figures are the unrooted trees, for which the figure-counting series is the as yet unknown series $T(x)$. The content of a figure is the number of points in the tree. Clearly the appropriate group for this problem is S_k, and the configuration-counting series is therefore given by the

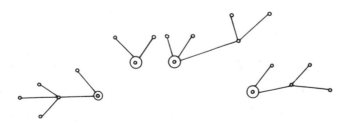

FIG. 5   A collection of rooted trees.

Hauptsatz, and is

$$Z(S_k; T(x)) \tag{B}$$

If we wish to find the number of rooted trees of all points, i.e., for all possible values of the root-degree k, we must sum (B) for all values of k, and we obtain

$$\sum_{k=1}^{\infty} Z(S_k; T(x)) \tag{C}$$

The coefficient of x^n in (C) is the number of configurations having content n, i.e., the number of collections of rooted trees of order n. Since, in reconstructing a single rooted tree from such a configuration, we add another point, it follows that the number of rooted trees with n points is the coefficient of x^{n-1} in (C) or, what amounts to the same thing, it is the coefficient of x^n in

$$x \sum_{k=1}^{\infty} Z(S_k; T(x)) \tag{D}$$

Thus (D) is almost the counting series for rooted trees, but it does not contain the term in x corresponding to the trivial rooted tree. If we supply this missing term we obtain the equation

$$T(x) = x + x \sum_{k=1}^{\infty} Z(S_k; T(x))$$

or, by writing conventionally $Z(S_0) = 1$,

$$T(x) = \sum_{k=0}^{\infty} Z(S_k; T(x)) \tag{E}$$

This result is not, as it stands, particularly convenient for computation. It can be converted to a more convenient form, but since we are more concerned here with chemical applications we shall not take space to show in detail how this is done. Suffice it to say that, by use of the general result,

$$\sum_{k=0}^{\infty} Z(S_k) t^k = \exp\left(\frac{f_1}{1} t + \frac{f_2}{2} t^2 + \frac{f_3}{3} t^3 + \dots\right)$$

which follows from the definition of $Z(S_k)$, equation (E) can be rewritten as

$$T_1 x + T_2 x^2 + T_3 x^3 + \dots = x(1-x)^{-T_1}(1-x^2)^{-T_2}(1-x^3)^{-T_3}\dots$$

— a form of the result already known to Cayley.[7]

Let us now consider a similar application of the Hauptsatz to a chemical problem, that of enumerating the alkyl radicals having the general formula $C_n H_{2n+1}$ ———. We first observe that we can remove all the hydrogen atoms from the structural formula of such a radical without losing any information — the hydrogen atoms can be uniquely restored any time we wish. Only carbon atoms are left and we can replace them by points to obtain a rooted tree; rooted

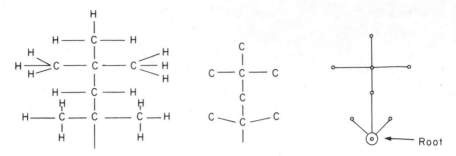

FIG. 6   Derivation of a rooted tree from an alkyl radical.

because one carbon atom is distinguished from the rest by being at the end of the 'free' bond of the radical. This process is illustrated in Fig. 6.

Thus these structures present an enumeration problem very similar to that of enumerating rooted trees. The difference is that (a) no point has degree greater than 4, and (b) the root has degree at most 3. Furthermore, these two statements are true of the rooted trees that are formed when the root is deleted.

Hence the only change that needs to be made is that the summation over k will be from $k = 0$ to $k = 3$. Let A_n be the number of different alkyl radicals having n carbon atoms, and let $\phi(x) = A_1 x + A_2 x^2 + A_3 x^3 + \ldots$ Then we obtain

$$\phi(x) = x \sum_{k=0}^{3} Z(S_k; \phi(x)). \tag{F}$$

This result (F) can be simplified by writing

$$A(x) = A_0 + A_1 x + A_2 x^2 + A_3 x^3 + \ldots$$

where $A_0 = 1$, so that

$$A(x) = 1 + \phi(x)$$

We then have

$$\phi(x) = x + x\phi(x) + 1/2 \, x \, [\phi^2(x) + \phi(x^2)]$$
$$+ 1/6 \, x \, [\phi^3(x) + 3\phi(x)\phi(x^2) + 2\phi(x^3)]$$

after making the appropriate substitutions in the cycle indices that occur in (F). On putting $\phi(x) = A(x) - 1$, we find that

$$A(x) = 1 + (1/6)x[A^3(x) + 3A(x)A(x^2) + 2A(x^3)]$$
$$= 1 + (1/6)xZ(S_3; A(x)) \tag{G}$$

Using hindsight we see that we could have obtained (G) directly, by regarding a hydrogen atom as a sort of 'degenerate' alkyl radical having zero carbon atoms.

The degree of the root can then always be taken as 3, and we have only one cycle index to deal with. This treatment of a hydrogen atom as if it were the first, or rather zero[th] member of the alkyl series is so convenient that we shall assume from now on that the term A_0 which represents it has been included in the counting series.

Equation (G) can be used recursively to find the values of A_n. If these are all known up to a given value N of n, then the right-hand side of (G) can be evaluated up to the term in x^{N+1}. Hence the value of A_{N+1} is found. Results equivalent to (G) were known to Cayley[7] and to Henze and Blair.[8] This result is the starting point for the enumeration of a wide variety of chemical compounds, expecially acyclic compounds. These acyclic enumerations will form the subject matter of the next chapter, and an application to a class of cyclic compounds will be given later in this chapter.

First, however, it is worth mentioning another, more interesting but more complicated, problem that Pólya dealt with in his paper. This is the graph theoretical problem of enumerating 'free' (unrooted) trees, and the similar chemical problem of enumerating the alkanes. It is easy to see that the alkanes, with general formula C_nH_{2n+2}, bear the same relationship to unrooted trees as the alkyl radicals do to rooted trees. Thus the enumeration problems for unrooted trees and for the alkanes are similar, and Pólya discussed both of them. His solutions, however, were effected by an indirect and cumbersome procedure, as were those of Cayley and Henze and Blair for similar problems. It was not until some time after the publication of Pólya's main paper[1] that a more direct and systematic method was discovered for tackling this problem. This method will be discussed in detail in chapter 4.

As a final simple example of the use of Pólya's Hauptsatz in chemical enumeration, let us consider the problem of enumerating the alkyl derivatives of benzene. A typical example is given in Fig. 7.

This problem is readily seen to be one to which the Hauptsatz can be applied. Around the benzene ring we have six 'boxes' into each of which we can place a single alkyl radical. Thus the figure counting series will be

$$A(x) = A_0 + A_1x + A_2x^3 + A_3x^3 + \ldots .$$

where A_0 (=1) is included since a box may contain just a hydrogen atom. The

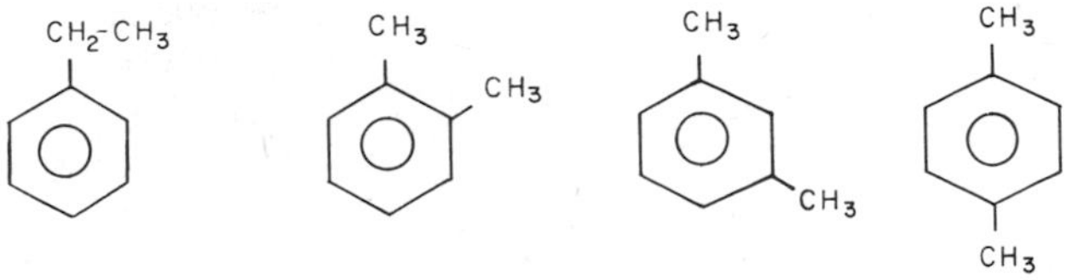

FIG. 7   All alkyl derivatives of C_8H_{16} of benzene.

TABLE I
Numbers of alkyl derivatives of benzene

Number of carbon atoms	6	7	8	9	10	11	12	13	14	15	16	17
Number of derivatives	1	1	4	8	22	51	136	335	871	2217	5749	14837

Number of carbon atoms	18	19	20	21	22	23	24	25
Number of derivatives	38636	100622	263381	690709	1817544	4793449	12675741	33592349

Number of carbon atoms	26	27	28	29	30
Number of derivatives	89223734	237455566	633176939	1691377956	4525792533

boxes can be permuted by any permutation of the dihedral group D_6, as in the example given in the previous section for the number of ways of substituting chlorine atoms around a benzene ring. Thus D_6 is the appropriate group for this problem and the configuration counting series is therefore

$Z(D_6; A(x))$.

Since

$$Z(D_6) = (1/12) (f_1^6 + 4f_2^3 + 2f_3^2 + 3f_1^2 f_2^2 + 2f_6)$$

we have

$$(1/12) [A^6(x) + 4A^3(x^2) + 2A^2(x^3) + 3A^2(x)A^2(x^2) + 2A(x^6)] \qquad \text{(H)}$$

as the counting series for these alkyl derivatives of benzene.

The general formula for these alkyl derivatives of benzene is $C_n H_{2n-6}$. Table 1 gives the number of isomers for each value of n from 6 to 30, calculated from the counting series (H).

3   ASYMPTOTIC FORMULAS IN CHEMICAL ENUMERATION

Pólya,[6] with his vast knowledge of power series, which he regarded as functions of a complex variable, succeeded in establishing asymptotic expressions for various families of chemical compounds. To explain these results we require quite a lot of notation, and in general we follow that of Pólya.

Let t_n be the number of trees of order n, and let T_n be the number of rooted ones. Let r_n be the number of trees having n points of degree 4 in which every point has degree 1 or 4, and let R_n be the number of planted ones. These represent paraffins (saturated hydrocarbons) and alcohols respectively. Similarly, s_n is the number of stereoisomeric paraffins (with n carbons) and S_n is the number of such alcohols (planted paraffins). Finally, q_n denotes the number of achiral paraffins and Q_{12} the number of achiral planted alcohols.

With this notation, we are almost ready to indicate without proofs the asymptotic results which Pólya obtained. But first we still need to introduce the radii of convergence of the four power series $\Sigma Q_n x^n$, $\Sigma R_n x^n$, $\Sigma S_n x^n$, and $\Sigma T_n x^n$. These are written κ, ρ, σ, and τ respectively. All that Pólya stated concerning these four numbers is:

$$1 > \kappa > \rho > \sigma; \quad \rho > \tau; \quad \rho = 0.35 \ldots; \quad \sigma = 0.30 \ldots;$$
$$\text{and} \quad 1/e > \tau > 1/4.$$

However, much more is now known. For example Otter[9] found that $\tau = 0.3383219$ and very recently Robinson and Schwenk[10] established, with the aid of a very clever computer program which he devised, that $\tau = 0.3383218568992$.

We first note the asymptotic expressions for R_n and r_n developed by Pólya.[6] In connection with these, he found the approximate value $\rho = 0.35 \ldots$ and indicated how the constants a and b are determined by ρ and the numbers R_1, $R_2, R_3, \ldots$:

$$R_n \sim \rho^{-n} n^{-3/2} \frac{b}{2\sqrt{\pi}}$$

$$r_n \sim \rho^{-n} n^{-5/2} \frac{\rho a b^3}{4\sqrt{\pi}}.$$

Pólya did not give the details of the remaining asymptotic expressions which now follow, but said that they are handled similarly, with K_i standing for constants:

$$Q_n \sim K_1 \kappa^{-n}$$
$$q_n \sim K_2 \kappa^{-n}$$
$$S_n \sim K_3 n^{-3/2} \sigma^{-n}$$
$$s_n \sim K_4 n^{-5/2} \sigma^{-n}$$
$$T_n \sim K_5 n^{-3/2} \tau^{-n}$$
$$t_n \sim K_6 n^{-5/2} \tau^{-n}$$

The last two of these expressions were specified more precisely by Otter[9] who calculated:

$$K_5 = 0.4399237$$
$$K_6 = 0.5349485.$$

Pólya[6] also developed asymptotic formulas for other families of chemical compounds. For example, the simplest case of unsaturated hydrocarbons, $C_n H_{2n}$, containing exactly one double bond, are counted asymptotically by

$$\frac{\rho^{-n}}{4n}.$$

However, for more general unsaturated hydrocarbons $C_n H_{2n+2-2b}$, containing b double bonds, the asymptotic number is only known up to a constant:

$$K_7 = \rho^{-n} \; m^{(3b-5)/2},$$

where K_7 depends only on b.

Consider substituted paraffins with m distinct radicals substituted for hydrogen atoms, of the form $C_n H_{2n+2-m} X' X'' \ldots X^{(m)}$. The asymptotic expression obtained by Pólya was

$$\rho^{-n} \quad n^{(2m-5)/2} \; \frac{\rho a b^3}{4\sqrt{\pi}} \left(\frac{2}{\rho a b^2} \right)^2,$$

where ρ, a and b are the same constants as in the expressions for R_n and r_n.

Finally, Pólya also showed that the number of hydrocarbons with a given cyclic structure is asymptotically proportional to R_n, the number of alcohols, and that the constant of proportionality is determined by the cycle index of the group of the graph having the given cyclic structure.

REFERENCES

1. Pólya, G. (1937). Kombinatorische Anzahlbestimmung für Gruppen, Graphan und chemische Verbindungen, *Acta. Math.* **68**, 145.
2. Harary, F. and Palmer, E. M., (1973). "Graphical Enumeration", Academic Press, New York, London and San Francisco.
3. Pólya, G., (1936). Algebraische Berechnung der Isomeren einiger organischer Verbindungen, *Z. Krystallogr.* (A), **93**, 414.
4. Pólya, G., (1935). Un problème combinatoire général sur les groupes des permutations et le calcul du nombre des composés organiques, *Compt. rend. Acad. Sci. Paris,* **201**, 1167.
5. Pólya, G., (1936). Sur le nombre des isomères de certains composés chimiques, *Compt. rend. Acad. Sci. Paris,* **202**, 1554.
6. Pólya, G., (1936). Über das Anwachsen des Isomerenzahlen in den homologen Reihen des organischen Chemie, *Vierteljschr. Naturforsch. Ges. Zürich,* **81**, 243.
7. Cayley, A., (1874). On the mathematical theory of isomers, *Philos. Mag.* **67**, 444.
8. Blair, C. M. and Henze, H. R., (1931). The number of structurally isomeric alcohols of the methanol series, *J. Amer. Chem. Soc.* **53**, 3042.
9. Otter, R., (1948). The number of trees, *Ann. of Math.* **49**, 583.
10. Robinson, R. W. and Schwenk, A. J. (1975). The distribution of degrees in a large random tree, *Discrete Math.* **12**, 359.

CHAPTER 4

The Enumeration of Acyclic Chemical Compounds*

Ronald C. Read

University of Waterloo,
Department of Combinatorics and Optimization,
Waterloo, Ontario, Canada

1 INTRODUCTION

In the last chapter we saw one or two ways in which combinatorial methods, and, in particular, Pólya's Theorem, can be used to enumerate chemical compounds of various kinds. We shall now follow up this idea in greater detail.

The most general problem of this type would be the following. Given an empirical formula, which merely states how many atoms of each kind there are in a compound, determine the number of distinct chemical structures that have those numbers of atoms. For example, we might ask how many structural isomers there are having the empirical formula $C_{30}H_{51}N_7O_{15}$. More generally, we ask for a mathematical formula, or some automatic procedure, whereby the number of isomers corresponding to *any* given empirical formula can be

*The original work described in this chapter was carried out under grant number A8142 from the National Research Council of Canada.

determined. A different, and somewhat more realistic, version of this problem is to ask for the number of stereo-isomers corresponding to the given empirical formula. Unfortunately, both these problems are notoriously intractable,* and it is most unlikely that any practical solution to them will be found in the near future. However, if we lower our sights a little, and confine our attentions to some restricted class of chemical compounds, then a mathematical treatment of the problem becomes possible. In particular, the acyclic hydrocarbons form an important class of compounds for which an explicit enumeration is feasible, and it is to the counting of these compounds, and some of their derivatives, that we shall address ourselves in this chapter.

The enumeration of acyclic hydrocarbons containing at most one double or triple bond has been well studied in the past by Cayley, Blair and Henze, Pólya and others (see the references given below). Later, new techniques were discovered whereby these same enumerations could be obtained more easily, but which were not, apparently, used to derive any new results. This is surprising, since all the apparatus was then at hand for the solution of more difficult enumerative problems, including the one already alluded to of counting the general acyclic hydrocarbons, i.e. having any numbers of double or triple bonds. As it happened, this problem does not appear to have been tackled at all prior to the publication of my paper,[7] the results of which will be extended here.

Thus the aim of this chapter will be threefold. First, it will present a short history of the early work in the counting of acyclic compounds; secondly, the more important of these enumerations will be derived, using (and thereby illustrating) the most up-to-date techniques for obtaining these results; and thirdly, the enumeration of both structural and stereo isomers of acyclic hydrocarbons in general (a hitherto neglected problem) will be given.

The compounds that we shall be enumerating fall naturally into families, within which they are ranked according to the number of carbon atoms that they have. This makes the generating function, or counting series, in which the power of x is the number of carbon atoms, the most suitable method of displaying our results. Our starting point will be the counting series that has already been derived in the previous chapter, namely the counting series $A(x)$ for the numbers, A_n, of structural isomers of the alkyl radicals, having the general formula $-C_n H_{2n+1}$. A word about notation. The symbols Q, R, X and Y will be used to denote unspecified atoms or radicals. When we need more symbols we shall use the letters Q and R with subscripts ($Q_1, Q_2, \ldots, R_1, R_2, \ldots$) even though, strictly speaking, this conflicts with the chemical usage of subscripts to indicate the number of atoms. This will be preferable to the use of a more complicated notation such as would be required to avoid this conflict. In any case, this conflict is more apparent than real, since the letters Q and R are not atom symbols.

---

*Substantial progress has recently been achieved by Masinter and his colleagues.[8]

2   SOME STRAIGHT-FORWARD USES OF THE COUNTING SERIES FOR ALKYL RADICALS

As we saw in Chapter 3, the recursion formula for this counting series $A(x) = A_0 + A_1 x + A_2 x^2 + \ldots$ is

$$A(x) = 1 + xZ(S_3; A(x))$$

$$= 1 + \tfrac{1}{6} x [A^3(x) + 3A(x)A(x^2) + 2A(x^3)] \tag{2.1}$$

and from this equation the coefficients in $A(x)$ can be calculated to as many terms as one likes.

The calculation by hand of these coefficients by means of (2.1) is tedious, however, and their computation up to A_{20} by Blair and Henze[1] in 1931, and by Perry[5] up to A_{30} in 1932, was a remarkable *tour de force*. A modern electronic computer can be persuaded to perform these calculations in a very short time, and the results of such a calculation are given in Table I. We shall now see how the numbers A_n and the series $A(x)$ can be used to enumerate several families of acyclic compounds.

We first note that if we join an alkyl radical to a hydroxyl group $-$OH we get an alkyl alcohol of the methanol series $C_n H_{2n+1}OH$. The number of structural isomers of the alcohols with this formula is thus A_n. If $n = 0$ we get the compound H$-$OH, which, of course, is not an alcohol in the chemical sense (or in any other sense!); but its inclusion in the series will do no harm, and will give some advantages.

Cayley[3] and Blair and Henze[1] derived their general results by considering separately the number of primary, secondary and tertiary alcohols. We shall find it more convenient to derive these from the general result.

(a) Primary alcohols. The primary alcohols with n carbons are of the form

$$
\text{R}-\overset{\displaystyle \text{H}}{\underset{\displaystyle \text{H}}{\text{C}}}-\text{OH}
$$

where R denotes an alkyl radical with $n - 1$ carbon atoms. Hence there is a one-to-one correspondence between the primary alcohols with n carbons and the general alcohols with $n - 1$ carbons. Thus the counting series for primary alcohols is

$$xA(x) \tag{2.2}$$

the multiplied x accounting for the difference of 1 in the number of carbon atoms. Since H$-$OH is catered for in the counting series $A(x)$, the counting series $xA(x)$ takes account of the corresponding alcohol, methanol (CH$_3$OH). In a sense this is not a primary alcohol, but is best included here since it is certainly not a secondary or tertiary alcohol.

(b) Secondary alcohols. The secondary alcohols with n carbon atoms are of

TABLE I

The numbers of structural isomers of some acyclic compounds

n	1	2	3	4	5	6	7	8	9	10	11	12	13
Alkyl radicals $A(x)$	1	1	2	4	8	17	39	89	211	507	1238	3057	7639
Primary alcohols	1	1	1	2	4	8	17	39	89	211	507	1238	3057
Secondary alcohols			1	1	3	6	15	33	82	194	482	1188	2988
Tertiary alcohols				1	1	3	7	17	40	102	249	631	1594
$B(x)$	1		2	3	7	14	32	72	171	405	989	2426	6045
Esters		1	2	4	9	20	45	105	249	599	1463	3614	9016
Ethylene derivs.		1	1	3	5	13	27	66	153	377	914	2281	5690
$C_nH_{2n}XY$	1	2	5	12	31	80	210	555	1479	3959	10652	28760	77910
$C_nH_{2n}X_2$	1	2	4	9	21	52	129	332	859	2261	5983	15976	42836
Glycols		1	2	6	14	38	97	260	688	1856	4994	13550	36791
$P(x)$	1	1	2	4	9	18	42	96	229	549	1347	3326	8330
$Q(x) - A(x^2)$			1	2	6	13	33	78	194	474	1188	2971	7528
Alkanes	1	1	1	2	3	5	9	18	35	75	159	355	802

n	14	15	16	17	18	19	20
Alkyl radicals $A(x)$	19241	48865	124906	321198	830219	2156010	5622109
Primary alcohols	7639	19241	48865	124906	321198	830219	2156010
Secondary alcohols	7528	19181	49060	126369	326863	849650	2216862
Tertiary alcohols	4074	10443	26981	69923	182158	476141	1249237
$B(x)$	15167	38422	97925	251275	648061	1679869	4372872
Esters	22695	57564	146985	377555	974924	2529308	6589734
Ethylene derivs.	14397	36564	93650	240916	623338	1619346	4224993

$C_nH_{2n}XY$	211624	576221	1572210	4297733	11767328	32266801	88594626
$C_nH_{2n}X_2$	115469	312246	847241	2304522	6283327	17164401	46972357
Glycols	100302	273824	749316	2053247	5635266	15484532	42599485
$P(x)$	21000	53407	136639	351757	909962	2365146	6172068
$Q(x) - A(x^2)$	19142	49060	126280	326863	849439	2216862	5805749
Alkanes	1858	4347	10359	24894	60523	148284	366319

n	21	22	23	24	25
Alkyl radicals $A(x)$	14715813	38649152	101821927	269010485	712566567
Primary alcohols	5622109	14715813	38649152	101821927	269010485
Secondary alcohols	5806256	15256265	40210657	106273050	281593237
Tertiary alcohols	3287448	8677074	22962118	60915508	161962845
$B(x)$	11428365	29972078	78859809	208094977	550603722
Esters	17234114	45228343	119069228	314368027	832193902
Ethylene derivs.	11062046	29062341	76581151	202365823	536113477
$C_nH_{2n}XY$	243544919	670228623	1846283937	5090605118	14047668068
$C_nH_{2n}X_2$	128741107	353345434	970999198	2671347292	7356752678
Glycols	42599485	117312742	323373356	892139389	2463252315
$P(x)$	16166991	42488077	112004630	296080425	784688263
$Q(x) - A(x^2)$	15256265	40209419	106273050	281590180	747890675
Alkanes	910726	2278658	5731580	14490245	36797588

the form

$$\begin{array}{c} R_1 \\ | \\ H{-}C{-}OH \\ | \\ R_2 \end{array} \qquad (2.3)$$

where R_1 and R_2 denote alkyl radicals that are not hydrogen atoms, having $n-1$ carbon atoms between them. To enumerate these isomers we apply Pólya's theorem. We have two boxes (R_1 and R_2) which can be interchanged (since we are counting *structural* isomers), and in each we must put an alkyl radical (not hydrogen). Hence the figure-counting series is $A(x)-1$, and the group is S_2. Hence the numbers of ways of choosing R_1 and R_2 are given by the configuration-counting series

$$Z(S_2; A(x)-1).$$

Taking into account the extra carbon atoms in (2.3) we see that the counting series for the secondary alcohols is

$$xZ(S_2; A(x)-1) = \tfrac{1}{2}x[A^2(x) - 2A(x) + A(x^2)]. \qquad (2.4)$$

 (c) Tertiary alcohols. The tertiary alcohols have the formula

$$\begin{array}{c} R_1 \\ | \\ R_2{-}C{-}OH \\ | \\ R_3 \end{array} \qquad (2.5)$$

where R_1, R_2, R_3 are alkyl radicals (not hydrogen) having $n-1$ carbon atoms between them. Using Pólya's theorem with three boxes, permutable by S_3, we obtain the counting series

$$xZ(S_3; A(x)-1)$$

in much the same way as before. This reduces to

$$\tfrac{1}{6}x[A^3(x) - 3A^2(x) + 3A(x)A(x^2) - 3A(x^2) + 2A(x^3)] \qquad (2.6)$$

 The reader can verify that; by adding the right-hand sides of (2.2), (2.4) and (2.6) we obtain the right-hand side of (2.1) except for the constant term which, corresponding to H–OH, is not included.

 There are many other things that we can attach to an alkyl radical besides a hydroxyl group, and we see therefore that the counting series (2.1) will enumerate compounds of the form $C_nH_{2n+1}X$, i.e. monosubstituted saturated hydrocarbons, where X is a monovalent atom (other than hydrogen) or a monovalent radical (provided it can be clearly delineated from the rest of the molecule — it could not be an alkyl radical for example). If X contains carbon atoms then we need to juggle with the counting series slightly if we wish the power of x to be the *total* number of carbon atoms in the molecule. Thus if X is the carboxyl group —COOH, the coefficient A_n will enumerate the isomers of

$$C_n H_{2n+1} \cdot COOH$$

in which the total number of carbon atoms is $n + 1$. The counting series for these acids by total number of carbon atoms is therefore $xA(x)$, in which A_n is the coefficient of x^{n+1} (Cf. the enumeration of primary alcohols).

The enumeration of compounds of the form $C_n H_{2n+1} X$ is therefore a trivial variation on that of enumerating the alcohols, and we quickly turn to some less trivial deductions from the basic result (2.1).

Aldehydes and ketones. It will be convenient to consider aldehydes and ketones together, to begin with, since then the general formula is

$$\begin{matrix} R_1 \\ \\ R_2 \end{matrix} \!\!> C = O \qquad (2.7)$$

where R_1 and R_2 are alkyl radicals — possibly hydrogen atoms. We have here two interchangeable boxes in which to place a figure enumerated by $A(x)$. Hence, applying Pólya's theorem we obtain the configuration counting series $Z(S_2; A(x))$. This is similar to our treatment of secondary alcohols, and (as there) we must multiply by x to allow for the extra carbon in (2.7). Thus the counting series for aldehydes and ketones together is

$$B(x) = xZ(S_2; A(x))$$
$$= \tfrac{1}{2}x[A^2(x) + A(x^2)] \qquad (2.8)$$

The coefficients in $B(x)$ are tabulated in Table I.

The numbers of aldehydes are easily found, since their general formula $C_n H_{2n+1} \cdot CHO$ is of the form $C_n H_{2n+1} X$. The counting series is therefore $xA(x)$, the same as, for example, that of the organic acids discussed above, which is hardly surprising considering the chemical affinity between these two types of compound. Subtracting this counting series from $B(x)$ we obtain the counting series for the ketones alone. The ketones are therefore the difference between $B(x)$ and the number of primary alcohols.

Another simple enumeration is that of the hydrocarbons of the acetylene series, having the general formula.

$$R_1 - C \equiv C - R_2 \qquad (2.9)$$

where R_1 and R_2 are alkyl radicals or hydrogen atoms. The combinatorial situation is the same as for aldehydes and ketones (two interchangeable boxes etc.) but there are now two extra carbon atoms. The counting series is therefore $xB(x)$, with $B(x)$ defined by (2.8).

A different situation arises if we consider the esters with general formula

$$R_1 - C \!\!\begin{matrix} \nearrow O \\ \searrow OR_2 \end{matrix} \qquad (2.10)$$

where R_1 and R_2 are alkyl radicals, and where R_1 can be a hydrogen atom, whereas R_2 must not (since this would give an acid). If R_1 has r_1 carbon atoms and R_2 has r_2 carbon atoms, then the number of ways of choosing R_1 and R_2 in (2.10) to get a total of carbon atoms is

$$\Sigma A_{r_1} A_{r_2} \tag{2.11}$$

where the summation is for $r_1 + r_2 = n - 1$, $0 \leqslant r_1 \leqslant n - 2$ and $1 \leqslant r_2 \leqslant n - 1$. It is easily seen that (2.11) is the coefficient of x^n in

$$x(A_0 + A_1 x + A_2 x^2 + \ldots)(A_1 x + A_2 x^2 + \ldots).$$

Hence the counting series for these esters is

$$xA(x)[A(x) - 1] \tag{2.12}$$

The numbers of esters are also given in Table I.

3   DISUBSTITUTED HYDROCARBONS

The previous results were fairly immediate deductions from the basic result (2.1), using Pólya's Theorem. We now turn to the problem of enumerating the disubstituted hydrocarbons with the formula $C_n H_{2n} XY$. This can also be achieved using Pólya's theorem, but the method is less immediate and more difficult. We shall first assume that X and Y are *different* radicals or atoms.

Let us imagine that we pick up the molecule $C_n H_{2n} XY$ by the radicals X and Y, and pull them apart as far as we can. We see that we can write the structural formula as

$$X-\underset{\underset{R_1}{|}}{\overset{\overset{Q_1}{|}}{C}}-\underset{\underset{R_2}{|}}{\overset{\overset{Q_2}{|}}{C}}- \cdots -\underset{\underset{R_p}{|}}{\overset{\overset{Q_p}{|}}{C}}-Y \tag{3.1}$$

where the Q_i and R_i are alkyl radicals (or hydrogen atoms). Concentrating on those molecules with a given value of p, we have the problem of placing alkyl radicals having a total carbon content of $n - p$ atoms, into $2p$ boxes. Thus we have an enumeration problem of the Pólya type; our remaining task is to discover the appropriate permutation group, and its cycle index.

Consider any pair of 'boxes' Q_i and R_i. Since we are counting structural isomers these boxes are interchangeable, i.e. can be permuted independently of the others, and it follows that the group we require is

$$S_2 \times S_2 \times \ldots \times S_2 \quad (p \text{ factors})$$

whose cycle index is

$$\{Z(S_2)\}^p = \frac{1}{2^p} (s_1^2 + s_2)^p. \tag{3.2}$$

The number of ways of allocating alkyl radicals to these $2p$ boxes is therefore given by the coefficient of x^{n-p} in

$$\{½[A^2(x) + A(x^2)]\}^p$$

i.e. by the coefficient of x^n in

$$x^p\{½[A^2(x) + A(x^2)]\}^p = B^p(x)$$

with $B(x)$ defined as in (2.8).

To find the total number of compounds of this type we must sum for $p = 1$, 2, 3, ... We obtain

$$\sum_{p=1}^{\infty} B^p(x) = \frac{B(x)}{1 - B(x)} \tag{3.3}$$

Our next main task will be to enumerate the disubstituted hydrocarbons with formula $C_nH_{2n}X_2$, that is, the problem like the one just completed, but with X and Y the same. To make the solution of this problem a little simpler, however, we shall digress to consider first the enumeration of the alkyl derivatives of ethylene.

These are the compounds having the general formula

$$\begin{array}{cc} R_1 & R_2 \\ \diagdown \ /\ \\ C \\ \| \\ C \\ \diagup \ \diagdown \\ R_3 & R_4 \end{array} \tag{3.4}$$

where the R_i are alkyl radicals. From (3.4) we see that we have 4 boxes into which to put alkyl radicals – a straightforward Pólya problem provided we can determine the relevant permutation group.

We can interchange R_1 and R_2, or R_3 with R_4, or both of these, and have the same compound. In addition, we can turn the diagram (3.4) upside-down. If we think of R_1, R_2, R_3 and R_4 as elements of a matrix

$$\begin{pmatrix} R_1 & R_2 \\ R_3 & R_4 \end{pmatrix} \tag{3.5}$$

then we can permute each row of (3.5) by a permutation in S_2; and in addition we can permute the two rows by a permutation in S_2. The resulting permutation group is a special case of what Pólya called the 'wreath product' of two groups, a concept that we must briefly discuss.

Let **M** be an $m \times n$ matrix, and let G and H be two permutation groups of degrees m and n respectively. Now consider all permutations of the mn elements of **M** that can be obtained in the following way:

(a) For each row of **M** choose a permutation belonging to H, and permute the elements of that row by that permutation. The permutations for the various rows are chosen independently; they need not be all the same, nor need they all be different.

(b) Having permuted each row according to the permuatation chosen for it, now permute the rows among themselves by a permutation belonging to G.

The set of all permutations of the mn elements of **M** that are obtainable in this way form a group. It is called the 'wreath product' of G and H and is denoted by $G[H]$.

Pólya,[6] gave a formula for the cycle index of a wreath product in terms of those of the two groups. Suppose, for brevity, we write

$$Z(G) = g(s_1, s_2, s_3, \ldots, s_m)$$

and

$$Z(H) = h(s_1, s_2, s_3, \ldots, s_n)$$

Then the cyclic index of the wreath product $G[H]$ is obtained by replacing each occurrence of s_i in $g(s_1, s_2, s_3, \ldots, s_m)$ by $h(s_i, s_{2i}, s_{3i}, \ldots, s_{mi})$. In the original notation for cycle indices this can be written as

$$Z(G[H]) = Z(G; Z(H; s_1, s_2, s_3, \ldots, s_m), Z(H; s_2, s_4, s_6, \ldots, s_{2m}),$$
$$\ldots, Z(H; s_n, s_{2n}, s_{3n}, \ldots, s_{mn})) \qquad (3.6)$$

It will be seen that the group of permutations of the elements R_1, R_2, R_3, R_4 of the matrix (3.5) is obtained in just this way, taking G and H to be S_2. Hence it is the wreath product $S_2[S_2]$. We could easily have found the cycle index of this group by simply listing its 8 permutations, but we shall make use of (3.6) instead. This is akin to using a steam-hammer to crack a nut, but it will serve to demonstrate the mechanism of the hammer.

In $Z(G) = \frac{1}{2}(s_1^2 + s_2)$ we have to replace each occurrence of s_1 by $Z(H) = \frac{1}{2}(s_1^2 + s_2)$, and each occurrence of s_2 by $\frac{1}{2}(s_2^2 + s_4)$. We then have

$$Z(S_2[S_2]) = \frac{1}{2}\{ [\frac{1}{2}(s_1^2 + s_2)]^2 + [\frac{1}{2}(s_2^2 + s_4)] \}$$
$$= \frac{1}{8}\{ s_1^4 + 2s_1^2 + 3s_2^2 + 2s_4 \}.$$

We can now resume the enumeration that we interrupted in order to define the wreath product. We have the appropriate cycle index, and since the figure-counting series is $A(x)$ we derive the configuration-counting series

$$\frac{1}{8}\{ A^4(x) + 2A^2(x) A(x^2) + 3A^2(x^2) + 2A(x^4) \}. \qquad (3.8)$$

This is the first example where we have used the wreath product, but we shall need it again for the enumeration of the compounds $C_n H_{2n} X_2$ to which we now return.

Let us first suppose that the number p of carbon atoms between the two X's is an even number, $p = 2k$ say. Then we can write the general formula as

$$
X-\overset{\overset{\displaystyle Q_1}{|}}{\underset{\underset{\displaystyle R_1}{|}}{C}}-\overset{\overset{\displaystyle Q_2}{|}}{\underset{\underset{\displaystyle R_2}{|}}{C}}-\cdots-\overset{\overset{\displaystyle Q_{2k}}{|}}{\underset{\underset{\displaystyle R_{2k}}{|}}{C}}-X
\tag{3.9}
$$

much as before (3.1). This time, however, the group will be different, because we can now reverse the diagram (3.9) end-for-end and have an equivalent scheme for the molecule. The group of permutations for the alkyl radicals in the left-hand half of the molecules (i.e. Q_i and R_i for $i = 1, 2, \ldots, k$) is

$$
S_2 \times S_2 \times \ldots \times S_2 \quad (k \text{ factors})
$$

[Compare (3.2)]. The same is true for the right-hand half. Since the two sides can be permuted, the group for the whole diagram is the wreath product

$$
S_2 [S_2 \times S_2 \times \ldots \times S_2]
\tag{3.10}
$$

To see the analogy with the ethylene example, and with the definition of wreath product, think of the $2k$ boxes on the left as forming the first row of a matrix while those on the right form the second row. Then each row can be permuted by $S_2 \times S_2 \times \ldots \times S_2$; while the rows can be permuted bodily by S_2.

The cycle index of (3.10) is

$$
\tfrac{1}{2}\{ [\tfrac{1}{2}(s_1^2 + s_2)]^{2k} + [\tfrac{1}{2}(s_2^2 + s_4)]^{k}\}
$$

and on substituting the figure-counting series $A(x)$, and multiplying by x^{2k} to accommodate the $2k$ carbon atoms between the X's, we obtain the series

$$
\frac{1}{2}\left\{ \frac{x}{2}(A^2(x) + A(x^2)]^{2k} + \left[\frac{x^2}{2}(A^2(x^2) + A(x^4)\right]^{k}\right\}
$$

$$
= \frac{1}{2}\{B^{2k}(x) + B^k(x^2)\}.
\tag{3.11}
$$

When p is odd, say $p = 2k + 1$, then the above remarks apply to all the 'boxes' except the two attached to the middle carbon atom of the chain. These latter can be permuted among themselves independently of what happens to the others. Hence our group is now

$$
S_2 \times S_2 [S_2 \times S_2 \times \ldots \times S_2]
$$

which eventually gives us the counting series

$$
\tfrac{1}{2}B(x)\{B^{2k}(x) + B^k(x^2)\},
\tag{3.12}
$$

as the analogue of (3.11).

We now sum (3.11) and (3.12) over all values of k. Since we need $k = 0$ when p is odd (to include CH_2X_2) we shall include it also for p even. This will introduce a constant term, corresponding to the compound $X - X$! This inclusion does no harm, and simplifies the result. We have

$$\frac{1}{2} \sum_{k=0}^{\infty} \{B^{2^k}(x) + B^k(x^2)\} + \frac{1}{2} \sum_{k=0}^{\infty} B(x)\{B^{2^k}(x) + B^k(x^2)\}$$

which reduces to

$$\frac{1}{2} \cdot \frac{1}{1 - B(x)} + \frac{1}{2} \cdot \frac{1 + B(x)}{1 - B(x^2)} \tag{3.13}$$

By substituting specific radicals or atoms (Cl, OH, COOH etc.) for X and Y in the preceding enumerations, we can obtain the numbers of compounds of various specific types. One important family of compounds that can *almost* be enumerated in this way is that of the glycols. These are of the form $C_2H_{2n}(OH)_2$, i.e. they are of the form $C_2H_{2n}X_2$ just considered, except that the two hydroxyl groups must not be attached to the same carbon atom.

Hence to enumerate the glycols we must subtract from the numbers given by (3.13) the numbers of 'invalid' glycols. These will have the general formula

$$\begin{array}{ccc} R_1 & & OH \\ & \diagdown \diagup & \\ & C & \\ & \diagup \diagdown & \\ R_2 & & OH \end{array}$$

and are therefore equinumerous with the compounds

$$\begin{array}{c} R_1 \\ \diagdown \\ \diagup\quad C = 0 \\ R_2 \end{array}$$

(aldehydes and ketones) which have already been enumerated (2.8). Thus the numbers of glycols can be found. They are given in Table I, along with the numbers of the two kinds of disubstituted hydrocarbons that we have just considered. The numbers of ethylene derivatives, first found by Blair and Henze,[2] are also included.

4   UNSUBSTITUTED ALKANES

It might be thought that the enumeration of the alkanes, C_nH_{2n+2}, would present less of a problem than those considered hitherto, in so far as the alkanes are, so to speak, the compounds from which the others that we have considered can be derived; but this is not the case. The reason is that, for example, the X in $C_nH_{2n+1}X$ provides a convenient distinguishable point at which to attack the problem, while for the alkanes there is no such distinguished atom to pick up the problem by. In graph-theoretical terms the previous enumerations have been of trees that were *rooted* in some way; what we shall now consider is a problem in the enumeration of unrooted trees.

The early researchers in this field (Cayley, Blair and Henze, and even Pólya) had a great deal of difficulty with this problem, and were forced to solve it by complex methods that made use of the centre and bicentre of the tree structure given by the structural formula. More recent developments have provided a powerful tool by which the enumeration of unrooted trees can be performed with comparative ease. We shall examine this tool shortly, but first we need some preliminary results.

Some of the advantages of having a carbon atom at which a substitution has taken place can be regained if, without substituting for any of the hydrogen atoms in an alkane molecule, we contrive to distinguish one carbon atom from all the rest. This corresponds to a feasible chemical procedure – the labelling of a molecule by replacing one carbon atom by a radio-isotope. We shall call such a molecule an 'alkane with a labelled carbon atom', and we shall denote the labelled carbon atom by C^*.

We shall now enumerate these labelled alkanes, whose general formula is

$$R_4 - \overset{\displaystyle R_1}{\underset{\displaystyle R_3}{\overset{|}{\underset{|}{C^*}}}} - R_2$$

the R_i's being alkyl radicals. Since we are considering structural isomers only, the four 'boxes' into which these radicals go can be permuted in any way, i.e. by any permutation of S_4.

We therefore apply Pólya's theorem with $A(x)$ as the figure-counting series, and S_4 as the group. Since the cycle index of the group is

$$\frac{1}{24}(s_1^4 + 6s_1^2 s_2 + 3s_2^2 + 8s_1 s_3 + 6s_4)$$

and since we must multiply by x to accommodate the extra (labelled) carbon atom, our configuration counting series takes the form

$$P(x) = \sum_{n=1}^{\infty} P_n x^n$$

$$= xZ(S_4; A(x))$$

$$= \frac{1}{24} x\{A^4(x) + 6A^2(x)A(x^2) + 3A^2(x^2) + 8A(x)A(x^3) + 6A(x^4)\}$$

$$(4.1)$$

where P_n is the number of these compounds with n carbon atoms in all.

The other preliminary result that we require is the enumeration of alkanes in which one carbon-to-carbon valency bond has been labelled, that is, distin-

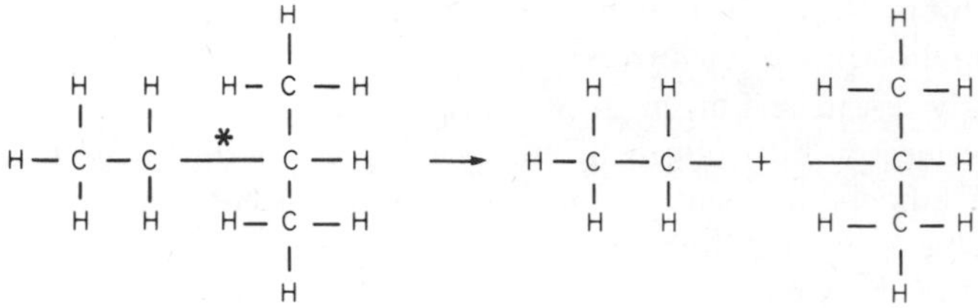

```
                    H                                         H
                    |                                         |
   H    H    H — C — H              H    H    H — C — H
   |    |    |                      |    |    |
H — C — C ——*—— C — H    ⟶    H — C — C —  +  —— C — H
   |    |    |                      |    |    |
   H    H    H — C — H              H    H    H — C — H
                    |                                         |
                    H                                         H
```

FIG. 1

guished from the other bonds. This labelling does not correspond in any direct way to a feasible chemical procedure, but is only a means to an end. As with the labelling of a carbon atom, this labelling of a bond gives us a point to start from. If we 'break' the labelled bond in two, we get two alkyl radicals as shown in Fig. 1 (in which the labelled bond is indicated by the asterisk).

Conversely, by joining two alkyl radicals, in an obvious way, we obtain an alkane in which one carbon-to-carbon bond is distinguished from the rest by virtue of being the common 'free bond' of the two radicals.

Hence the enumeration of these 'bond-labelled' molecules is a Pólya type problem. There are two boxes, permutable by S_2, into each of which we must place an alkyl radical, but not just a hydrogen atom (since the labelled bond must be a carbon-to-carbon bond). Pólya's theorem then gives the configuration counting series

$$Q(x) = \sum_{n=1}^{\infty} Q_n x^n$$

$$= Z(S_2; A(x) - 1)$$

$$= \frac{1}{2} \{(A(x) - 1)^2 + A(x^2) - 1\} \tag{4.2}$$

We now look at the comparatively recent advance (alluded to above) which enables us to derive the counting series for unsubstituted (and unlabelled) alkanes from the two series $P(x)$ and $Q(x)$ just given (see the paper by Harary and Norman[4]). We shall not give the proof here, but discuss the theorem in the context of trees in general.

Let T be any tree. There will be certain one-to-one mappings (if only the identity) of the vertex set of T onto itself which leave the tree invariant, i.e. adjacent vertices remain adjacent, etc. These mappings (called automorphisms) form a group, the automorphism group of T. If there is an automorphism which maps a vertex u onto a vertex v we say that u and v are equivalent. This relation between vertices is an equivalence relation, as its name suggests, and divides the vertex set of T into equivalence classes. We shall let p^* denote the number of equivalence classes.

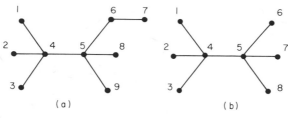

(a)   (b)

FIG. 2

Clearly any automorphism induces, in an obvious way, a mapping of the set of edges of T onto itself. If an edge e maps onto an edge f under some automorphism of T we say e and f are equivalent, and in this way we define equivalence classes of edges. Let q^* be the number of equivalence classes.

The result that we shall use, and which is given by Harary and Norman[4] is that

$$p^* - q^* + s = 1 \qquad\qquad (4.3)$$

Here $s = 1$ if the tree has a symmetric edge (i.e. an edge uv which maps onto itself, in the sense that u and v map onto v and u respectively, under some automorphism of T), and $s = 0$ otherwise. A symmetric edge must clearly have the same 'half-tree' at its two ends, and hence there can be at most one such edge in any tree. The following two examples should make this clear.

In the tree of Fig. 2(a) vertices 1, 2, 3 are equivalent, and so are 8 and 9. All other vertices are in classes by themselves. Hence $p^* = 6$. The edges $(1, 4), (2, 4)$ and $(3, 4)$ are equivalent; so are $(5, 8)$ and $(5, 9)$. Hence $q^* = 5$. There is no symmetric edge, so $s = 0$ and (4.3) is satisfied. In Fig. 2(b) vertices 1, 2, 3, 6, 7, 8 are all equivalent; so are 4 and 5. Hence $p^* = 2$. As for the edges, edge $(4, 5)$ is a symmetric edge and is in a class by itself. All other edges make up a second equivalence class. Hence $q^* = 2$. Since now $s = 1$, equation (4.3) is again satisfied.

This result is true for the alkanes, which are trees in the usual graph-theoretical sense if we ignore the hydrogen atoms. The vertices are then the carbon atoms and the edges are the carbon-to-carbon bonds. We do not alter the problem by ignoring the hydrogen atoms since they can be uniquely restored when the rest of the molecule is known. Suppose we take two replicas of an alkane molecule and label a carbon atom in each. If the two carbon atoms so labelled are equivalent vertices then the two labelled molecules are indistinguishable; if they are in different equivalence classes then we shall have two differently labelled molecules. Hence the number of *different* labelled molecules that can be obtained by labelling one carbon atom in an unlabelled alkane is p^*, the number of vertex-equivalence classes. In an exactly similar way we show that by labelling a carbon-to-carbon bond in an alkane we can get exactly q^* different alkanes with a labelled bond.

Now let us sum equation (4.3) over all the alkanes having a given number n of carbon atoms. We have

$$\Sigma p^* - \Sigma q^* + \Sigma s = \Sigma 1. \tag{4.5}$$

Now, by what has just been said, Σp^* will be the total number of alkanes with a labelled carbon atom; Σq^* will be the number with a labelled bond; Σs will be the number of alkanes having a symmetric bond; and $\Sigma 1$ will be just the number of alkanes — the number we are looking for. Thus Σp^* is the number P_n given by (4.1) and Σq^* is the number Q_n given by (4.2). If we can determine the number Σs of alkanes with a symmetric bond we shall know everything in (4.5) except the number we wish to find.

A molecule with a symmetric bond must have an even number (say $2k$) of carbon atoms, and, as already remarked, the 'half-trees' at the end of the symmetric bond must be the same. Hence, if we split the symmetric bond (as in the determination of Q_n) we get two alkyl radicals as before, but now they are identical. Thus, having chosen one of the A_k alkyl radicals for one half, we have determined the whole molecule. Hence the number of alkanes having a symmetric edge is $A_{n/2}$ (which we interpret to be 0 if n is odd). Denoting the required number of alkanes by C_n we thus rewrite (4.5) as

$$P_n - Q_n + A_{n/2} = C_n. \tag{4.6}$$

If we now multiply (4.6) by x^n, sum from $n = 1$ upwards, and rearrange, we have

$$C(x) = \sum_{n=1}^{\infty} C_n x^n$$
$$= P(x) - Q(x) + A(x^2) - 1 \tag{4.7}$$

Since $P(x)$, $Q(x)$ and $A(x)$ are known we can calculate the series $C(x)$ and thus determine the number of alkanes. They are given in the last row of Table I. If we write $A_1(x)$ for $A(x) - 1$, then (4.7) can be put in the more elegant form

$$C(x) = P(x) - \tfrac{1}{2}\{A_1^2(x) - A_1(x^2)\}. \tag{4.8}$$

5   STEREO-ISOMERS

So far we have confined ourselves to problems of counting the *structural* isomers of the compounds that we have considered. This is chemically rather unrealistic, the enumeration of *stereo-isomers* being rather more to the point. However, both problems are of interest, and once one has been solved, the solution of the other follows readily. It therefore makes little difference which we tackle first.

We start by considering the monosubstituted alkanes (or alkyl radicals) as before, and we observe that the simplest such compound for which there are

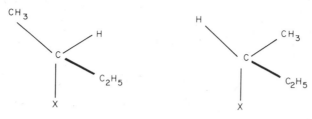

FIG. 3

stereo-isomers has the formula C_4H_9X, and can exist in the two enantiomorphic forms given in Fig. 3.

The X, H, CH_3 and C_2H_5 are at the vertices of a regular tetrahedron, and in both drawings in Fig. 3 the ethyl radical is to be thought of as above the plane of the paper. At once we see the difference that stereo-isomerism produces. As before, our general formula is

$$R_1-\underset{\underset{X}{|}}{\overset{\overset{R_2}{|}}{C}}-R_3$$

but instead of our being able to allow *any* permutation of the three 'boxes' into which R_1, R_2 and R_3 are to be put, we can allow only those corresponding to a cyclic permutation.

Thus the problem is basically the same as before; only the group is different. We shall use lower-case letters in the stereo-isomer problems to correspond to the upper-case letters that we used in the enumeration of structural isomers, so that, for the monosubstituted alkanes, we shall write the generating function as

$$a(x) = a_0 + a_1 x + a_2 x^2 + \dots . \tag{5.1}$$

The equation analogous to (2.1) is then

$$a(x) = 1 + xZ(C_3 ; a(x))$$
$$= 1 + \tfrac{1}{3}x\{a^3(x) + 2a(x^3)\} \tag{5.2}$$

since $Z(S_3) = \tfrac{1}{3}(s_1^3 + 2s_3)$. This enumerates for example, the stereo-isomers of the alcohols of the methanol series. Primary alcohols are enumerated exactly as before. Secondary alcohols have the form (2.3), but now the radicals R_1 and R_2 are not interchangeable. Tertiary alcohols have the form (2.5), but R_1, R_2 and R_3 may only be permuted cyclically. From these remarks we deduce the following counting series:

Primary alcohols	$xa(x)$
Secondary alcohols	$\tfrac{1}{2}x\,[a(x) - 1]^2$
Tertiary alcohols	$\tfrac{1}{3}x\{\,[a(x) - 1]^3 + 2\,[a(x^3) - 1]\,\}.$

We can also readily verify the following results:

Ketones and aldehydes	$b(x) = \tfrac{1}{2}x\,[a^2(x) + a(x^2)]$
Aldehydes alone Alkanoic acids	$xa(x)$
Acetylene derivatives	$xb(x)$
Esters	$a(x)\{a(x) - 1\}$

The enumeration of the stereo-isomers of the ethylene derivatives is somewhat different. The basic formula is

$$R_1 \diagdown \quad \diagup R_2$$
$$C$$
$$\|$$
$$C$$
$$R_2 \diagup \quad \diagdown R_4$$

as before, but now we cannot interchange R_1 and R_2 unless we also interchange R_3 and R_4. Bearing this in mind we see that the appropriate group now contains only the following permutations (with an obvious notation)

$$\begin{pmatrix} 1 & 2 & 3 & 4 \\ 1 & 2 & 3 & 4 \end{pmatrix} \begin{pmatrix} 1 & 2 & 3 & 4 \\ 2 & 1 & 4 & 3 \end{pmatrix} \begin{pmatrix} 1 & 2 & 3 & 4 \\ 3 & 4 & 1 & 2 \end{pmatrix} \begin{pmatrix} 1 & 2 & 3 & 4 \\ 4 & 3 & 2 & 1 \end{pmatrix}$$

and its cycle index is therefore $\tfrac{1}{4}(s_1^4 + 3s_2^2)$. Hence the counting series for ethylene derivatives is

$$\tfrac{1}{4}x^2\,[a^4(x) + 3a^2(x^2)]. \tag{5.3}$$

Turning now to the disubstituted alkanes we see that when $X \neq Y$ we can draw the molecule in a standard way, so as to make X and Y and the chain of p carbon atoms joining them lie in the plane of the paper, as in Fig. 4.

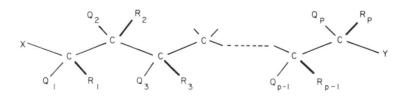

FIG. 4

(The bonds drawn with a heavy line in Fig. 4 are those which project above the plane of the paper.) We see that no permutation of the alkyl radicals Q_i and R_i is possible. Hence the group for this problem is the identity group E_{2p} of degree $2p$, whose cycle index is s_1^{2p}. The required counting series is thus $x^p a^{2p}(x)$ for a given value of p, or

$$\sum_{p=0}^{\infty} x^p a^{2p}(x) = \frac{1}{1 - xa^2(x)}$$

for all compounds of this type.

When X and Y are the same, the molecule can be turned end-for-end and the 'boxes' change places in pairs. By the definition of the wreath product the group in question is now $S_2[E_p]$. (Note: this is irrespective of the parity of p, since if p is odd the 'boxes' attached to the central carbon atom in the chain change places along with the others.) The cycle index of this group is $\frac{1}{2}(s_1^{2p} + s_2^p)$, and hence the counting series for all compounds of this type is

$$\sum_{p=0}^{\infty} \frac{x^p}{2} \{a^{2p}(x) + a^p(x^2)\} = \frac{1}{2}\left\{\frac{1}{1 - xa^2(x)} + \frac{1}{1 - xa(x^2)}\right\}$$

The numbers of the compounds considered so far in this section are tabulated in Table II.

To enumerate the alkanes, having regard to stereo-isomers, we use the same method as before, and first enumerate alkanes with a labelled carbon atom. It is readily verified that the 'boxes' in the diagram

$$R_3 - C* \underset{\underset{R_4}{\vert}}{\overset{\nearrow R_2}{\searrow R_1}}$$

can be permuted by any permutation of the alternating group A_4, whose cycle index is

$$\frac{1}{12}(s_1^4 + 3s_2^2 + 8s_1 s_3)$$

Applying Pólya's theorem, with $a(x)$ as the figure-counting series, we obtain

$$p(x) = \sum_{n=1}^{\infty} \frac{1}{12} x\{a^4(x) + 3a^2(x^2) + 8a(x)a(x^3)\} \tag{5.5}$$

as the stereo-analogue of (4.1).

The counting series for alkanes in which a carbon-carbon bond has been labelled is

$$q(x) = Z(S_2; a(x) - 1)$$
$$= \frac{1}{2}\{[a(x) - 1]^2 + a(x^2) - 1\}, \tag{5.6}$$

the analogue of (4.2).

We now use the theorem, described and used in Section 4, to the effect that $p^* - q^* + s = 1$, and obtain the result

$$c(x) = p(x) - q(x) + a(x^2) - 1 \tag{5.7}$$

TABLE II

The numbers of stereo-isomers of some acyclic compounds

n	1	2	3	4	5	6	7	8	9	10	11	12
Alkyl radicals $A(x)$	1	1	2	5	11	28	74	199	551	1553	4436	12832
Primary alcohols	1	1	1	2	5	11	28	74	199	551	1553	4436
Secondary alcohols			1	2	5	14	36	98	273	768	2197	6360
Tertiary alcohols				1	1	3	10	27	79	234	686	2036
$b(x)$	1	1	2	3	8	18	47	123	338	935	2657	7616
Esters	1	1	2	4	10	25	64	172	472	1319	3750	10796
Ethylene derivs.	1	1	1	4	6	18	42	118	314	895	2521	7307
$C_n H_m$ XY	1	3	8	23	69	208	636	1963	6099	19059	59836	188576
$C_n H_m$ X_2	1	2	5	13	37	108	325	993	3070	9564	29979	94392
$p(x)$	1	1	2	10	22	60	158	439	1229	3525	10178	29802
Alkanes	1	1	1	2	3	5	11	24	55	136	345	900

n	13	14	15	16	17	18	19
Alkyl radicals $A(x)$	37496	110500	328092	980491	2946889	8901891	27012286
Primary alcohols	12832	37496	110500	328092	980491	2946889	8901891
Secondary alcohols	18584	54780	162672	486154	1461197	4413988	13393855
Tertiary alcohols	6080	18224	54920	166245	505201	1541014	4716540
$b(x)$	22138	64886	191873	571169	1711189	5153883	15599094
Esters	31416	92276	273172	814246	2441688	7360877	22295746

		20	21	22	23	24	25
Ethylene derivs.	21238	62566	185310	553288	1660490	5011299	15190665
C_nH_mXY	596252	1890548	6008908	19139155	61074583	195217253	624913284
$C_nH_mX_2$	298311	945592	3005021	9570559	30539044	97611676	312462096
$p(x)$	87862	261204	781198	2350249	7105081	21577415	65787902
Alkanes	2412	6563	18127	50699	143255	408429	1173770

n	20	21	22	23	24	25
Alkyl radicals $A(x)$	82300275	251670563	772160922	2376294040	7333282754	22688455980
Primary alcohols	27012286	82300275	251670563	772160922	2376294040	7333282754
Secondary alcohols	40807290	124783669	382842018	1178140280	3635626680	11247841224
Tertiary alcohols	14480699	44586619	137648341	425992838	1321362034	4107332002
$b(x)$	47415931	144692886	443091572	1361233280	4194107380	12957209782
Esters	67819576	207083944	634512581	1950301202	6011920720	18581123978
Ethylene derivs.	46244031	141296042	433204573	1332261200	4108833222	12704949506
C_nH_mXY	2003090071	6428430129	20653101216	66420162952	213802390264	688796847976
$C_nH_mX_2$	1001554565	3214232129	10326580526	33210135104	106901289420	344398593149
$p(x)$	201313311	618040002	1903102730	5876174472	18189503139	56435742554
Alkanes	3396844	9892302	28972080	85289390	252260276	749329719

where $c(x) = \sum\limits_{n=1}^{\infty} c_n x^n$, and c_n is the number of stereo-isomers with the formula $C_n H_{2n+2}$. This is the analogue of (4.7). The values of c_n are tabulated in the last row of Table II.

6   GENERAL ACYCLIC HYDROCARBONS–STRUCTURAL ISOMERS

The results derived so far in this chapter have all been results that have been known for some time; though, as already remarked, the methods used to obtain them here were not usually those by which they were first derived. We now apply the methods and techniques of this chapter to a new and more general enumeration – one that will contain many of the previous results as special cases. We shall study the enumeration of acyclic hydrocarbons having a given number of carbon atoms and given numbers of double and triple bonds, an enumeration that has been briefly described by myself.[7] As before, we start by looking at the enumeration of monosubstituted hydrocarbons or, what amounts to the same thing, monovalent hydrocarbon radicals. This problem, though used here as a means to an end, is of some interest in its own right.

Let $G_{n,d,t}$ be the number of structural isomers of the monovalent acyclic radicals having n carbon atoms, d double bonds and t triple bonds. The formula of such a radical is $C_n H_{2n+1-2d-4t}$. We shall obtain a counting series of the form

$$G(x, y, z) = \sum_{n=0}^{\infty} \sum_{d=0}^{\infty} \sum_{t=0}^{\infty} G_{n,d,t} x^n y^d z^t. \tag{6.1}$$

Now such a radical, occurring as part of a molecule, will be connected to the rest of the molecule by a single bond. We shall need to consider also the number of divalent radicals connected to the rest of the molecule by a double bond – let the number having parameters n, d and t defined as above be $H_{n,d,t}$ – and the numbers of trivalent radicals connected to the rest of the molecule by a triple bond – let the corresponding number be $I_{n,d,t}$. Diagrammatically they are as shown in Fig. 5.

Thus $G_{n,d,t}$ is the number of radicals of the form shown in Fig. 5(a); $H_{n,d,t}$ is the number of the form shown in Fig. 5(b), and $I_{n,d,t}$ is the number of the form shown in Fig. 5(c). We shall call these three types of radicals 'G-radicals', 'H-radicals', and 'I-radicals' respectively. The shaded 'balloons' here represent the rest of the radical. Note that the double and triple bonds of attachment to the rest of the molecule are included in the count for d and t.

In conformity with (6.1) we define

$$H(x,y,z) = \sum_{n=0}^{\infty} \sum_{d=0}^{\infty} \sum_{t=0}^{\infty} H_{n,d,t} x^n y^d z^t \tag{6.2}$$

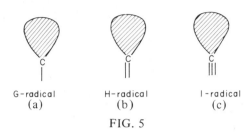

G-radical   H-radical   I-radical
(a)   (b)   (c)

FIG. 5

$$I(x,y,z) = \sum_{n=0}^{\infty} \sum_{d=0}^{\infty} \sum_{t=0}^{\infty} I_{n,d,t} x^n y^d z^t \tag{6.3}$$

When examined in more detail, the radicals enumerated by $G(x, y, z)$ are seen to be of the three types shown in Fig. 6.

In Fig. 6(a) we have three boxes, permutable among themselves, into each of which we must put a G-radical. Since a box can contain a hydrogen atom alone we shall regard a hydrogen atom as a G-radical. The numbers of ways of forming G-radicals of this type are therefore enumerated by the series

$$xZ(S_3; G(x,y,z))$$

by Pólya's theorem. Note that there is a multiplied x since an extra carbon atom is present over and above the carbon content of the radicals in the boxes.

In Fig. 6(b) we have two dissimilar boxes. In one we must put a G-radical while in the other we must put an H-radical. Having regard to the extra carbon atom we see that the G-radicals of this type are enumerated by

$$xG(x,y,z)H(x,y,z).$$

In Fig. 6(c) we have only one box, in which to place an I-radical. The corresponding counting series here is therefore just $xI(x,y,z)$.

Putting these results together, and adding a constant term ($= 1$) to represent the G-radical consisting of a single hydrogen atom (which would not otherwise be included) we obtain the following

$$G(x,y,z) = 1 + x\{Z(S_3; G(x,y,z)) + G(x,y,z)H(x,y,z) + I(x,y,z)\} \tag{6.4}$$

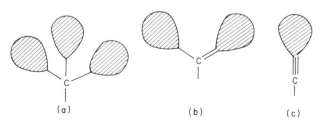

(a)   (b)   (c)

FIG. 6

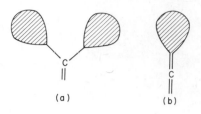

(a)   (b)

FIG. 7

On closer inspection the H-radicals are seen to be of the two types shown in Fig. 7.

In Fig. 7(a) we have two interchangeable boxes into which G-radicals must be put. This gives us the counting series

$$xyZ(S_2 ; G(x,y,z))$$

the factor xy being present because of the extra carbon and extra double bond.

In Fig. 7(b) we have a single box into which we must put an H-radical. This gives us $xyH(x,y,z)$. Taking account of the extra double bond and the extra carbon atom introduced we derive the following equation

$$H(x,y,z) = xyZ(S_2 ; G(x,y,z)) + xyH(x,y,z) \qquad (6.5)$$

whence

$$H(x,y,z) = \frac{xy}{1 - xy}\, Z(S_2 ; G(x,y,z)). \qquad (6.6)$$

The I-radicals are of just one type, shown in Fig. 8 and they contribute the counting series $xzG(x,y,z)$. Hence

$$I(x,y,z) = xzG(x,y,z). \qquad (6.7)$$

From (6.4), (6.6) and (6.7) we derive

$$G(x,y,z) = 1 + x\,\{Z(S_3 ; G(x,y,z))$$

$$+ \frac{xy}{1 - xy}\, G(x,y,z) \cdot Z(S_2 ; G(x,y,z)) + xzG(x,y,z)\} \qquad (6.8)$$

FIG. 8

or

$$G(x,y,z) = 1 + x \left\{ \frac{1}{6} \left[G^3(x,y,z) + 3G(x,y,z)G(x^2,y^2,z^2) + 2G(x^3,y^3,z^3) \right] \right.$$

$$\left. + \frac{1}{2} \frac{xy}{1-xy} \left[G^3(x,y,z) + G(x,y,z)G(x^2,y^2,z^2) \right] + xzG(x,y,z) \right\}$$

$$(6.9)$$

Given $G(x,y,z)$ up to the terms in x^n we can evaluate the right-hand side of (6.9), and hence $G(x,y,z)$, up to the terms in x^{n+1}. Hence $G(x,y,z)$ can be calculated to whatever extent we wish. The calculation is extremely tedious by hand, but has been programmed on a computer, and the results given in Table III were obtained.

TABLE III

Monovalent acyclic radicals (d = double bonds and t = triple bonds)

$n = 2$	t 0	1	$n = 3$	t 0	1	$n = 4$	t 0	1	2	$n = 5$	t 0	1	2
0	1	1	0	2	2	0	4	4	1	0	8	10	4
d 1	1	1	d 1	3		1	8	3		1	21	14	
			2	1		d 2	5			d 2	20	3	
						3	1			3	7		
										4	1		

$n = 6$	t 0	1	2	3	$n = 7$	t 0	1	2	3	$n = 8$	t 0	1	2	3	4
0	17	25	12	1	0	39	64	38	7	0	89	166	115	29	1
1	56	50	7		1	149	166	45		1	398	531	206	13	
2	69	25			2	228	134	7		2	725	587	84		
d 3	37	3			d 3	165	36			d 3	664	261	7		
4	9				4	60	3			4	326	47			
5	1				5	11				5	88	3			
					6	1				6	13				
										7	1				

$n = 9$	t 0	1	2	3	4	$n = 10$	t 0	1	2	3	4	5
0	211	437	348	114	11	0	507	1157	1040	417	62	1
1	1068	1656	829	115		1	2876	5076	3103	661	22	
2	2261	2325	577	13		2	6932	8639	3066	222		
3	2505	1470	123			3	9032	7121	1155	13		
d 4	1570	433	7			d 4	6909	2998	162			
5	570	58				5	3204	648	7			
6	122	3				6	915	69				
7	15					7	161	3				
8	1					8	17					
						9	1					

As a check on this result we see that if we put $y = z = 0$, thus eliminating any radicals with double or triple bonds, we obtain

$$G(x,0,0) = 1 + \tfrac{1}{6}x[G^3(x,0,0) + 3G(x,0,0)G(x^2,0,0) + 2G(x^3,0,0)]$$

(6.10)

which is simply equation (2.1), thinly disguised.

By putting $y = z = 1$ we obtain a recursion formula for the generating function $T(x) = G(x,1,1)$ for all hydrocarbon radicals, irrespective of how many multiple bonds they have. Other special results can be obtained in similar ways, but we must hurry on to consider the problem of enumerating the *unsubstituted* hydrocarbons.

As before, we now consider compounds with a labelled carbon atom. These are of 4 types, as shown in Fig. 9. It is easily verified that the type I compounds are enumerated by

$$xZ(S_4; G(x,y,z));$$

(6.11)

the type II compounds by

$$xZ(S_2; G(x,y,z)) \cdot H(x,y,z);$$

(6.12)

the type III compounds by

$$xZ(S_2; H(x,y,z));$$

(6.13)

and the type IV compounds by

$$xG(x,y,z) \cdot I(x,y,z).$$

(6.14)

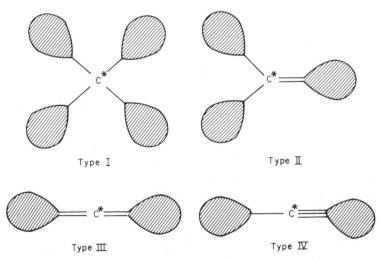

Type I   Type II

Type III   Type IV

FIG. 9

(a) (b)

FIG. 10

All these results follow by application of Pólya's theorem. On adding these four series (6.11–14) we get the counting series for compounds with a labelled carbon atom; we shall denote this series by $K(x,y,z)$.

We now turn to compounds with a labelled carbon–carbon bond. If this bond is single, then our compound is as shown in Fig. 10(a) and we have two interchangeable boxes into which to put a G-radical (other than just hydrogen). The result is

$$Z(S_2 ; G(x,y,z) - 1). \tag{6.15}$$

However, following our previous method, we shall eventually have to subtract from this the numbers of compounds that are symmetrical about the labelled bond, and now is as good a time as any to do this. These symmetrical compounds are enumerated by $G(x^2,y^2,z^2) - 1$. Subtracting this from (6.15) and, for convenience, writing $G_1(x,y,z)$ in place of $G(x,y,z) - 1$ we obtain

$$Z(S_2 ; G_1(x,y,z)) - G_1(x^2,y^2,z^2)$$
$$= \tfrac{1}{2}[G_1^2(x,y,z) - G_1(x^2,y^2,z^2)] \tag{6.16}$$

This corresponds to the portion $\tfrac{1}{2}[A_1^2(x) - A_1(x^2)]$ of (4.8).

If the labelled bond is a double bond, as in Fig. 10(b) we proceed similarly but note that in joining two H-radicals in this way we lose a double bond from the total of double bonds that the two radicals had between them; for two double bonds have become one. Hence the analogue of (6.16) is

$$\frac{1}{2y} \{H^2(x,y,z) - H(x^2,y^2,z^2)\} \tag{6.17}$$

where the factor $1/y$ reduces the double bond count by 1 for each term.

Those compounds for which the labelled bond is a triple bond are treated similarly to give

$$\frac{1}{2z} \{I^2(x,y,z) - I(x^2,y^2,z^2)\} \tag{6.18}$$

We now apply the same theorem as before, and we find that the unsubstituted acyclic hydrocarbons are enumerated according to carbon content and numbers of double and triple bonds by the counting series $L(x,y,z)$, where

$$L(x,y,z) = K(x,y,z) - \tfrac{1}{2}\{G_1^2(x,y,z) - G_1(x^2,y^2,z^2)\}$$

$$- \frac{1}{2y}\{H^2(x,y,z) - H(x^2,y^2,x^2)\} \qquad (6.19)$$

$$- \frac{1}{2z}\{I^2(x,y,z) - I(x^2,y^2,z^2)\}$$

and everything on the right-hand side of (6.19) is known.

All the coefficients $L_{n,d,t}$ in $L(x,y,z)$ have been computed for $n \leqslant 10$ and are given in Table IV.

TABLE IV

Acyclic hydrocarbons (d = double bonds, t = triple bonds)

d	$n=2$, $t=0$	1	$n=3$, $t=0$	1	$n=4$, $t=0$	1	2	$n=5$, $t=0$	1	2
0	1	1	1	1	2	2	1	3	3	2
1	1		1	1	3	1		5	4	
2			1		2			6	1	
3					1			2		
4								1		

d	$n=6$, $t=0$	1	2	3	$n=7$, $t=0$	1	2	3	$n=8$, $t=0$	1	2	3	4
0	5	7	5	1	9	14	11	3	18	32	28	10	1
1	13	12	3		27	34	12		66	95	48	4	
2	16	7			44	29	3		120	110	22		
2'					(44)	(29)	(3)						
3	10	1			32	9			115	53	3		
4	3				15	1			62	12			
5	1				3				21	1			
6					1				4				
7									1				

d	$n=9$, $t=0$	1	2	3	4	$n=10$, $t=0$	1	2	3	4	5
0	35	72	69	28	5	75	171	179	88	20	1
1	153	262	157	29		377	718	518	138	8	
2	328	376	120	4		901	1245	537	53		
3	367	354	29			1196	1074	226	4		
4	253	85	3			964	498	39			
5	100	14				491	124	3			
6	28	1				160	17				
7	4					36	1				
8	1					5					
9						1					

7   ENUMERATION OF UNLABELLED STEREO-HYDROCARBONS

As a final exercise we shall solve the problem for stereo-isomers analogous to that for structural isomers that we have just solved. The reasoning is, for the most part, much the same as before, so that not much need be said about it. There are however a few differences which will be pointed out as occasion arises.

When we examine the monovalent radicals, we see that they are of the same three types as shown in Fig. 6. But in Fig. 6(a) the three boxes may only be permuted cyclically, so that the group of permutations is C_3, not S_3. There are not other differences at this stage, and we thus easily obtain the stereo-analogue of (6.4). It is

$$g(x,y,z) = 1 + x\{Z(C_3 ; g(x,y,z))$$
$$+ g(x,y,z)h(x,y,z) + i(x,y,z)\} \qquad (7.1)$$

where the lower case letters g, h and i correspond, in the case of stereo-isomers, to G, H and I for structural isomers.

The H-radicals are again of the two types shown in Fig. 7, but now the two boxes in Fig. 7(a) cannot be permuted. Thus we have the group E_2 instead of S_2, and the stereo-analogue of (6.5) becomes

$$h(x,y,z) = xyg^2(x,y,z) + xyh(x,y,z) \qquad (7.2)$$

whence

$$h(x,y,z) = \frac{xy}{1 - xy} g^2(x,y,z). \qquad (7.3)$$

A word of clarification is in order here. We have glibly stated that the double bond of the H-radical in Fig. 7(a) prevents the two boxes from being interchanged (by rotation). But in deriving our final result we shall have situations in which this H-radical will be a part of a larger radical. What if the rest of this larger radical has rotational symmetry? The argument would then break down. Fortunately this cannot happen. On the other side of the carbon to which our H-radical is attached there may be another double bond, and beyond that possibly another, but if we follow this chain of double bonds we must eventually reach a carbon atom at which there are two single bonds. Thus the larger radical, of which the radical of Fig. 6(b) is a part, will appear as in Fig. 11.

Now one of the two portions A and B contains the free single bonds for this larger radical; the other does not. Hence there can be no rotational symmetry about the double bond in question.

For the I-radicals we have the same situation as before, and we derive the equation

$$i(x,y,z) = xzg(x,y,z). \qquad (7.4)$$

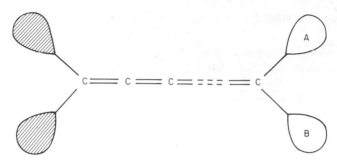

FIG. 11

Substituting from (7.3) and (7.4) in (7.1) we obtain

$$g(x,y,z) = 1 + x\left\{ \frac{1}{3} \left[g^3(x,y,z) + 2g(x^3,y^3,z^3) \right] \right.$$

$$\left. + \frac{xy}{1-xy} g^3(x,y,z) + xzg(x,y,z) \right\} \tag{7.5}$$

the analogue of (6.9). Table V gives the numbers of these radicals for $n \leqslant 11$.

We now enumerate hydrocarbons with a labelled carbon atom, and the method is slightly different from that of the structural isomer enumeration. We distinguish three types, given in Fig. 12.

The enumeration of types V and VII is the same as that of types I and IV in section 6, except that for the type V enumeration we must use the alternating group A_4 instead of S_4. These types therefore together yield the counting series

$$xZ(A_4; g(x,y,z)) + xg(x,y,z) \cdot i(x,y,z) \tag{7.6}$$

The method previously used for types II and III in section 6 will not work here, for in Fig. 9 (Type II) there is no way of telling whether the two boxes can be permuted or not — it will depend on what is at the other end of the double bond shown in the figure. Accordingly we adopt a different approach, and subsume the analogues of types II and III under a single type, type VI of Fig. 12, in which one of the p carbon atoms in the chain of atoms linked by double bonds is the labelled atom. It is easily verified that this description includes all the previous type II and III compounds.

If p is even, $= 2k$, we can orient the figure so that the labelled carbon atom is in the left-hand portion of the chain. There is then no symmetry about a 'vertical' axis, and there are k ways of choosing which carbon atom to label.

Since there is still possible symmetry about the chain itself, i.e. interchanging Q_1 with Q_2 and R_1 with R_2, the permutation group that we need is $S_2[E_2]$ whose cycle index is $\frac{1}{2}(s_1^4 + s_2^2)$. Thus the numbers of ways of allocating the G-radicals Q_1, Q_2, R_1 and R_2 is given by $Z(S_2[E_2]; g(x,y,z))$. But we must

TABLE V

The numbers of stereo-isomers of monovalent acyclic radicals (d = double bonds and t = triple bonds)

$n=2$	t 0	1		$n=3$	t 0	1		$n=4$	t 0	1	2		$n=5$	t 0	1	2
0	1	1		0	2	2		0	5	5	1		0	11	14	4
d 1	1	1		d 1	4			1	12	4			1	36	22	
				2	1			d 2	7				d 2	34	4	
								3	1				3	10		
													4	1		

$n=6$	t 0	1	2	3	$n=7$	t 0	1	2	3	$n=8$	t 0	1	2	3	4
0	28	41	16	1	0	74	122	60	8	0	199	370	219	42	1
1	111	92	10		1	342	356	76		1	1060	1328	420	20	
2	142	43			2	553	282	10		2	2072	1486	160		
d 3	68	4			3	378	64			3	1851	592	10		
4	13				4	114	4			4	797	85			
5	1				5	16				5	172	4			
					6	1				6	19				
										7	1				

$n=9$	0	t 1	2	3	4	$n=10$	0	1	2	t 3	4	5
0	551	1134	786	202	13	0	1553	3505	2784	899	96	1
1	3306	4828	2020	210		1	10357	17264	8980	1476	35	
2	7572	6988	1360	20		2	27194	30678	8810	462		
3	8384	4130	244			3	35988	24344	2960	20		
d 4	4790	1022	10			4	25878	9001	328			
5	1456	106				5	10426	1572	10			
6	242	4				6	2410	127				
7	22					7	324	4				
8	1					8	25					
						9	1					

$n=11$	0	1	t 2	3	4	5
0	4436	10908	9766	3796	578	20
1	32576	61000	37920	8604	498	
2	96439	128536	49160	4966	35	
3	148588	128726	25354	714		
4	129488	64966	5220	20		
d 5	66006	16816	412			
6	20092	2242	10			
7	3712	148				
8	418	4				
9	28					
10	1					

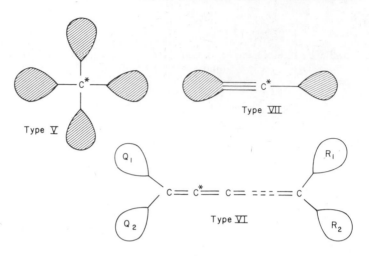

FIG. 12

allow for the addition of $2k$ extra carbon atoms, $2k - 1$ extra double bonds, and the k choices of carbon atom to label. Altogther we have the counting series

$$kx^{2k}y^{2k-1} \cdot \tfrac{1}{2}\{g^4(x,y,z) + g^2(x^2,y^2,z^2)\}. \tag{7.7}$$

Summing (7.7) over all k we have

$$\sum_{k=1}^{\infty} kx^{2k}y^{2k-1} \cdot \tfrac{1}{2}\{g^4(x,y,z) + g^2(x^2,y^2,z^2)\} \tag{7.8}$$

$$= \frac{x^2 y}{2(1-x^2 y^2)^2}\{g^4(x,y,z) + g^2(x^2,y^2,z^2)\}.$$

If p is odd ($= 2k + 1$) and the atom that is labelled is not the centre one of the chain, then we can take it to be on the left-hand side, and obtain an overall counting series of

$$\sum_{k=1}^{\infty} kx^{2k+1}y^{2k} \cdot \tfrac{1}{2}\{g^4(x,y,z) + g^2(x^2,y^2,z^2)\} \tag{7.9}$$

$$= \frac{x^3 y^2}{2(1-x^2 y^2)^2}\{g^4(x,y,z) + g^2(x^2,y^2,z^2)\}$$

since now $2k + 1$ extra carbons and $2k$ extra double bonds are introduced.

If $p = 2k + 1$ and we label the central carbon atom, then we open up the possibility of symmetry about a 'vertical' axis (i.e. interchange of the left and right ends). The permutations shown in Fig. 13 are then possible.

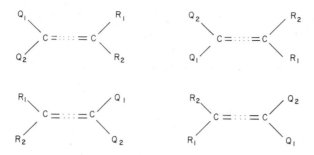

FIG. 13

The cycle index of this group of permutations is $\frac{1}{4}(s_1^4 + 3s_2^2)$. [Compare the problem of ethylene derivatives, and equation (5.3).] Thus in this case we get an additional counting series of

$$\sum_{k=1}^{\infty} x^{2k+1} y^{2k} \cdot \frac{1}{4} \{g^4(x,y,z) + 3g^2(x^2,y^2,z^2)\} \qquad (7.10)$$

$$= \frac{x^3 y^2}{4(1 - x^2 y^2)} \{g^4(x,y,z) + 3g^2(x^2,y^2,z^2)\}$$

By adding up the contributions (7.6), (7.8), (7.9) and (7.10) we obtain the counting series for all labelled stereo-isomers, this being the analogue of the series $K(x,y,z)$ for the structural isomer problem. However, since some simplification can be carried out later, we shall not collect together these several series at present.

As before, we now count the stereo-isomers of hydrocarbons with a labelled bond. Those compounds for which the labelled bond is single or triple give no trouble, and if we subtract the numbers of compounds for which the bond in question is symmetric, we obtain

$$\frac{1}{2}[g_1^2(x,y,z) - g_1(x^2,y^2,z^2)] \qquad (7.11)$$

where $g_1(x,y,z) = g(x,y,z) - 1$, and

$$\frac{1}{2z} [i^2(x,y,z) - i(x^2,y^2,z^2)]$$

as the analogues of (6.16) and (6.18) respectively. The compounds for which the labelled bond is double present more difficulty, and we consider again the diagram of Fig. 12 (Type VI).

If $p = 2k + 1$, arrange to have the labelled double bond on the left. There are

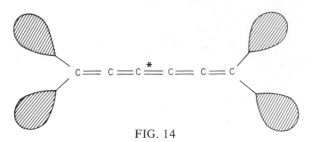

FIG. 14

k ways of choosing it. The group is $S_2[E_2]$ and as before we obtain

$$\sum_{k=1}^{\infty} kx^{2k+1}y^{2k} \cdot \tfrac{1}{2}\{g^4(x,y,z) + g^2(x^2,y^2,z^2)\}$$

$$= \frac{x^3y^2}{2(1-x^2y^2)^2}\{g^4(x,y,z) + g^2(x^2,y^2,z^2)\}.$$

Note that this term will cancel with (7.9) when we eventually collect the terms together.

If $p = 2k$ and the labelled bond is not the centre one, arrange for it to be on the left. There are now $k-1$ choices and we obtain

$$\sum_{k=1}^{\infty} (k-1)x^{2k}y^{2k-1} \cdot \tfrac{1}{2}\{g^4(x,y,z) + g^2(x^2,y^2,z^2)\} \tag{7.14}$$

$$= \frac{x^4y^3}{2(1-x^2y^2)^2}\{g^4(x,y,z) + g^2(x^2,y^2,z^2)\}$$

If the labelled bond is the centre one (Fig. 14) the requisite group is now that having cycle-index $\tfrac{1}{4}(s_1^4 + 3s_2^2)$ (compare Fig. 13), and we obtain the series

$$\sum_{k=1}^{\infty} x^{2k}y^{2k-1} \tfrac{1}{4}\{g^4(x,y,z) + 3g^2(x^2,y^2,z^2)\}$$

$$= \frac{x^2y}{4(1-x^2y^2)}\{g^4(x,y,z) + 3g^3(x^2,y^2,z^2)\} \tag{7.15}$$

We now consider those compounds having a labelled double bond which is symmetric, which can occur only if $p = 2k$ and the centre bond is labelled. Given two different G-radicals Q and R to place at one end of the chain of carbons, we can construct *two* molecules of this type, as shown in Fig. 15.

These are different (*cis* and *trans* isomerism) but each has an automorphism group mapping the central bond end-for-end on itself. However if Q and R are the same there is clearly only one such molecule. Hence the problem is equivalent to that of placing a figure in each of two *distinguishable* boxes, since, for a given choice of figures we have two ways of placing the figures if they

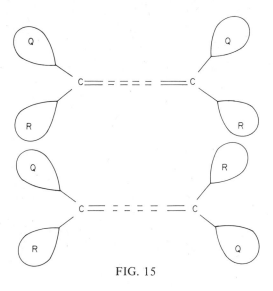

FIG. 15

are different, but only one if they are identical. Hence the number of compounds for which the four G-radicals have parameters n, d and t is the coefficient of $x^{n/2}y^{d/2}z^{t/2}$ in $g^2(x,y,z)$. Allowing for the $2k$ extra carbon atoms and the $2k-1$ double bonds we obtain the series

$$x^{2k}y^{2k-1}\{g^2(x^2,y^2,z^2)\}$$

On summing for $k=1$ upwards we get

$$\frac{x^2y}{1-x^2y^2}\{g^2(x^2,y^2,z^2)\} \tag{7.16}$$

To obtain the required counting series $l(x,y,z)$ – analogous to $L(x,y,z)$ in (6.19) – we must add (7.6), (7.8), (7.9) and (7.10), subtract (7.11), (7.12), (7.13), (7.14) and (7.15), and finally add (7.16). Two of these terms cancel, as already remarked, and the remainder, after some simplification, yield the following result:

$$l(x,y,z) = xZ(A_4; g(x,y,z)) + \tfrac{1}{2}x^2z\{g^2(x,y,z) - g(x^2,y^2,z^2)\}$$

$$+ \frac{x^2y}{2(1-x^2y^2)}\{g^4(x,y,z) + g^2(x^2,y^2,z^2)\}$$

$$- \frac{x^2y}{4(1-xy)}\{g^4(x,y,z) + 3g^2(x^2,y^2,z^2)\}$$

$$- \tfrac{1}{2}\{g_1^2(x,y,z) - g_1(x^2,y^2,z^2)\}$$

$$+ \frac{x^2y}{1-x^2y^2}\{g^2(x^2,y^2,z^2)\} \, .$$

TABLE VI

The numbers of stereo-isomers of acyclic hydrocarbons (d = double bonds, t = triple bonds)

$n = 2$

d \ t	0	1
0	1	1
1	1	1

$n = 3$

d \ t	0	1
0	1	1
1	1	1
2	1	

$n = 4$

d \ t	0	1	2
0	2	2	1
1	4	1	
2	2	2	
3	1		

$n = 5$

d \ t	0	1	2
0	3	3	2
1	6	5	
2	8	1	
3	2		
4	1		

$n = 6$

d \ t	0	1	2	3
0	5	8	5	1
1	18	17	4	
2	23	9		
3	14	1		
4	3			
5	1			

$n = 7$

d \ t	0	1	2	3
0	11	18	13	3
1	42	54	16	
2	75	46	4	
3	52	12		
4	21	1		
5	3			
6	1			

$n = 8$

d \ t	0	1	2	3	4
0	24	47	37	11	1
1	118	172	76	5	
2	236	207	32		
3	228	90	4		
4	106	16			
5	30	1			
6	4				
7	1				

$n = 9$

d \ t	0	1	2	3	4
0	55	123	108	36	5
1	314	544	286	42	
2	758	832	217	5	
3	868	532	44		
4	547	149	4		
5	180	19			
6	40	1			
7	4				
8	1				

$n = 10$

d \ t	0	1	2	3	4	5
0	136	338	325	135	23	1
1	895	1717	1102	238	11	
2	2418	3214	1164	86		
3	3356	2720	443	5		
4	5784	1108	60			
5	1135	223	4			
6	295	23				
7	52	1				
8	87					
9	11					

This, then, is the counting series for stereo-isomers of the acyclic hydrocarbons. Values of the coefficients in this series are given in Table VI.

From these values we can deduce some of the results that we had earlier. If we pick out the coefficients of the terms independent of y and z we get the series $a(x)$. The terms of the form $x^n y$ will give us the counting series for the ethylene derivatives, i.e. (5.3); those of the form $x^n z$ will give the counting series for the acetylene derivatives, i.e. $xb(x)$.

Other results can be obtained. Thus the counting series for the stereo-isomers of *all* acyclic hydrocarbons is

$$x + 3x^2 + 4x^3 + 13x^4 + 31x^5 + 109x^6 + 372x^7$$
$$+ 1446x^8 + 5714x^9 + 27100x^{10} + \ldots .$$

while that for those hydrocarbons having no triple bonds is

$$x + 2x^2 + 3x^3 + 9x^4 + 20x^5 + 64x^6 + 205x^7$$
$$+ 747x^8 + 2767x^9 + 14159x^{10} + \ldots.$$

and so on. It is also of interest to compare corresponding entries in Tables IV and VI, to see the effect that stereo-isomerism has on the numbers of isomers; but this, and other amusements, will be left to the reader.

REFERENCES

1. Blair, C. M. and Henze, H. R. (1931). The number of structurally isomeric alcohols of the methanol series. *J. Amer. Chem. Soc.* **53**, 3042–3046.
2. Blair, C. M. and Henze, H. R. (1933). The number of structurally isomeric hydrocarbons of the ethylene series. *J. Amer. Chem. Soc.* **55**, 680–686.
3. Cayley, A. (1874). On the mathematical theory of isomers. *Phil. Mag.* **47**, 444–446.
4. Harary, F. and Norman, R. (1960). Dissimilarity characteristic theorems for graphs. *Proc. Amer. Math. Soc.* **11**, 332–334.
5. Perry, D. (1932). The number of structural isomers of certain homologs of methane and methanol. *J. Amer. Chem. Soc.* **54**, 2918–2920.
6. Pólya, G. (1938). Kombinatorische Anzahlbestimmungen für Gruppen, Graphen und chemische Verbindungen. *Acta Math.* **68**, 145–254.
7. Read, R. C. (1972). Some recent results in chemical enumeration. *In*: Graph Theory and Applications (Eds. Y. Alavi, D. R. Lick, and A. T. White). Springer-Verlag, Berlin.
8. Masinter, L. M., Sridhavan, N. S., Lederburg, J. and Smith, D. H. (1974) *J. Amer. Chem. Soc.* **96**, 7702.

CHAPTER 5

Enumeration of Cyclic Graphs

Alexandru T. Balaban

Institute of Atomic Physics,
Bucharest, Roumania
and
Polytechnic University,
Chair of Organic Chemistry,
Bucharest, Roumania

In Chapter 3 several enumerations of cyclic graphs were examined, making use of Pólya's enumeration theorem.[1] There is no rigid distinction in applying this theorem to cyclic or acyclic graphs. In the present chapter we shall start by examining several problems associated with enumerations of cyclic graphs, which can be solved by using Pólya's theorem, and then pass on to other problems where different mathematical approaches have to be employed.

In several of the following cases, graph theory has served not only to count graphs, but also for definition, classification, systematization, coding and nomenclature of chemical species.

1   ISOTOPE-ISOMERISM OF LABELLED COMPOUNDS

In cyclic or acyclic isotopically labelled compounds, a new kind of isomerism
appears: isotope-isomerism. To encompass labelling by both radioactive and
stable isotopes, we shall use the term 'labelling' to mean 'changing the natural
isotopic composition'. One or several atoms of the same kind can be labelled in a
molecule by an isotopic species, this number of atoms being called 'isotopic
multiplicity', e.g. two for CHD_2-CH_3 and CH_2D-CH_2D. On the other hand,
one can label a molecule with several different isotopic atoms (either of the same
element, or of different isotopic atoms (either of the same element, or of
different elements) obtaining a 'multiply-labelled molecule', e.g. the deuterated
ethanes in the previous example were singly-labelled, but the following benzenes
are doubly-labelled: C_6H_4DT, $^{12}C_5{}^{13}CH_5D$.

Isotope-isomers are molecules with the same structure and isotopic composition, but differing in the position of the label(s), as in the above example of
ethane-d_2.

The first attempt to apply Pólya's counting theorem to enumerating
isotope-isomers of labelled compounds was made by Hübner,[2] who considered,
however, only singly-labelled compounds. Balaban *et al.*[3] generalized the
treatment to multiply-labelled molecules with any isotopic multiplicity.

Isotopic substitution of hydrogen in organic molecules leads to the same
numbers of isotope-isomers as, for instance, the numbers of isomers obtained by
chemical substitution using halogens. However, isotopic substitution of elements
with higher valencies, such as carbon, leads to numbers of isotope-isomers
different from those of chemical isomers, because one can label molecules with
carbon isotopes in positions where no chemical substitution can take place, e.g.
naphthalene at positions 9 and/or 10. Finally, isotope-isomerism in multiply-
labelled compounds necessitates a generalization of Pólya's counting theorem.

We shall start by discussing isotope-isomers of carbon-labelled benzene and
toluene. We recall that in Chapter 3 the number of singly-labelled benzenes or
toluenes with one hydrogen isotope and various isotopic multiplicities (which is
equal to the number of chemical isomers obtained by chlorination of these
molecules), was given by generating functions of the form $C(x) = \sum_i a_i x^i$:

for benzene : $C(x) = 1 + x + 3x^2 + 3x^3 + 3x^4 + x^5 + x^6$

for toluene : $C(x) = 1 + 4x + 10x^2 + 16x^3 + 18x^4 + 16x^5 + 10x^6 + 4x^7 + x^8$.

The coefficient a_i of any power x^i indicates the number of isotope-isomers of
benzene or toluene, deuterated (tritiated) with isotopic multiplicity i, e.g. there
exist three benzenes-d_2, four toluenes-d_1 and ten toluenes-d_2.

To obtain the numbers of carbon-labelled isotope-isomers, we apply Pólya's
counting theorem to the graphs G representing the given molecules when
ignoring hydrogen atoms. Graphs G whose lines represent covalent bonds, are
called 'reduced graphs'. The reduced graph of benzene is the cyclic graph with

six points C_6; the reduced graph for toluene is graph $K_2 \cdot C_6$ resulting by identifying one point of the complete graph K_2 of order 2, with one point of the cyclic graph C_6 of order 6. The symmetry mappings of these graphs are used to determine their point groups $\Gamma(G)$ and then their cycle indices Z.[4]

These cycle index formulas can be derived either by using operations on permutation groups, e.g. for toluene

$$\Gamma(K_2 \cdot C_6) = E_3 + S_2[E_2]$$
$$Z(E_3 + S_2[E_2]) = Z(E_3) \cdot Z(S_2[E_2]) = c_1^3 \cdot \tfrac{1}{2}(c_1^4 + c_2^2) = \tfrac{1}{2}(c_1^7 + c_1^3 c_2^2),$$

or by examining the symmetry elements (excluding inversion centres or symmetry planes), e.g. for toluene the identity operation where all seven points remain unchanged (c_1^7) and the rotation with $180°$ around a C_2 axis passing through three points and mapping two points on another two $(c_1^3 c_2^2)$.

Thus one obtains·

for benzene : $Z(D_{6h}) = \tfrac{1}{12}(c_1^6 + 3c_1^2 c_2^2 + 4c_2^3 + 2c_3^2 + 2c_6)$

for toluene : $Z(C_{2v}) = \tfrac{1}{2}(c_1^7 + c_1^3 c_2^2)$.

Then according to Pólya's theorem, in the cycle index Z, each variable c_k is substituted by a figure-counting series:

$$c_k = 1 + x^k$$

(for singly-labelled molecules);

$$c_k = 1 + x_1^k + x_2^k$$

(for doubly-labelled molecules with two isotopic labels which cannot be attached to the same point of the reduced graph);

$$c_k = 1 + x_1^k + x_2^k + x_1^k x_2^k$$

(for doubly-labelled molecules with two isotopic labels which can be attached to the same point of the reduced graph. This last type of figure-counting series represents the generalization of Pólya's theorem.)

The resulting polynomial configuration counting series $C(x)$ is the generating function whose coefficients are the numbers of isotope-isomers:

For carbon singly-labelled toluene, with any isotopic multiplicity,

$$C(x) = Z(C_{2v}, 1 + x) = 1 + 5x + 13x^2 + 21x^3 + 21x^4 + 13x^5 + 5x^6 + x^7,$$

indicating for example, by the coefficient of x, that there are five isotope-isomers of singly carbon-labelled toluene with isotopic multiplicity equal to one (Fig. 1).

For doubly-labelled benzene we obtain the following partition of coefficients in the configuration-counting series (to save space, we do not present the whole configuration-counting series since some terms are repeated for x_1 and x_2 by symmetry).

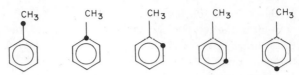

FIG. 1   Isotope-isomers of singly carbon-labelled toluene with isotopic multiplicity equal to one.

When the two labels cannot involve the same point of the reduced graph, the partition is

$$Z(D_{6h}, 1 + x_1 + x_2) : x_1^6, x_1^5 x_2, x_1^5, x_1, 3x_1 x_2, 3x_1^2, 3x_1^3, 3x_1^4, 3x_1^4 x_2,$$
$$3x_1^3 x_2^3, 3x_1^4 x_2^2, 6x_1^2 x_2, 6x_1^3 x_2, 6x_1^2 x_2^2, 11x_1^2 x_2^2$$

hence there are three isomers for $^{12}C_4 {}^{13}C^{14}CH_6$, as the coefficients of $x_1 x_2$ or of $x_1^4 x_2$ are 3.

When the two labels can involve the same point of the reduced graph, the partition is

$$Z(D_{6h}, 1 + x_1 + x_2 + x_1 x_2) : x_1^6, x_1^6 x_2^6, x_1, x_1^5, x_1^5 x_2, x_1^6 x_2^5, 3x_1^2, 3x_1^4,$$
$$3x_1^6 x_2^2, 3x_1^6 x_2^4, 3x_1^3, 3x_1^6 x_2^3, 4x_1 x_2, 4x_1^5 x_2,$$
$$4x_1^5 x_2^5, 9x_1^2 x_2, 9x_1^4 x_2, 9x_1^5 x_2^2, 9x_1^5 x_2^4, 12x_1^3 x_2,$$
$$12x_1^5 x_2^3, 24x_1^2 x_2^2, 24x_1^4 x_2^2, 24x_1^4 x_2^4, 28x_1^3 x_2^2,$$
$$28x_1^4 x_2^3, 38x_1^3 x_2^3$$

hence there are e.g. four isomers of $^{12}C_5 {}^{13}CH_5 D$ as the coefficients of $x_1 x_2$ or of $x_1^5 x_2$ are 4 (Fig. 2).

FIG. 2   Isotope-isomers of doubly carbon- and deuterium-labelled benzene with isotopic multiplicities equal to one.

Another problem which can be solved by straightforward use of Pólya's theorem is the enumeration of substituted porphyrins,[90] in agreement with results obtained by direct counting;[91] of substituted arenes;[92] of inorganic ring systems;[93,94] cf. also earlier references[95-97] and a more recent one.[98]

2   ENUMERATION OF ALL POSSIBLE MONOCYCLIC AROMATIC SYSTEMS

In 1959, the present author tried to enumerate all possible monocyclic aromatic systems formed from atoms of the first row of the periodic table.[5] The initial postulates were simple: for first-row atoms possessing sp^2 hybridization, the numbers of electrons in the non-hybridized atomic orbitals p_z (which give rise to the delocalized molecular orbitals in the Hückel approximation) can only be, according to Pauli's exclusion principle, 2, 1, or 0. These atoms can therefore be classified into three classes, X, Y, and Z, according to the above numbers of electrons in the p_z orbital, irrespective of the chemical nature of these atoms. Nor is the number m of ring atoms restricted.

The numbers x, y, and z of atoms of type X, Y, and Z, respectively, in such a monocyclic aromatic ring $X_x Y_y Z_z$ must satisfy two requirements for an m-membered ring which obeys the Hückel rule (stating that the number of π-electrons is $4n + 2$, where n is a non-negative integer):

$$x + y + z = m$$
$$2x + y \quad = 4n + 2. \tag{2.1}$$

To find integer values for x, y, and z from this set of two equations with three unknowns, monocyclic aromatic systems were separated into those with π-electron doublet ($n = 0$), sextet ($n = 1$), etc. Then the ring size (m) was varied systematically and the solutions were found by simple algebra: e.g. for the most interesting case of π-electron sextet, with $k = 0, 1, 2$, or 3, the solutions are: $x = k$, $y = 6 - 2k$, $z = m - 6 + k$, with the restriction $0 \leqslant x,y,z \leqslant m$.

For each of these solutions, the question then arises to find the numbers $R(x, y, z)$ of isomers for a molecule $X_x Y_y Z_z$ with given x, y, and z values. We will call these isomers 'basic monocyclic aromatic rings' because we do not yet specify the nature of the X, Y, or Z atoms. In collaboration with S. Teleman,[5] an algebraic formula was found for this problem, taking into account rotation and reflection symmetries of the $X_x Y_y Z_z$ ring, through the terms $K(x, y, z)$ and $L(x, y, z)$, respectively:

$$R(x,y,z) = \tfrac{1}{2}[K(x,y,z) + L(x,y,z)] \tag{2.2}$$

The term $K(x, y, z)$ is expressed as a sum depending on all common divisors d of the three integers x, y, and z:

$$K(x,y,z) = \frac{1}{m}\Sigma_d \, d \cdot T\left(\frac{x}{d},\frac{y}{d},\frac{z}{d}\right) \tag{2.3}$$

The function $T(r, s, t)$ can be calculated by the general recurrence formula, where d is a common divisor of r, s, and t:

$$\Sigma_d \, T\left(\frac{r}{d},\frac{s}{d},\frac{t}{d}\right) = \frac{(r+s+t)!}{r!\,s!\,t!} \tag{2.4}$$

To find out the values of the terms T in the sum, one applies the preceding relation until the numbers r, s, and t are coprime (e.g. until $d = 1$) when (2.4) becomes

$$T(r, s, t) = \frac{(r + s + t)!}{r!\, s!\, t!} \tag{2.5}$$

which can be calculated and then used to determine successive values of terms T from the recurrence (2.4).

The term $L(x, y, z)$ depends on the parity of the set of integers x, y, and z. Let the even numbers of this set be denoted by $p_1, p_2, \ldots, p_r$ and the odd ones by $q_1, q_2, \ldots, q_s$ (where $r + s = 3$). There are four possible cases:

$$s = 0, L = \frac{\left(\dfrac{m}{2}\right)!}{\left(\dfrac{p_1}{2}\right)!\left(\dfrac{p_2}{2}\right)!\left(\dfrac{p_3}{2}\right)!}$$

$$s = 1, L = \frac{\left(\dfrac{m-1}{2}\right)!}{\left(\dfrac{p_1}{2}\right)!\left(\dfrac{p_2}{2}\right)!\left(\dfrac{q_1-1}{2}\right)!}$$

$$s = 2, L = \frac{\left(\dfrac{m}{2} - 1\right)!}{\left(\dfrac{p_1}{2}\right)!\left(\dfrac{q_1-1}{2}\right)!\left(\dfrac{q_2-1}{2}\right)!}$$

$$s = 3, L = 0.$$

For example, let us calculate by this procedure the number $R(2, 2, 4)$. By applying relations (2.2) and (2.3) we find

$$R(2, 2, 4) = \tfrac{1}{2}[K(2, 2, 4) + L(2, 2, 4)] \tag{2.6}$$

$$K(2, 2, 4) = \frac{1}{8}\left[2T\left(\frac{2}{2}, \frac{2}{2}, \frac{4}{2}\right) + T\left(\frac{2}{1}, \frac{2}{1}, \frac{4}{1}\right)\right] \tag{2.7}$$

By virtue of (2.5),

$$T(1, 1, 2) = \frac{(1 + 1 + 2)!}{1!\, 1!\, 2!} = 12.$$

We then substitute this value in recurrence (2.4):

$$T(1, 1, 2) + T(2, 2, 4) = \frac{(2 + 2 + 4)!}{2!\, 2!\, 4!} = 420,$$

whence

$$T(2, 2, 4) = 420 - T(1, 1, 2) = 420 - 12 = 408.$$

Therefore from relation (2.7),

$$K(2, 2, 4) = \tfrac{1}{8}(2 \cdot 12 + 408) = 54.$$

The term $L(2, 2, 4)$ for three even numbers ($s = 0$) is

$$L(2, 2, 4) = \frac{4!}{1! \, 1! \, 2!} = 12.$$

Therefore by applying relation (2.6) we arrive at the desired number

$$R(2, 2, 4) = \frac{54 + 12}{2} = 33.$$

Several years later, in collaboration with F. Harary,[6] a simpler solution of the problem of determining the number of basic aromatic monocyclic aromatic rings was found by using Pólya's theorem. The problem of counting the isomers of a monocyclic aromatic system $X_x Y_y Z_z$ is analogous to the well-known problem in enumerative combinatorial analysis called 'the necklace problem': how many different necklaces can one make with m beads of three different colours $x + y + z = m$? The answer is obtained by substituting c_k in the cycle index $Z(D_m)$ for the respective m value by the figure-counting series $c_k = 1 + x_1 + x_2$ to obtain the configuration-counting series $C_m(x)$, i.e. the generating function. Thus one finds the following partitions of terms and coefficients:

$$C_3(x) : x_1^3, x_1^2 x_2, x_1 x_2$$
$$C_4(x) : x_1^4, x_1^3 x_2, 2x_1^2 x_2^2, 2x_1^2 x_2$$
$$C_5(x) : x_1^5, x_1^4 x_2, 2x_1^3 x_2^2, 2x_1^3 x_2, 4x_1^2 x_2^2$$
$$C_6(x) : x_1^6, x_1^5 x_2, 3x_1^4 x_2^2, 3x_1^3 x_2^3, 3x_1^4 x_2, 6x_1^3 x_2^2, 11x_1^2 x_2^2$$
$$C_7(x) : x_1^7, x_1^6 x_2, 3x_1^5 x_2^2, 4x_1^4 x_2^3, 3x_1^5 x_2, 9x_1^4 x_2^2, 10x_1^3 x_2^3, 18x_1^3 x_2^2$$
$$C_8(x) : x_1^8, x_1^7 x_2, 4x_1^6 x_2^2, 5x_1^5 x_2^3, 8x_1^4 x_2^4, 4x_1^6 x_2, 12x_1^5 x_2^2, 19x_1^4 x_2^3,$$
$$33x_1^4 x_2^2, 38x_1^3 x_2^3.$$

This indicates that there are three isomers of a ring $X_2 Y_4$ (the coefficient of $x_1^4 x_2^2$ in $C_6(x)$ is 3), eleven isomers of $X_2 Y_2 Z_2$ (the coefficient of $x_1^2 x_2^2$ in $C_6(x)$ is 11), and 33 isomers of $X_2 Y_4 Z_2$ in agreement with the calculation above of $R(2,2,4)$ since the coefficient of $x_1^4 x_2^2$ in $C_8(x)$ is 33. The extension of this procedure to bicyclic systems,[6] however, presents some still unexplained limitations.[99]

The solutions of the indeterminate set of equations (2.1) and of the isomerism problem are presented for the π-electron sextet case in Table I and Fig. 3 for ring sizes $4 \leqslant m \leqslant 8$ (for easy reading, in Fig. 3 atoms of type Y are not written explicitly, but are understood to exist in the unlabelled vertices of the polygons). The numbers in the column 'unrestricted' are the corresponding coefficients of the above partitions of $C_m(x)$ or the numbers $R(x, y, z)$ found by the formulas (2.2) to (2.5).

TABLE I

Numbers $R(x, y, z)$ of basic monocyclic aromatic rings $X_x Y_y Z_z$ with
$4 \leqslant x + y + z = m \leqslant 8$

| | | | | Number of basic monocyclic aromatic rings | | |
| | | | | | Without adjacent | Without adjacent |
m	x	y	z	Unrestricted	Z-type atoms	X- or Z-type atoms
4	2	2	0	2	2	1
	3	0	1	1	1	0
5	1	4	0	1	1	1
	2	2	1	4	4	2
	3	0	2	2	1	0
6	0	6	0	1	1	1
	1	4	1	3	3	3
	2	2	2	11	7	5
	3	0	3	3	1	1
7	0	6	1	1	1	1
	1	4	2	9	6	6
	2	2	3	18	4	3
	3	0	4	4	0	0
8	0	6	2	4	3	3
	1	4	3	19	6	6
	2	2	4	33	2	2
	3	0	5	5	0	0

The free electron model provides a theoretical explanation for the observation[5] that isomers with two or more adjacent X-type or Z-type atoms are less stable than isomers without such adjacencies. Counting isomers with adjacency restrictions is, however, a more complicated problem, which was solved very recently by Lloyd.[7] Previous calculations had been made empirically,[5] and the results are given in the last two columns of Table I, and illustrated in Fig. 3.

Another useful application of the classification into types X, Y, and Z is for defining *mesoionic* aromatic compounds as those systems which have chains of an odd number of Y-type atoms between chains of X- and/or Z-type atoms. Mesoionic systems are usually less stable than isomeric non-mesoionic aromatics, and the best known mesoionic compounds have five-membered rings. This difference in stability is in agreement with molecular orbital calculations.[8] Counting separately mesoionic and non-mesoionic systems is a still unsolved problem. In Fig. 3, mesoionic systems are denoted by letter M.

Chemically, first-row elements of X-, Y-, or Z-type can be classified according to the number of the group of the periodic system, and according to whether they have a bonding (shared), or an unshared electron pair (types b and u,

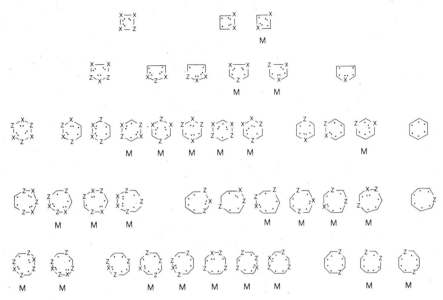

FIG. 3   Monocyclic aromatic systems with π-electron sextet and 4- to 8-membered rings having no adjacent Z-type atoms.

respectively) as presented in Table II. Atoms with more than one formal electric charge, as well as Be atoms for Z-type or F atoms for X-type atoms, were excluded.

Combination of Table I and II (i.e. of the basic rings presented in Fig. 3 with the atom types from Table II) gives, for the monocyclic π-electron sextet systems, a very large number of possible solutions (without counting the supplementary variation of substituents R for the atom types in Table II).

The central position of carbon atoms (column 4b) and nitrogen atoms (column 5u) explains why the most typical aromatics are benzene, azines and azoles. However, not all systems resulting from the combination of Table II with the solutions of equations (2.1), e.g. those from Fig. 3 for π-electron sextet, correspond to stable aromatic systems. The main restriction (besides mesoionic character and adjacency conditions) is that concerning electronegativities of atoms in the ring.

By means of formulas (2.8) and (2.9), each atom type from Table II (with $R = H$ for b-type atoms) was assigned an electronegativity value (relatively to carbon atom $Y4b$) termed 'aromaticity constant'[8,9] and denoted by k_H.

$$k_H = 100 \frac{0.478 \, Z^*}{r} - 1.01 - n_\pi \qquad (2.8)$$

$$Z^* = Z_n - 0.85 n_K - 0.525 n_{L,n} - 0.175 n_{L,b} \qquad (2.9)$$

TABLE II
First-row elements which may form aromatic rings

Atom type	Group							
	3u	3b	4u	4b	5u	5b	6u	6b
X	—	—	—	$-\overset{\cdot\cdot}{\underset{R}{C}}{}^{\ominus}$	$-\overset{\cdot\cdot}{\underset{\cdot\cdot}{N}}{}^{\ominus}$	$-\overset{\cdot\cdot}{\underset{R}{N}}-$	$-\overset{\cdot\cdot}{\underset{\cdot\cdot}{O}}-$	$-\overset{\cdot\cdot}{\underset{R}{O}}{}^{\oplus}$
Y	—	$-\overset{\cdot}{\underset{R}{B}}{}^{\ominus}$	$-\overset{\cdot}{C}{}^{\ominus}$	$-\overset{\cdot}{\underset{R}{C}}-$	$-\overset{\cdot}{N}-$	$-\overset{\cdot}{\underset{R}{N}}{}^{\oplus}$	$-\overset{\cdot}{O}{}^{\oplus}$	—
Z	$-\overset{\cdot\cdot}{B}{}^{\ominus}$	$-\underset{R}{B}-$	$-\overset{\cdot\cdot}{C}-$	$-\underset{R}{C}{}^{\oplus}$	$-\overset{\cdot\cdot}{N}{}^{\oplus}$	—	—	—

where Z^* = effective charge; Z_n = nuclear charge; r = covalent radius (in Ångströms); $n_K = 2$ (number of electrons in the K-shell); $n_{L,n}$ = number of non-bonding L electrons (2 for u-type atoms, zero for b-type atoms); $n_{L,b}$ = number of bonding L-electrons (4 for u-type atoms, 6 for b-type atoms); n_π = number of π-electrons (0, 1, or 2 for Z-, Y-, and X-type atoms, respectively).

The sum of these constants over all atoms in the ring gives a value $K_{arom} = \Sigma k_R$ which for all known aromatic compounds lies in the range -200 to $+200$. Formula (2.8) gives $K_{arom} = 0$ for benzene, $+100$ for the tropylium cation, and -100 for the cyclopentadienide anion. For comparison, taking rings formed from atoms of type $Y4b$ for one heteroatom, the K_{arom} values are $+23$ for pyridine, $+74$ for the conjugated acid of pyridine (the N-protonated pyridinium cation), $+97$ for pyrylium, and -26 for pyrrole. By the relation

$$k_R = k_H + 20\sigma_{p,R}$$

values of aromaticity constants for b-type atoms bearing various substituents R can be calculated (p,R is Hammett's sigma constant for the *para* R substituent).

Finally, there are also other restrictions such as:

(i) formation of a closed electronic configuration, i.e. all bonding molecular orbitals occupied, all antibonding molecular orbitals vacant, no non-bonding molecular orbitals. (Therefore we did not include the three-membered ring X_3 in Fig. 3 since it cannot accommodate six π-electrons in bonding orbitals; for the same reason it is debatable whether there is significant aromatic stabilization in four-membered rings X_2Y_2 or X_3Z.)

(ii) the possibility of forming very stable decomposition products by thermally-allowed pericyclic reactions, explaining the non-existence of hexazine, in opposition to the isolable pentazoles.

The row of Table II with *Y*-type atoms leads by pairwise combination of two such atoms to all known molecules with a triple bond (homo- or hetero-diatomic).[10] After the prediction made on this basis that $O_2^{2\oplus}$ should be detectable, it was indeed found mass-spectrometrically.[11]

A similar classification of atom types occurring in free radicals,[12] has nitrogen atoms as the central and most important atom types.

Second-row atoms like phosphorus or sulphur can take part in aromatic systems, but the presence of *d*-orbitals introduces complications. Polycyclic systems are examined in the next section.

3   *cata*-CONDENSED BENZENOID POLYCYCLIC AROMATIC SYSTEMS

The usual definition of a *cata*-condensed benzenoid polycyclic system is that in such a molecule no carbon atom is common to more than two benzenoid rings, e.g. molecules I–VI.* When there exist carbon atoms common to three benzenoid rings, the system is considered to be *peri*-condensed (e.g. VII). Borderline cases like coronaphenes (VIII) could be considered according to the usual definition to be *cata*-condensed.

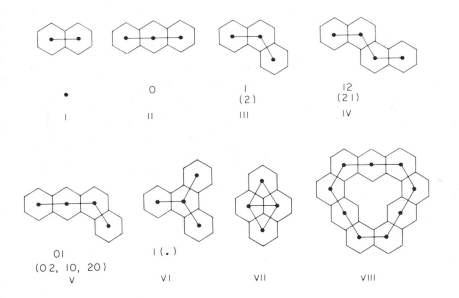

By considering the 'characteristic graph' whose points are the centres of hexagons, and whose lines are obtained by linking the centres of hexagons

*In these and subsequent formulas, alternating double bonds are implied on the periphery.

whenever these are condensed, a better definition is obtained:[13] *cata*-condensed systems have characteristic graphs which are trees, *peri*-condensed systems have characteristic graphs which contain cycles. Only by this definition, all *cata*-condensed systems with the same number n of condensed benzenoid rings are isomeric. Coronaphenes like VIII are *peri*-condensed according to this definition (in this case, the characteristic graph has a larger ring, whereas in usual *peri*-condensed systems like VII, the characteristic graph has three-membered rings).

For brevity, we will call benzenoid condensed polycyclic systems 'polyhexes': these will be classified according to the new definition into *cata*-condensed (catafusenes) and *peri*-condensed systems (perifusenes). Perifusenes with a given number n of hexagons are not isomeric with each other; they can be neutral molecules (e.g. VII), monoradicals (IX) or polyradicals (X). The Hückel rule is certainly not applicable to perifusenes. Since perifusenes are not yet amenable to a rigorous mathematical treatment, the following discussion will deal only with catafusenes.

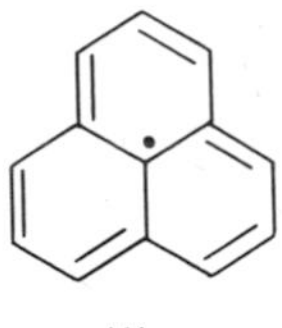

IX 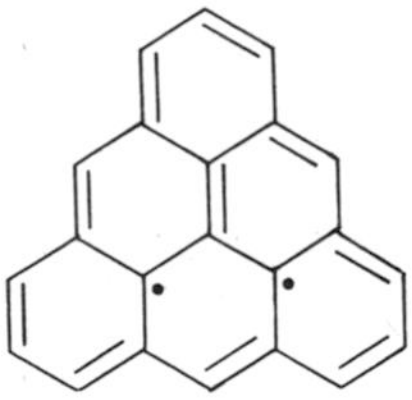   X

The characteristic graph differs from normal graphs in that its angles cannot be varied. Since a marginal hexagon, corresponding to an endpoint of the characteristic graph, can be annelated in three directions as shown in Fig. 4, we can code these three directions by digits 0 (for an angle of $180°$), 1, and 2 (for angles of $120°$ or $240°$). Thus a non-branched catafusene with n hexagons (points of the characteristic graph) can be coded by a string of $n - 2$ digits 0, 1, and 2; to make the notation unique and unambiguous, whenever more than one notation is possible, the smallest number resulting from this string of digits is selected: thus under formulas II–V the selected notation is given, with the

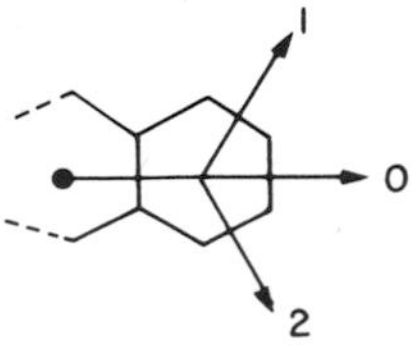

FIG. 4   The three possible annelations of marginal benzenoid rings in catafusenes, and corresponding notations.

alternative possible notations corresponding to larger numbers formed from these digits given in brackets. For naphthalene (I) the notation is a dot. It can be seen that for acenes (e.g. I, II) only one notation is possible, a string of digits 0; for symmetrical non-branched catafusenes (either relatively to an axis, e.g. III, or to a centre, e.g. IV) two notations are possible; for unsymmetrical non-branched catafusenes (e.g. V) four notations are possible. However, in all cases only one notation is selected.

By simple algebra it can be deduced that the number C_n of non-branched catafusenes with n condensed hexagons is:[13]

$$4C_n = 3^{n-2} + 4.3^{(n-3)/2} + 1 \qquad \text{(for } n \text{ odd)}$$
$$4C_n = 3^{n-2} + 2.3^{(n-2)/2} + 1 \qquad \text{(for } n \text{ even).}$$

(3.1)

For branched catafusenes, the formula is more complicated but it gives the possibility to enumerate all catafusenes.[14] The proposed[15] nomenclature for branched catafusenes makes use of brackets, e.g. for triphenylene (VI) the longest path in the characteristic graph corresponds to phenanthrene (with notation 1), and at the branching position there is just one hexagon (denoted by a dot in brackets).

Thus for catafusenes a definition, an enumeration, and a coding-nomenclature proposal can be devised on the basis of the characteristic graph. Numbers of polyhexes are presented in Table III. The problem of enumerating polyhexes is an old unsolved one in graph theory, and is generally known as the 'cell-growth problem': how many aggregates or 'animals' are there for a given number n of cells? In this case the cell is hexagonal, but analogous problems exist for square- and triangular-cell animals.

TABLE III

Numbers of polyhexes with n condensed hexagons

| | Catafusenes | | | | Polyhexes |
n	Non-branched[a]	Branched[b]	Total	Perifusenes	(total)[c]
1	1	0	1	0	1
2	1	0	1	0	1
3	2	0	2	1	3
4	4	1	5	2	7
5	10	2	12	10	22
6	25	12	37	45	82
7	70	53	123	210	333
8	196	250	446	1002	1448

[a]Calculated after formulas (3.1).[13]   [b]Calculated after reference 14.   [c]Results of computer calculations.[16]

Two further implications of the preceding problem of catafusenes will be discussed. The first is the nomenclature of *cata*-condensed benzenoid polycyclic compounds. There exist in use many trivial names for such compounds (chrysene, picene, etc.) and the systematic nomenclature is rather cumbersome. By using the preceding code, all catafusenes can be specified simply in the following form: [12] tetracatafusene for chrysene, and [121] pentacatafusene for picene. An alternative to the above proposed nomenclature for non-branched[13] or branched[15] catafusenes is possible, also making use of the characteristic graph; the above nomenclature is based on the longest path of the characteristic tree, after the model of the current alkane nomenclature. Any tree has a centre or a centroid, which is either a point, or a pair of adjacent points; this property can be used for a systematic nomenclature of acyclic systems as proposed by Lederberg[17] and by Read.[18,19] Along this line, a corresponding nomenclature of catafusenes can be devised, based on the centre of the characteristic graph.

The second implication concerns properties of catafusenes such as resonance energy, electronic absorption spectra, and polarographic reduction potentials, which depend on the frontier orbitals. As pointed out repeatedly (cf. Chapter 7), the Hückel matrix has the same eigenvalues as the adjacency matrix which is purely topological in nature. In other words, the Hückel M.O. method is a very clever way of relating the topology of a conjugated π-electron system (as expressed by the adjacency matrix of its reduced graph) to the Schrödinger equation for that system. This explains, *inter alia*, the equivalence between the H.M.O. method and the valence-bond (V.B.) method whose topological (graph-theoretical) nature is readily apparent.

Sahini proposed a linear correlation between the resonance energy E,[20,21] or the electrochemical reduction potential[22] of catafusenes, and structural parameters: n, the number of condensed benzenoid rings, and a, the number of 'imperfectly aromatic rings', i.e. those which accommodate less than three double bonds in the Kekulé structure with the maximum number of 'perfectly aromatic rings'. Thus $a = 0$ for I, III or IV, but $a = 1$ for II and V. It can be demonstrated[23] that a is the number of zero digits in the code of the characteristic graph of the catafusene, so that correlation (3.2) is purely topological in nature:

$$E = 1.75\ n - 0.15\ a + 0.25 \quad \text{(in beta units)} \tag{3.2}$$

Another correlation can be made with Baird's procedure[24] for calculating resonance energies of catafusenes; the original reference should be consulted for this correlation, also topological, since it is more elaborate.[23]

Interestingly, the numbers of diamondoid hydrocarbons[100] and of their 'characteristic graphs', which are the carbon skeletons of staggered alkane rotamers,[101] can be calculated by formulas similar to (3.1); for non-branched systems, the formulas are identical, not similar.

4 NON-BENZENOID *cata*-CONDENSED POLYCYCLIC CONJUGATED COMPOUNDS

We shall discuss a similar problem to that in the preceding section (namely to enumerate and code *cata*-condensed polycyclic systems), but this time for any size of the condensed rings, not only six-membered. Ring sizes will be denoted by z_i (irrespective of the parity), x_i (for even-membered rings) or y_i (for odd-membered rings).

For non-branched systems, we designate by 'class' all polycyclic condensed molecules with a given set of unordered ring sizes. A 'subclass' contains polycyclic compounds possessing a given set of ring sizes in a specified order.[25] A marginal z-membered ring can be annelated in $z-3$ ways; the mode of annelation and the characteristic graph will be coded as previously, with digit 0 for an angle of 180° (only for even-membered rings), using odd and even digits for angles larger and smaller than 180°, respectively, as shown in Fig. 5. These digits are larger for angles further apart from 180°. To code the characteristic graph, these digits are exponents of the respective rings sizes z_i. In the preceding section, it had been understood that $z_i = 6$ throughout, therefore the ring size could be left out. This is no longer possible here.

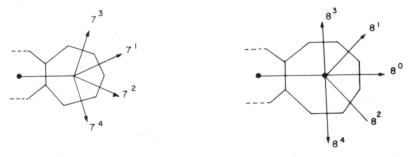

FIG. 5   The possible annelations of marginal rings with ring sizes $z = 7$ or 8 in *cata*-condensed non-benzenoid polycyclic conjugated compounds, and corresponding notations.

Examples of such polycyclic systems with their notation (again, selecting from possible alternative notations the one corresponding to the smallest number formed by the ordered exponents, ignoring the z_i values) are presented in formulas XI–XIV.

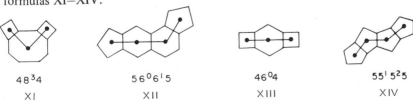

48^34 56^06^15 46^04 $55^1 5^25$
XI   XII   XIII   XIV

A subclass of unsymmetrical non-branched *cata*-condensed systems will be noted by the ordered sequence $z_1 z_2 \ldots z_{n-1} z_n$; a subclass of symmetrical systems (relatively to the order of ring sizes) containing an even number $(n = 2m)$ of rings will be denoted $z_1 z_2 \ldots z_{m-1} z_m z_{m-1} \ldots z_2 z_1$; a subclass of symmetrical systems with an odd number $(n = 2m + 1)$ of condensed rings will be denoted by $z_1 z_2 \ldots z_{m-1} z_m z_0 z_m z_{m-1} \ldots z_2 z_1$.

The numbers of isomers within a subclass of such non-branched systems can be calculated by the following formulas:

Unsymmetrical subclass:
When at least one non-marginal ring is odd-membered,

$$C = \frac{1}{2} \prod_{i=2}^{n-1} (z_i - 3).$$

When all non-marginal rings are even-membered,

$$C = \frac{1}{2} \left[\prod_{i=2}^{n-1} (x_i - 3) + 1 \right]$$

Symmetrical subclass:
Odd number of rings:
   The central ring is odd-membered:

$$C = \frac{1}{4} (y_0 - 3) \prod_{i=2}^{m} (z_i - 3) \left[\prod_{i=2}^{m} (z_i - 3) + 1 \right]$$

The central ring is even-membered:
Some non-marginal rings are odd-membered:

$$C = \frac{1}{4} (x_0 - 3) \prod_{i=2}^{m} (z_i - 3)^2 + \frac{1}{4} (x_0 - 2) \prod_{i=2}^{m} (z_i - 3)$$

All non-marginal rings are even-membered:

$$C = \frac{1}{4} \left[(x_0 - 3) \prod_{i=2}^{m} (x_i - 3)^2 + (x_0 - 2) \prod_{i=2}^{m} (x_i - 3) + 1 \right]$$

Even number of rings
   Some non-marginal rings are odd-membered:

$$C = \frac{1}{4} \prod_{i=2}^{m} (z_i - 3) \left[\prod_{i=2}^{m} (z_i - 3) + 2 \right]$$

All non-marginal rings are even-membered:

$$C = \frac{1}{4} \left[\prod_{i=2}^{m} (x_i - 3) + 1 \right]^2$$

The question arises whether some of these *cata*-condensed systems are aromatic, like their benzenoid analogues. The answer in the general case is negative: the Hückel rule is generally valid only for monocyclic systems. It can be extended to benzenoid *cata*-condensed systems, but not to non-benzenoid *cata*-condensed, or to any *peri*-condensed systems.

Indeed, even though they comply with the Hückel rule, many non-benzenoid *cata*-condensed systems with non-bonding orbitals, with occupied antibonding orbitals, or with empty bonding orbitals can be found, starting with the isolated cases noted by Wilcox[26] (XIII) and by Bochvar[27] (XIV). It was shown[28] that this is a general phenomenon in *cata*-condensed non-benzenoid polycyclic systems (branched or non-branched) possessing certain structural features. Since the Hückel rule is modulo 4, it is convenient to have the following notation: $u \equiv 0 \pmod 4$, $v \equiv 1 \pmod 4$, $w \equiv 2 \pmod 4$, $t \equiv 3 \pmod 4$. Alternant non-branched *cata*-condensed systems have non-bonding levels when they contain an even number of u-membered (e.g. XV) or u- and w-membered rings (e.g. XIII) linked in a certain fashion. Non-alternant non-branched systems with a symmetry centre have:

(i) vacant bonding orbitals when they contain $2k$ pairs of v_i-membered and any number of w_i-membered or of pairs of u_i-membered rings, e.g. XV;

(ii) occupied antibonding orbitals when they contain $2k$ pairs of t_i-membered and any number of w_i-membered or of pairs of u_i-membered rings, e.g. XVI.[28]

Branched *cata*-condensed non-benzenoid polycyclics obeying Hückel's rule can also have anomalous frontier orbitals.[28]

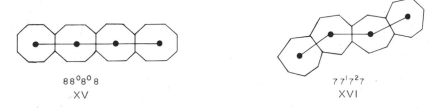

$8\,8^{\circ}8^{\circ}\,8$
XV

$7\,7^{\prime}7^{2}7$
XVI

5   COUNTING CONFIGURATIONS OF ANNULENES

The characteristic graphs of polyhexes were coded by specifying with three digits in a string the directions of annelation differing by angles of $60°$ as shown in Fig. 4. One may code with three digits: 1, 2, and 3 as in formula XVII the three directions[29] on a graphite lattice (honeycomb lattice) differing by angles of $120°$. If necessary, one can also specify the sense along these three directions, as shown in formulas XVIII. Thus one may code any circuit superimposable on the graphite lattice, in particular the periphery of any polyhex. From all possible alternative notations, again that one is selected which corresponds to the smallest resulting number. Accordingly, the periphery of benzene XIX can be described by the code $1\bar{2}3\bar{1}2\bar{3}$ after XVIII, or simpler, since the signs for a circuit always alternate, 123123 after XVII. The perimeter of any polyhex is a configuration of an annulene, superimposable on the graphite lattice. We will count how many such possible configurations exist for an $[n]$ annulene. A similar problem, the determination of circuits superimposable on the diamond lattice, was solved by means of a computer program by Saunders.[30]

A circuit may have single edges (e.g. XIX), or if an edge of the graphite lattice is used more than once, multiple-edges, e.g. XX. In the latter case, the notation is enclosed in brackets.[31] No multiple-edge configuration has yet been experimentally observed.

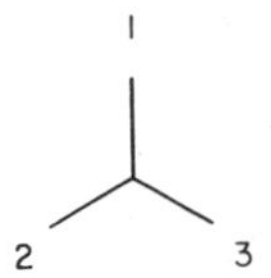

XVII

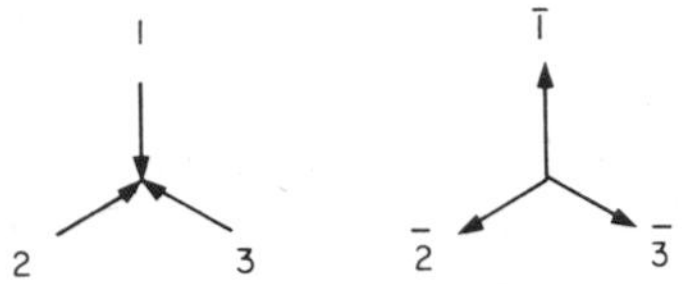

XVIII

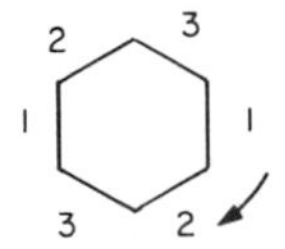

123123

XIX

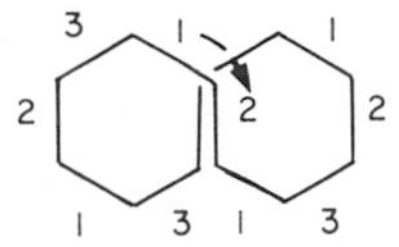

(121321323123)

XX

This procedure enables us to code and to enumerate all circuits super-imposable on the graphite lattice, corresponding to configurations of annulenes (cf. Table IV). This is an example where graphs can assist in solving structural-stereochemical problems.

Not all these configurations are stable, owing to steric and electronic factors. Thus, [16] annulene was shown by low-temperature NMR spectroscopy[32,33] to assume two stable configurations, one of which, XXI, is superimposable on the graphite lattice, hence has no angle strain, but has steric repulsions of inner hydrogens (this is the least crowded configuration, corresponding to the circuit with the maximum area in Table IV). The other, XXII, is not superimposable on the graphite lattice, hence it does not have 120° angles, leading to angle strain, but has less steric crowding of inner hydrogens.

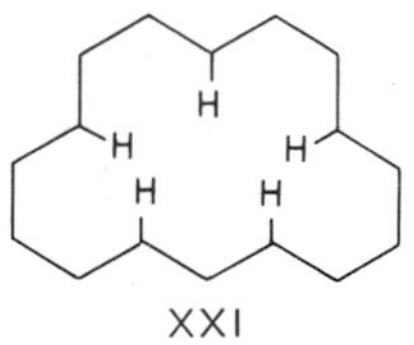

XXI

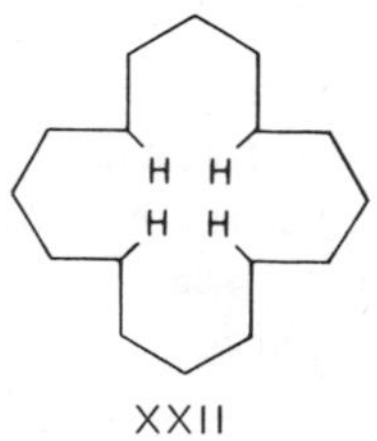

XXII

TABLE IV

Configurations and notations of [n]annulenes

n	Total	Single edge	Multiple edge	Configurations
6	1	1	0	123123
8	0	0	0	—
10	1	1	0	1212312123
12	3	1	2	121231312323 (121321323123) (123123123123)
14	5	3	2	12121231212123 12123121312313 12123131212313 (12121321313123) (12131231213123)
16	8	2	6	1212123121312323 (12121231232313213)(12121321212323123) 1212123131212323 (1212123123231323) (1212312132132123) (1212312321213213) (1212312123123123)

In the higher annulenes, the only stable configurations are predicted[31] to be superimposable on the graphite lattice. It can be shown[31] that starting with [22]annulene the steric crowding of inner hydrogens in these circuits with largest areas remains constant.

The coding of configurations with three digits presented above[31] is an alternative to that using two symbols for *cis* and *trans* configuration of each double bond (as the smallest such possible binary number, converted into a decimal one).[32,33]

6   VALENCE ISOMERS OF ANNULENES, AND CUBIC MULTIGRAPHS

Graph theory can provide an elegant and concise definition of valence isomerism. Given a set of isomeric molecules, their reduced graphs symbolize by points the atoms of valence $\geqslant 2$ (univalent atoms are ignored in the reduced graphs) and by lines the covalent bonds between these atoms. The subsets of molecules whose reduced graphs have the same partition of degrees of their points (vertices) constitute classes of valence isomers.

Let us take for illustration isomers with molecular formula $C_4 H_4$. Ignoring stereoisomerism, we consider only constitutional isomers and depict their reduced graphs. By a procedure involving the construction of all possible graphs of degree $\leqslant 4$ to be discussed in the last section of this chapter, one can[34] draw the pictures of all eleven possible graphs corresponding to molecular formula $C_4 H_4$, presented in Table V.*

Of course, not all graphs from this table correspond to stable molecules (see, however, the six classes of valence isomers totalling 26 isomers with molecular formula $C_5 H_8$, all corresponding to stable molecules[34]). In fact, only two of the eleven graphs correspond to molecules which can be isolated (vinylacetylene and allene), and two more afford stable substituted derivatives (methylene-cyclopropene, cyclobutadiene).

The sums of degrees (cf. Table V, second column) are always equal, for any isomeric ensemble of molecules, to $2q$, i.e. twice the number of lines. The four classes of valence isomers of Table V include the class $(CH)_4$ with two valence isomers. This is a particular case of the general problem of enumerating valence isomers of $[p]$ annulenes, $(CH)_p$ (for short, n-valenes).

The problem, as stated, is equivalent to the long-time unsolved problem of enumerating trivalent (cubic) multigraphs, i.e. regular graphs of degree 3, which must be connected and may have single or multiple edges. Robinson[35] has recently enumerated simple cubic graphs; the enumeration of cubic multigraphs can also be treated mathematically by a procedure which, however, does not allow an easy numerical computation, and can by no means provide the geometrical realization (i.e. the drawing) of the multigraphs.[36]

Starting with a paper published in 1966 (part I in the series 'Chemical Graphs'),[37] an algorithm for constructing all cubic multigraphs was imagined.[37,38] It can be easily demonstrated[4] that the order p of a cubic graph must be an even number: this results *inter alia* from the fact that the number of edges, $q = 3p/2$, must be an integer.

The algorithm will be discussed in the next section, because it involves general cubic graphs. By building up the table of cubic multigraphs (valence isomers of

---

*For this chapter we will adopt the convention: points of degree 3 by white dots; points of degree 4 by black dots; and points of degree 1 or 2 without special symbols.

TABLE V

The isomers with molecular formula $C_4 H_4$ grouped into four classes of valence isomers

Partition of degrees		Valence isomers	Number of valence isomers
Formula	Degrees		
$(CH)_4$	$3+3+3+3 = 12$		2
$(CH_2)(CH)_2 C$	$4+3+3+2 = 12$		4
$(CH_2) C_2$	$4+4+2+2 = 12$		3
$(CH_3)(CH)C_2$	$4+4+3+1 = 12$		2

annulenes), all multigraphs with $p = 4 - 10$,[37] and, later, those with $p = 12$,[38] were enumerated and depicted. Only the former valence isomers are presented in Tables VI and VII. They are grouped according to the number of double bonds (cf. the division into similar groups by van Tamelen[39]), and each has a notation including (in the order from left to right):

— the order p of the graph (the number of CH groups);
— the number of double bonds, β (circuits with two points);
— the number of three-membered circuits, t;*
— the number of four-membered circuits, q;
— a serial number, s, which is 1 unless the preceding numbers are the same for r graphs, where $r > 1$; in this case, s discriminates among these graphs by taking all integer values between 1 and r. In order to have an objective assignment of the serial numbers to the r graphs, s values are assigned in the order of increasing numbers of 5-circuits (or when this fails to discriminate, 6-circuits, and so on). Notations of cubic multigraphs in the present

---

*A circuit is characterized by a set of vertices in a given order, rather than by its lines, therefore a cyclopropenyl group has one, not two, 3-circuits; similarly, cyclobutadiene has only one four-circuit.

TABLE VI

Valence isomers of [4]-, [6]-, and [8]annulene (cubic multigraphs with 4, 6, and 8 points). Non-planar graphs have notations enclosed in brackets.

p\\β	0	1	2	3	4
4	4-0-4-3-1		4-2-0-1-1		
6	(6-0-0-9-1) 6-0-2-3-1	6-1-2-1-1	6-2-0-2-1 6-2-2-0-1	6-3-0-1-1	
8	(8-0-0-4-1) 8-0-0-6-1 (8-0-1-3-1) 8-0-2-2-1 8-0-4-2-1	(8-1-0-5-1) 8-1-1-2-1 8-1-2-1-1 8-1-2-2-1 8-1-3-3-1	8-2-0-1-1 8-2-0-3-1 8-2-1-0-1 8-2-2-1-1 8-2-2-1-2	8-3-0-0-1 8-3-0-1-1 8-3-1-0-1 8-3-2-0-1	8-4-0-0-1

chapter conform to this last requirement, but notations of references 37 and 38 do not.

According to a well-known theorem by Kuratowski,[4] graphs are classified into planar (those which can be drawn on a plane so that no two lines meet except at a vertex), and non-planar (when this cannot be done, e.g. for $K_{3,3}$ XXIII, or K_5, XXIV, or any graph containing these as subgraphs).

XXIII

XXIV

β↑ \ t	0	1	2	3	4
0	10-0-0-5-1	10-0-1-3-1	10-0-2-3-1, 10-0-2-2-1, 10-0-2-0-1	10-0-3-1-1, 10-0-3-0-1	100-4-21 100-422 100-46, 10-0-4-21
1	10-1-0-4-1	10-1-1-3-1, 10-1-1-2-1, 10-1-1-1-1	10-1-2-2-2, 10-1-2-2-1, 10-1-2-3-1, 10-1-2-0-1, 10-1-2-1-1, 10-1-2-3-1, 10-1-2-2-3	10-1-3-1-1, 10-1-3-1-2, 10-1-3-4-1	10-1-1-4-1-1, 10-1-1-4-2-1
2	10-2-0-2-1, 10-2-0-4-1, 10-2-0-2-2, 10-2-0-1-1	10-2-1-1-2, 10-2-1-0-1, 10-2-1-1-3, 10-2-1-1-1, 10-2-1-2-1, 10-2-1-1-3-1	10-2-2-1-2, 10-2-2-2-1, 10-2-2-1-1, 10-2-2-1-4, 10-2-2-0-1, 10-2-2-1-3, 10-2-2-2-3, 10-2-2-2-2, 10-2-2-3-1	10-2-3-3-1, 10-2-3-1-1	10-2-4-1-1
3	10-3-0-0-2, 10-3-0-2-1, 10-3-0-1-2, 10-3-0-0-1	10-3-1-0-2, 10-3-1-0-1, 10-3-1-1-2, 10-3-1-1-1, 10-3-1-0-3	10-3-2-1-1, 10-3-2-0-1, 10-3-2-1-2, 10-3-2-1-3	10-3-3-0-1	
4	10-4-0-0-2, 10-4-0-0-1, 10-4-0-0-3, 10-4-0-1	10-4-1-0-2, 10-4-1-0-1	10-4-2-0-1		
5	10-5-0-0-1				

85

No molecule corresponding to a non-planar graph is yet known, as may be expected; however, if a line could represent a chain of atoms instead of a covalent bond, it should be possible to prepare molecules corresponding to non-planar graphs. Notations of non-planar graphs are enclosed in brackets.

These notations (cf. also later sections of this chapter where the notations will be extended to include the number of loops in general cubic graphs, and the number of vertices of various degrees in irregular graphs) allow a simple ordering of cubic graphs in Tables VI and VII. Whereas for $p = 4$, 6, and 8 both planar and non-planar graphs are shown in Table VI, only planar graphs with $p = 10$ are presented in Table VII (and likewise only planar graphs with $p = 12$ were depicted in ref. 38). Since these planar graphs represent only constitutional, not geometrical isomerism, they were drawn so as to emphasize their planarity, avoiding crossing lines, rather than the geometrical perspective as in actual molecules.

It can be seen that there are two possible valence isomers of $(CH)_4$ as already known from Table V, both of which have proved to be very reactive molecules; six possible cubic multigraphs of order six, one of which is XXIII, non-planar and chemically impossible (but all molecules corresponding to the remaining five planar cubic multigraphs with 6 points are now known); from the 17 planar multigraphs of order 8, 10 are known as chemical counterparts; from the 71 planar multigraphs of order 10, 23 are known so far; and only 10 compounds are so far known out of the 357 possible planar multigraphs of order 12.*

Valence isomers of annulenes have become interesting in the last ten years because of the many unusual features they present:[40] the numerous thermal or photochemical rearrangements, some of them degenerate, the relative simplicity of their thermodynamic relationships enabling chemists to consider all $(CH)_p$ valence isomers with the same p value as 'points on an energy surface'.[41-43] Since the topic of valence isomers was reviewed recently,[37,38,40,44-49] we shall only make a few comments concerning some outstanding molecules in Tables VI and VII.

Both valence isomers of $(CH)_4$, namely cyclobutadiene $(4-2-0-1-1)$[50,51] and tetrahedrane $(4-0-4-3-1)$[52] are unstable, reactive molecules. The valence isomers $(CH)_6$ of benzene $(6-3-0-0-1)$ are comparatively more stable, though the thermally forbidden valence isomerization into benzene is very exothermal as one may see in the explosive conversion of benzvalene $(6-1-2-1-1)$ into benzene.[53] The photochemical rearrangement of carbon atoms in benzene (XXV–XXVII) discovered by Wilzbach and Kaplan by labelling experiments,[54] was the incentive to isolate the intermediately formed benzvalene XXVI. Historically, the first valence isomer of benzene to be isolated was Dewar-benzene $(6-2-0-2-1)$.[47-55] Earlier chemists (e.g. Dewar and Ladenburg) had a

---

*Only constitutional isomers are taken into account in the above figures; if stereoisomers are considered, larger numbers result. For instance, several configurations of [10] annulene were obtained; two geometrical isomers of 9,10-dihydronaphthalene and tricyclo-[4.2.0.0^{2,5}] octa-3,7-diene (cyclobutadiene dimers) are known.

XXV   XXVI   XXVII

long argument with Kekulé on the benzene formula, and proposed alternative formulas called nowadays Dewar-benzene and benzprismane (6-0-2-3-1).* However, bicyclopropenyl (6-2-2-0-1)[56] was not considered by earlier chemists, and was omitted from the first reviews on valence isomers of benzene,[48,49] possibly because it lacks a Hamiltonian circuit. From the valence isomers $(CH)_8$ of cyclooctatetraene (8-4-0-0-1),[57] we mention bicyclo[4.2.0]-octatriene-2,4,7 (8-3-0-1-1) which is in dynamic equilibrium with cyclooctatetraene by a thermally-allowed pericyclic reaction,[58] the *syn* and *anti* dimers of cyclo-butadiene (tricyclo-[4.2.0.0^{2,5}]octa-3,7-diene, 8-2-0-3-1),[59,60] and semi-bullvalene (8-2-1-0-1) which is a molecule rearranging rapidly by thermally allowed Cope rearrangements, even at the lowest attainable temperatures in NMR studies, in a degenerate process.[43] From the valence isomers of $(CH)_{10}$ we mention the first valence isomer of an annulene to be prepared in 1963, Nenitzescu's hydrocarbon 10-3-0-1-1;[61,62] the very interesting molecule, bullvalene (10-3-1-0-1) predicted by Doering,[62] and shortly afterwards found by Schröder[63] to rearrange at slightly elevated temperatures so that every CH group becomes equivalent; and the bridged [10]annulenes studied by Vogel,[64] as well as the unsubstituted [10]annulene (10-5-0-0-1) which, despite Hückel's rule, is unstable owing to steric factors.[45,47]

The numerical values for the numbers of planar and total cubic multigraphs and simple graphs are given in Table XI (see p. 94). Between 1963 and 1975, chemists dealt extensively with the valence isomers of the lower annulenes. There is a considerable challenge to organic chemists to look at the much larger numbers of possible cubic multigraphs with more than 10 points, most of which correspond to unstrained, hence fairly stable, molecules. There are many interesting unanswered questions, e.g. will some of the benzene dimers be isolated?

7   VALENCE ISOMERS OF OTHER CYCLIC SYSTEMS, AND GENERAL CUBIC GRAPHS

Following the definition of valence isomers given in the previous section, we want to examine valence isomers of other cyclic systems, when not all points of the reduced graph have degree three, such as heterocycles, carbenium ions, etc.

*By applying Pólya's theorem to n-valenes $(CH)_n$, several cases of 'coisomeric cubic multigraphs' were found, possessing the same cycle index: the smallest coisomeric n-valenes are benzene and benzprismane (ignoring the stereoisomerism of the latter).[102] The historical controversy between Kekule and Ladenburg is based partly on this coincidence of isomer numbers.

Thus for instance we may ask how many valence isomers have the five-membered aromatic heterocycles such as furan and thiophene, i.e. how many non-isomorphic graphs exist with four points of degree 3, and one point of degree 2. There are two possibilities for solving this problem: (1) counting and displaying general cubic graphs, or (2) enumerating all graphs with points of degree equal to, or smaller than, four. The latter approach will be examined in the last section, since it involves irregular graphs. In the present section we will continue to examine regular graphs, where all points have the same degree.

General graphs can have *loops*, i.e. circuits with one point, in addition to multiple edges (circuits with two points) like the multigraphs discussed in the preceding section. Therefore if we can draw an exhaustive table of general cubic graphs, we may obtain from it the table of graphs whose points have degrees equal to, or lower than, three. Indeed, by eliminating a point on a loop and all the lines incident with it from a general cubic graph (XXVIII), we obtain a point of degree two (XXIX); by eliminating two points on a geminal pair of loops and all lines incident with them (XXX), we obtain a point of degree one (XXXI).

XXVIII   XXIX   XXX   XXXI

To construct the table of general cubic graphs,[65] we start from the two simplest general graphs of order two and use the following algorithm. We apply operations 1 and 2 to increase by two the order of the graphs, progressively:

Operation 1: mark two new vertices on any line(s) or loop(s) and join them by a new line (XXXII).
Operation 2: mark on any line or loop a new vertex and join it to another new point on a loop (XXXIII).

By applying these two operations exhaustively, all general cubic graphs of order $p + 2$ can be found from those of order p.

XXXII   XXXIII

To obtain a complete notation of general cubic graphs, we will note by λ the number of loops, so that the notation described in the preceding section now becomes p-λ-β-t-q-s (i.e. we denote as the first symbol the order of the cubic

TABLE VIIIa
General cubic graphs of orders p = 2, 4, and 6

β \ λ	0	1
0	—	2-0-2-0-0-1
1	—	—
2	2-2-0-0-0-1	—

$p = 2$

β \ λ	0	1	2
0	4-0-0-4-3-1	—	4-0-2-0-1-1
1	—	4-1-1-1-0-1	—
2	—	4-2-1-0-0-1	—
3	4-3-0-0-0-1	—	—

$p = 4$

β \ λ	0	1	2	3
0	(600091) 600231	6-0-1-2-1-1	602021 60220	6-0-3-0-0-1
1	6-1-0-2-3-1	6-1-1-1-1-1	6 12001 612101	—
2	6-2-0-2-1-1	621011 621101	6-2-2-0-01	—
3	6-3-0-1-01	6-3-1-0-0-1	—	—
4	6-4-00-01	—	—	—

$p = 6$

graph, then indicate the numbers of circuits with 1, 2, 3, and 4 points, and add at the end a serial number, s). Thus the notation of general graphs has an extra symbol as compared to multigraphs, or, stated otherwise, multigraphs are general graphs with $\lambda = 0$.

To make the assignment of s values non-arbitrary, the following supplementary priority rules apply for general cubic graphs when the criteria discussed

TABLE VIIIb

General cubic graphs of order $p = 8$. Notations of non-planar graphs are enclosed in brackets.

for multigraphs fail:

- The s value is smaller for general graphs having loops at a smaller distance.
- The s value is smaller for general graphs which have double bonds at a smaller distance.
- The s value is smaller for general graphs which have loops closer to double bonds.
- The s value is smaller for general graphs which have loops bonded to vertices of higher degree (this applies to the irregular graphs discussed in Section 9).

If we thus build up the table of general cubic graphs with orders 2–6 we obtain Table VIIIa, where these graphs are classified according to the numbers λ of loops and β of double bonds. Similarly for order $p = 8$ we obtain Table VIIIb. Reference 65 should be consulted for the table of general cubic graphs of order 10.

Two properties of this table are apparent if we denote by $G(p, \lambda, \beta)$ the number of general cubic graphs of order p with λ loops and β double bonds:

$$G(p, 0, 1) = G(p, 2, 0)$$
$$\sum_{\beta} G(p, 0, \beta) = \sum_{\lambda} G(p, \lambda, 0)$$

The first property is a consequence of the one-to-one correspondence between general cubic graphs of the same order with only one double bond, and those with exactly two loops (XXXIV–XXXV). The second property (which states that for any p, the number of cubic multigraphs without loops equals the number of cubic loop-graphs without double bonds) is the consequence of the one-to-one correspondence between multigraphs (XXXIV or XXXVII) and loop-graphs (XXXVI and XXXVIII). The latter correspondence (XXXVII–XXXVIII) is in fact the result of the former correspondence (XXXIV–XXXVI) applied twice in succession.[65]

XXXV   XXXIV   XXXVI   XXXVII   XXXVIII

The numbers of general cubic graphs of orders 2, 4, 6, 8, and 10 may be seen in Table IX. One can note an interesting coincidence of numbers for the planar multigraphs of order $p + 2$ and the general graphs of order p for $p = 2, 4, 6$, and 8. The numbers of planar cubic multigraphs with exactly β double bonds are

TABLE IX

Numbers of cubic (trivalent) graphs

Order p	2	4	6	8	10	12
Simple graphs	0	1	2	5	19	85
Simple planar graphs	0	1	1	3	9	32
Multigraphs	1	2	6	20	91	506
Planar multigraphs	1	2	5	17	71	357
General graphs	2	5	17	71	388	—

presented in Table X, which will be necessary in the next section. Both Table IX and Table X are obtained by the algorithm discussed above, constructing the full tables of non-isomorphic cubic graphs (multigraphs, general graphs) up to the required order, and then counting these graphs.

The table of general cubic graphs thus obtained can serve for obtaining the table of cubic multigraphs from the preceding section, by ignoring loop-graphs. It should be emphasized that it is not possible to obtain the table of cubic multigraphs by an algorithm involving only multigraphs, but that it is necessary to have general graphs.

Let us now use Table VIII for solving the initial problem — to find the valence isomers of furan or thiophene. Since there exist four general cubic graphs of order six with one loop, 6-1-0-2-3-1, 6-1-1-1-1-1, 6-1-2-1-0-1 and 6-1-2-0-0-1, by the correspondence XXVIII → XXIX we introduce a point of degree two replacing the loop, and we obtain the four possible valence isomers required (Fig. 6). Chemical evidence shows that the last two valence isomers of this figure are involved in the photochemical rearrangement of five-membered heterocycles[66] such as furan,[67] thiophene[68] and the azoles,[69] as shown in

TABLE X

Number of planar cubic multigraphs
(valence isomers of [p] annulenes)
having exactly β double bonds

			p		
β	4	6	8	10	12
0	1	1	3	9	32
1	0	1	4	16	75
2	1	2	5	22	112
3	—	1	4	16	86
4	—	—	1	7	41
5	—	—	—	1	10
6	—	—	—	—	1
Total	2	5	17	71	357

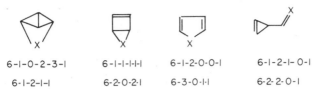

6-1-0-2-3-| 6-1-1-1-1-| 6-1-2-0-0-| 6-1-2-1- 0-|

6-1-2-1-| 6-2-0-2-| 6-3-0-1-| 6-2-2-0-|

FIG. 6   The possible valence isomers of furan ($X = O$) or $(CH)_5^{\oplus}$ ($X = CH^{\oplus}$) derived from general cubic graphs (upper notations) or cubic multigraphs (lower notations).

formulas XXXIX–XLI where the carbon atoms were traced by labelling experiments.

An alternative approach to producing the table of graphs with $p - b$ points of degree 3 and b points of degree 2 is to start from cubic multigraphs of order $p + b$ with at least b double bonds, and replace b double bonds by points of degree 2. As seen in Fig. 6 (lower notations), this procedure leads to the same results. When the double bonds in the cubic multigraph are not equivalent, the same cubic multigraph gives rise in this alternative approach to more than one graph, which will be denoted by letters A, B, etc. in Fig. 7.

An interesting problem is that of finding the valence isomers of the odd-membered cations or anions $(CH)_n^{\pm}$. For $n = 5$, the solutions are the same four graphs from Fig. 6. For $n = 7$, either from the Table VIIIb of general cubic graphs of order 8, or by the alternative procedure outlined above starting from cubic multigraphs of order 8 (Table VI), we obtain the graphs presented in Fig. 7.

Chemically, the most stable $(CH)_5^{\ominus}$ anion is the cyclopentadienide anion. Calculations for the $(CH)_5^{\oplus}$ cation have so far taken into account only three of the four graphs from Fig. 6, omitting the less stable vinylic cation derived from the general graph 6-1-2-1-0-1.[70,71] However, a non-classical species related to boranes (a tetragonal pyramid) is the most stable cation $(CH)_5^{\oplus}$: in agreement with theoretical predictions by Hoffmann,[72] and with theoretical calculations,[70-72] there is experimental evidence for such a structure.[73]

In the case of $(CH)_7^{\oplus}$ cations, the most stable valence isomer is tropylium, but the structure of the analogous species produced in the mass spectrometer by electron impact from benzyl derivatives is still uncertain owing to considerable atom scrambling.[74,75] Schleyer reviewed the rearrangements of such odd-membered cations $(CH)_n^{\oplus}$ discussing only the most stable, chemically significant, structures.[76]

XXXIX   XL   XLI

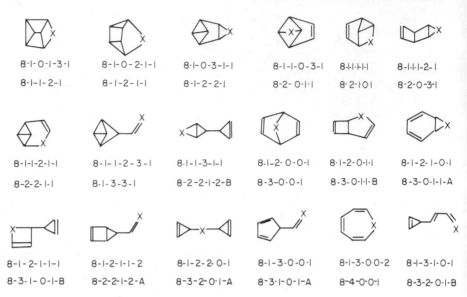

FIG. 7   The possible valence isomers of tropylium ($X = CH^{\oplus}$), tropone ($X = CO$) or oxepin ($X = O$) derived from general cubic graphs (upper notations) or cubic multigraphs (lower notations).

TABLE XI
Numbers of planar general
cubic graphs with one loop,
or of classical valence isomers
of $(CH)_n^{\oplus}$ cations

		n		
β	3	5	7	9
0	0	1	3	13
1	1	1	6	29
2	—	2	6	32
3	—	—	3	18
4	—	—	—	4
Total	1	4	18	96

The numbers of possible classical valence isomers of such ions are presented in Table XI. This table does not include non-classical structures like that discussed above for $(CH)_5^{\oplus}$ nor does its extension to higher orders allow the inclusion of such exotic structures like that found by Goldstein for a $(CH)_{11}^{\oplus}$ cation, namely a 'sandwich structure' from two allyl cations and a cyclopentadienide anion.[77]

8   VALENCE ISOMERS OF BENZOANNULENES

Significant progress has been made in the last few years in synthesizing not only valence isomers of annulenes, but also of benzoannulenes. Benzoannulenes are hydrocarbons which can be theoretically derived from an [n] annulene by replacing b double bonds with *ortho*-fused benzenoid rings ($n \geqslant 2b$). We will call these systems [b] benzo-[n] annulenes.

The simplest such compound is the elusive benzocyclobutadiene which dimerizes instantly,[78] but in the presence of transition metal catalysts another dimer is formed.[79] Naphthalene can be considered to be a monobenzo-[6] annulene; phenanthrene and anthracene are dibenzo-[6] annulenes; and triphenylene is tribenzo-[6] annulene. It is apparent that all [b] benzo-[n] annulenes are *cata*-condensed systems. A considerable collection of monobenzo-$(CH)_{10}$ valence isomers was obtained by Paquett.[80] Dibenzo-$(CH)_{10}$ valence isomers have also been prepared.[81]

Valence isomers of [b] benzo-[n] annulenes are to be found in the table of valence isomers of [n + 4b] annulenes possessing b dihydro benzenoid rings. Thus benzocyclobutadiene corresponds to graph 8-3-0-1-1 in Table VI containing valence isomers of [8] annulene. With the available tables of valence isomers of annulenes, this approach allows us to find only valence isomers of monobenzo-derivatives of [4] - to [8] annulene and dibenzo-[4] annulene (biphenylene).

A better procedure involves replacing double bonds in valence isomers of annulenes by benzenoid rings. The valence isomers of [b] benzo-[n] annulenes can be obtained from valence isomers of [n] annulenes with β double bonds, where $\beta \geqslant b$, by replacing b double bonds with *ortho*-fused benzenoid rings. This makes it necessary to compute Table XII giving the number of planar valence isomers of [n] annulenes (cubic multigraphs of order n) with at least β double bonds. Only planar graphs are considered, since only these are chemically

TABLE XII
Numbers of planar valence isomers of [n] annulenes (cubic multigraphs of order n) having at least β double bonds

			n		
β	4	6	8	10	12
0	2	5	17	71	357
1	1	4	14	62	325
2	1	3	10	46	250
3	—	1	5	24	138
4	—	—	1	8	52
5	—	—	—	1	11
6	—	—	—	—	1

significant. Each entry (n, β) from Table XII is the sum of entries $\Sigma(n, \beta_i)$ from Table X with $\beta_i = \beta, \beta + 1, \ldots, n/2$.

To find all possible valence isomers of $[b]$ benzo-$[n]$ annulenes,[82] we have to consider the symmetry of each cubic multigraph with n points, i.e. to see whether the double bonds are equivalent or not. For $n \geqslant 8$ points, there exist cubic multigraphs with two or more non-equivalent double bonds, e.g. for $n = 8$ (Table VI), 8-2-2-1-2, 8-3-0-1-1, 8-3-1-0-1, and 8-3-2-0-1. All these graphs give rise each to more than one monobenzoderivative; the last three, with 3 double bonds which may be annelated, give rise each to more than one dibenzo-derivative.

Whenever in a valence isomer of an annulene (cubic multigraph) two or more double bonds are conjugated (excluding cyclobutadiene), supplementary benzo-derivatives are possible, where a benzenoid ring does not replace a double bond, but a single bond between two conjugated double bonds: these benzoderivatives are *ortho*-quinonoid benzoderivatives: although the three double bonds in the Kekulé formula of benzene are equivalent, there exist two, not one, dibenzo-derivatives, namely phenanthrene and anthracene.

Figure 8 which represents the valence isomers of $[b]$ benzo-$[n]$ annulenes with $n = 4$, 6, and 8 (whose numbers may be seen in Table XIII, giving also the figures for $n = 10$; the structures in the latter case can be seen in the original reference[82]). The valence isomers of benzoannulenes without *o*-quinonoid structures are indicated with a circle inscribed in the annelated benzenoid ring, according to the usual convention. No supplementary significance according to Clar[83] is, however, attached to this circle. *o*-Quinonoid structures have, by contrast, no such inscribed circle, but a Kekulé formula with more localized double bonds. It is expected that *o*-quinonoid structures should be less stable than their isomers, therefore in their notation an asterisk is included, and their numbers are presented separately in Table XIII.

TABLE XIII

Numbers of possible planar valence
isomers of $[b]$ benzo-$[n]$ annulenes
without (but in brackets including also)
o-quinonoid structures

| | | | n | |
b	4	6	8	10
0	2	5	17	71
1	1	4(4)	18(21)	97(114)
2	1	3(4)	14(17)	80(103)
3	—	1	5 (6)	33 (42)
4	—	—	1	8 (10)
5	—	—	—	1

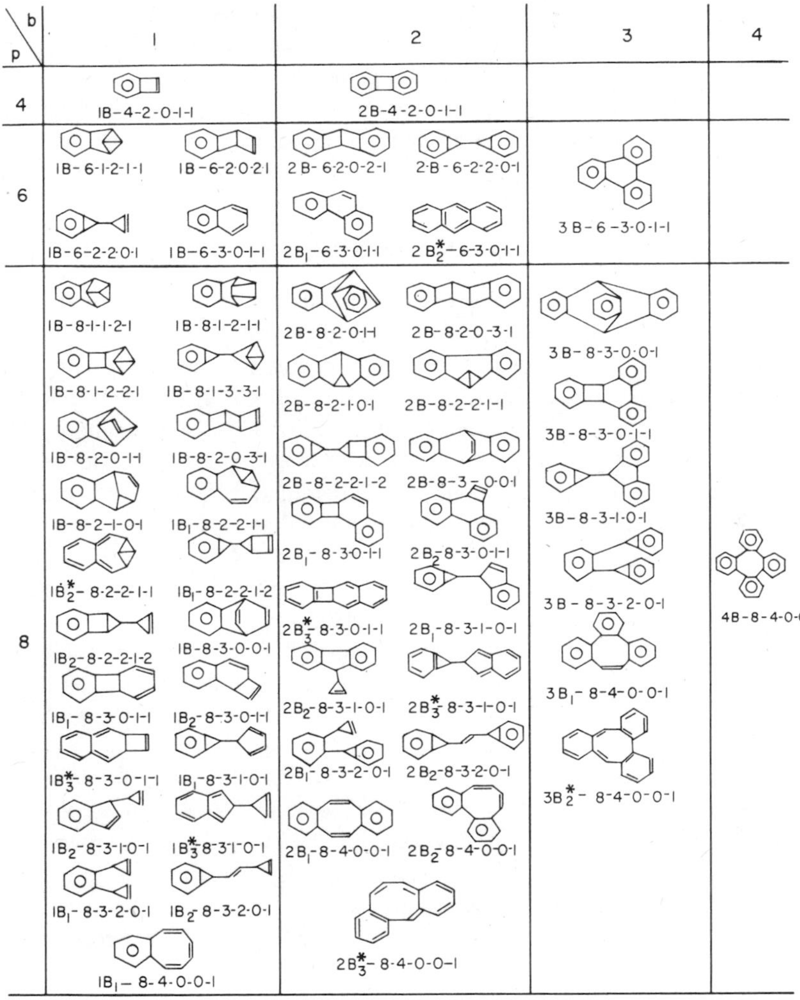

	1	2	3	4
4	1B-4-2-0-1-1	2B-4-2-0-1-1		
6	1B-6-1-2-1-1 1B-6-2·0·2·1 1B-6-2-2-0-1 1B-6-3-0-1-1	2B-6·2·0-2-1 2·B·6-2-2·0-1 2B₁-6-3·0·1·1 2B₂*-6-3·0·1·1	3B-6-3-0-1-1	
8	1B-8-1-1-2-1 1B-8-1-2-1-1 1B-8·1-2-2-1 1B-8-1-3-3-1 1B-8-2-0-1-1 1B-8-2-0-3-1 1B-8-2-1-0-1 1B₁-8-2-2-1-1 1B₂*-8-2-2-1-1 1B₁-8-2-2-1-2 1B₂-8-2-2-1-2 1B-8-3-0-0-1 1B₁-8-3·0-1-1 1B₂-8-3-0-1-1 1B₃*8-3-0-1-1 1B₁-8-3-1-0-1 1B₂-8-3-1-0-1 1B₃*8-3-1-0-1 1B₁-8-3-2-0-1 1B₂-8-3-2-0-1 1B₁-8-4-0-0-1	2B-8-2-0-1-1 2B-8-2-0-3-1 2B-8-2-1-0-1 2B-8-2-2-1-1 2B-8-2-2-1-2 2B-8-3·0-0-1 2B₁-8-3·0-1-1 2B₂-8-3-0-1-1 2B₃*8-3-0-1-1 2B₁-8-3-1-0-1 2B₂-8-3-1-0-1 2B₃*8-3-1-0-1 2B₁-8-3-2-0-1 2B₂-8-3-2-0-1 2B₁-8-4-0-0-1 2B₂-8-4-0-0-1 2B₃*-8-4-0-0-1	3B-8-3-0·0-1 3B-8-3-0-1-1 3B-8-3-1-0-1 3B-8-3-2-0-1 3B₁-8-4-0·0-1 3B₂*-8-4-0-0-1	4B-8-4-0-0-1

FIG. 8 Valence isomers of [b]benzo-[n]annulenes with n = 4-8.

As notation for benzoderivatives, we will use the notation of the corresponding valence isomer of annulene preceded by a symbol of the form bB_k, where the two numbers b and k indicate the number of annelated double bonds, and the serial number of benzoderivatives formed from the same valence isomer of an annulene, respectively. When all double bonds to be annelated are equivalent, and only one polybenzoderivative is formed from the given valence isomer of an annulene, the index k is absent. When one cubic multigraph gives rise to more than one polybenzoderivative, the serial index discriminates among these by taking values starting with 1. A set of precedence rules makes the assignment of these k values non-arbitrary.[82] Essentially, these rules state that acyclic double bonds to be annelated take precedence over cyclic ones, double bonds in smaller rings take precedence over those in larger rings, annelation of clustered double bonds takes precedence over annelation of remote double bonds, and non-o-quinonoid structures take precedence over o-quinonoid ones.

9   ISOMERS OF ANNULENES, AND IRREGULAR GRAPHS

The process of obtaining an exhaustive list of constitutional isomers with a given molecular formula is important both for chemistry and for mass spectroscopy. Molecular formulas for peaks in mass spectrograms can nowadays be obtained with a high degree of certainty from computerized double-focussing high-resolution mass spectrometers, since only a certain isotopic composition is compatible with the exact mass of a peak. Lederberg, Djerassi, Duffield and Feigenbaum[84-88] developed a computer program (DENDRAL) for obtaining isomers of *acyclic* compounds with a given molecular formula. Very recently an extension to cyclic systems has begun to be elaborated.

For cyclic graphs the situation is more complicated. A 'head-on' approach is to develop an algorithm and a table of all possible graphs of degrees four or less; since the degrees of vertices are no longer identical, the graphs are irregular. This was done in reference 34, which presented this table for degrees 2, 3, and 4 (and combinations thereof), and orders 1–5. In the present section, we give in Table XIV a portion of that table for orders 1–4 only, but expand it to include degrees 1, 2, 3, and 4.

Let, in a connected irregular graph, p' vertices have degree d', p'' vertices have degree d'', etc. General cubic graphs were noted p-λ-β-t-q-s; in the present case the notation must be expanded, and will have the degrees as indices, in the form: $p'_{d'}, p''_{d''} \ldots$ -λ-β-t-q-s. Since in this case we can have triple bonds between atoms, these triple bonds are counted each as two double bonds in β. In agreement with the priority rules discussed for the notation of general graphs, when all other indices fail to discriminate among two or more graphs, increasing values of s are assigned in the order: a triple bond – two cumulated double bonds – two conjugated double bonds – two isolated double bonds.

The loop-free graphs with four points from Table XIV namely with degrees 2/3/4 (notation $1_2 2_3 1_4$-0-...), 3 (notation 4_3-0-...), 2/4 (notation $2_2 2_4$-0-....) and 1/3/4 (notation $1_1 1_3 2_4$-0-...) constitute the eleven isomers of $C_4 H_4$ presented earlier in Table V. Indeed, these are the only graphical partitions of the integer representing the sum of vertex degrees of the reduced graph, which is equal to twice the number of edges (each edge having two endpoints), namely 12 in this case.

Each partition of this sum corresponds to a class of valence isomers, and the collection of all classes of valence isomers gives the desired isomers. Let us find

TABLE XIV(A)

Degrees \ p	1	1/2	2	1/2/3	1/3			2/3	3	1/2/3/4	1/2/4
					1, p-1	2, p-2	3, p-3				
1	1_1-0-0-0-0-I										
2	2_1-0-0-0-I		2_2-1-0-0-0-I								
			2_2-0-1-0-0-I								
3		$2_1 2_2$-0-0-0-0-I 3_2-0-1-0-0-I		$1_1 2_1 3$-0-1-0-0-I	$1_1 1_3$-1-0-0-0-I			$1_2 2_3$-0-1-0-I $1_2 2_3$-1-1-0-0-I $1_2 3_2$-0-0-0-0-I	2_3-0-2-0-0-I 2_3-2-0-0-0-I		
				$1_1 2_1 3$-1-0-0-0-I							
4		$2_1 2_2$-0-0-0-0-I 4_2-0-0-0-I		$1_1 2_2 3$-0-0-0-I $1_1 2_4 3$-1-0-0-0-I	3_3-0-1-1-0-I $1_3 5$-1-0-0-0-I $1_3 5$-2-0-0-0-I	$2_2 3$-0-1-0-0-I $2_2 3$-1-0-0-I	$3_1 3$-0-0-0-0-I	$2_2 2_3$-0-2-I-I $2_2 2_3$-0-1-0-0-I $2_2 2_3$-0-2-0-0-I $2_2 2_3$-1-0-0-I $2_2 2_3$-1-0-0-I $2_2 3_2$-0-0-0-0-I	4_3-0-4-3-I 4_3-0-2-0-I 4_3-1-1-0-I 4_3-2-1-0-0-I 4_3-3-0-0-0-I	$1_1 1_1 2 3 4$-0-1-1-0-1-I $1_1 1_1 2 3 4$-0-2-0-0-2-I $1_1 2 3 4$-0-2-0-0-2-I $1_1 2 3 4$-1-0-1-0-I $1_1 2 3 4$-1-1-0-0-2 $1_1 2 3 4$-1-1-0-0-3 $1_1 2 3 4$-2-0-0-0-I $1_1 2 3 4$ -2-0-0-0-2	$2_1 2_4$-0-I-0-0-I $2_1 2_4$-I-0-0-0-I

TABLE XIV(B)

TABLE XV

The numbers of isomers and possible partitions of vertex degrees corresponding to classes of valence isomers for C_6H_6, resulting by assigning integer non-negative values to indices j and k in the expanded formula $C_6H_6 \equiv (CH_3)_{2-k}(CH_2)_{2k-j}(CH)_{2j-k}(C)_{4-j}$

k	0	1	2	3	4
			j		
0	$7(CH_3)_2C_4$		—	—	—
1	–	$46 (CH_3)(CH_2)(CH)C_3$	$34 (CH_3)(CH)_3C_2$	—	—
2	–	$16 (CH_2)_3C_3$	$76 (CH_2)(CH)_2C_2$	$32 (CH_2)(CH)_4C$	$6 (CH)_6$

all isomers of an $[n]$annulene. The classes of valence isomers result by assigning integer non-negative values to both indices j and k in the following expanded formula[34] giving the partitions of vertex degrees:

$$C_nH_n \equiv (CH_3)_{n/2-k-1} (CH_2)_{2k-j} (CH)_{2j-k-n/2+3} (C)_{n-j-2}$$

It is easy to work out from the above formula that there exist exactly four classes of valence isomers for C_4H_4, as seen in Table V. For C_6H_6, it is found from the above formula that there exist seven classes of valence isomers as presented in Table XV. To find out all isomers in each class of valence isomers, one has to look in Table XIV (or its expansion to the required order) at the respective partition of degrees. Thus one finds that there exist a total of 217 possible graphs (one of which is non-planar XXIII) corresponding to the isomers of C_6H_6, as seen in Table XV. The structures of these graphs (most of which are too strained to be chemically possible) were given in reference 34.

Two rings may have 0, 1, 2, or more atoms in common. In the first case with no common points, the two rings are called isolated; the second case (one common point) is of spiranic compounds; in the third case with two adjacent common points, the rings are called condensed (*ortho*-fused); in the last case

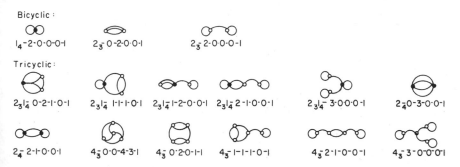

Bicyclic:

I_4-2·0·0·0-I $2_{\bar{3}}$ 0-2·0·0-I $2_{\bar{3}}$ 2·0·0·0-I

Tricyclic:

$2_3|_{\bar{4}}$ 0-2·1·0-I $2_3|_{\bar{4}}$ I-I·I·0-I $2_3|_{\bar{4}}$ I-2·0·0-I $2_3|_{\bar{4}}$ 2·1·0·0-I $2_{3|4}$ 3·0·0·0-I $2_{\bar{4}}$0-3·0·0-I

$2_{\bar{4}}$ 2-I·0·0-I $4_{\bar{3}}$0-0-4-3-I $4_{\bar{3}}$ 0·2·0-I-I $4_{\bar{3}}$ I-I-I·0-I $4_{\bar{3}}$2·I·0·0-I $4_{\bar{3}}$ 3·0·0·0-I

FIG. 9   Topologies of bi- and tricyclic systems containing atoms of valencies ≤ 4.

TABLE XVI

Numbers of possible topologies of bi-, tri-, and tetracyclic
systems of atoms with valence $\leqslant 4$

System	Bicyclic	Tricyclic	Tetracyclic
Including isolated rings	3	12	73
Excluding isolated rings	2	6	26

with more than two common points, or with two non-adjacent common points, the structure is called bridged.

In a reduced graph (not showing univalent atoms) of a molecule with p_i atoms of valence i ($1 \leqslant i \leqslant 4$), the number q of edges is:[34]

$$q = \frac{4p_4 + 3p_3 + 2p_2 - p_1}{2}$$

The cyclomatic number μ of a graph indicates the number of its cycles and is equal to the minimal number of cuts (removal of edges) necessary to convert the graph into an acyclic one, i.e. a tree:

$$\mu = q - p + 1$$

The topologies of polycyclic graphs or molecules (where a line no longer represents a covalent bond, but a sequence of covalent bonds) form a finite set which can be enumerated systematically making use of Table XIV, namely of graphs with degrees 3, 3/4, and 4. In reference 34 the pictures of topologies of bi-, tri-, and tetracyclic systems were given. Figure 9 presents only bi- and tricyclic topologies (i.e. graphs with $\mu = 2$ or 3).

Table XVI presents the numbers of possible topologies of bi-, tri-, and tetracyclic systems. If systems with isolated rings are to be left out, we have to omit graphs with line-connectivity equal to one, which have a 'bridge' or 'cut-edge', i.e. a line whose removal results in a disconnected graph. The corresponding numbers can be read in Table XVI, and increase sharply for larger μ values.

The current nomenclature of bridged polycyclic molecules was devised by Baeyer when the most complicated such molecule known was bicyclic. Its application to more complex topologies is rather cumbersome.[34] It is possible that in the future an improved approach to this problem could be found on the basis of graph theory.

Two recent reviews[103,89] include further applications of graph theory to the enumeration of cyclic systems.

REFERENCES

1. Pólya, G. (1937). *Acta Math.*, 68, 145.
2. Hübner, H. (1964). *Abh. Dtsch. Akad. Wiss. Berlin, Kl. Chem. Geol. Biol.*, Sonderband Stabile Isotope, 701.

3. Balaban, A. T., Fărcaşiu, D. and Harary, F. (1970). *J. Labelled Comp.*, 6, 211.
4. Harary, F. (1967). 'Graph Theory', Addison—Wesley, Reading, Mass.
5. Balaban, A. T. *Studii şi Cercetări Chim. Acad. R. P. România*, 7, 257 (1959); Appendix I, in collaboration with S. Teleman.
6. Balaban, A. T. and Harary, F. (1967). *Rev. Roumaine Chim.* 12, 1511.
7. Lloyd, D. (1973). Proceedings of British Combinatorial Conference, Aberystwyth.
8. Balaban, A. T. and Simon, Z. (1965). *Rev. Roumaine Chim.* 10, 1059.
9. Balaban, A. T. and Simon, Z. (1962). *Tetrahedron*, 18, 315.
10. Balaban, A. T. (1960). *Izv. Akad. Nauk, Otdel. Khim. Nauk*, 2260.
11. Meyerson, S. and Ihrig, P. J. (1973). *Internat. J. Mass Spectr. Ion Phys.* 10, 497.
12. Balaban, A. T. (1971). *Rev. Roumaine Chim.* 16, 725.
13. Balaban, A. T. and Harary, F. (1968). *Tetrahedron*, 24, 2505.
14. Harary, F. and Read, R. C. (1970).*Proc. Edinburgh Math. Soc.* Ser. II, 17, 1.
15. Balaban, A. T. (1969). *Tetrahedron*, 25, 2949.
16. Klarner, D. (1965). *Fibonacci Quart.* 3, 9.
17. Lederberg, J. (1968). *Proc. Natl. Acad. Sci. U.S.A.* 53, 134. 'The Mathematical Sciences' (G. A. W. Boehm, editor), MIT Press, Cambridge, Mass. 1969, p. 37.
18. Read, R. C. 'The coding of trees and tree-like graphs', Univ. of the West Indies, Jamaica, 1969.
19. Read, R. C. and Milner, R. S. 'A new system for the designation of chemical compounds for the purposes of data retrieval. I. Acyclic compounds', Univ. of the West Indies, Jamaica, 1969.
20. Sahini, V. E. (1962). *J. chim. phys.* 59, 177.
21. Sahini, V. E. (1962). *Rev. Chim. Acad. R. P. Romania*, 7, 1265.
22. Sahini, V. E. (1964). *Rev. Chim. (Bucuresti)*, 15, 551.
23. Balaban, A. T. (1970). *Rev. Roumaine Chim.* 15, 1243.
24. Baird, N. C. (1969). *Can. J. Chem.* 47, 3535.
25. Balaban, A. T. (1970). *Rev. Roumaine Chim.* 15, 1251.
26. Wilcox, C. F. Jr. (1968). *Tetrahedron Letters*, 795.
27. Bochvar, D. A., Stankevich, I. V. and Tutkevich, A. V. (1969). *Izv. Akad. Nauk S.S.S.R., Otdel. Khim. Nauk*, 1185.
28. Balaban, A. T. (1972). *Rev. Roumaine Chim.* 17, 1531.
29. Gordon, M. and Davison, W. H. T. (1952). *J. Chem. Phys.* 20, 428.
30. Saunders, M. (1967). *Tetrahedron*, 23, 2105.
31. Balaban, A. T. (1971). *Tetrahedron*, 27, 6115.
32. Oth, J. F. M., Antoine, G. and Gilles, J. M. (1968). *Tetrahedron Letters*, 6259, 6265.
33. Sondheimer, F. (1972). *Accounts Chem. Res.* 5, 81; *Also* 'Aromaticity', *Chem. Soc. Spec. Publ.* 21, 1967, p. 75.
34. Balaban, A. T. (1973). *Rev. Roumaine Chim.* 18, 635.
35. Robinson, R. W. (in press).
36. Harary, F. and Palmer, E. M. (1973). 'Graphical Enumeration', Academic Press, New York, London and San Francisco.
37. Balaban, A. T. (1966). *Rev. Roumaine Chim.* 11, 1097.
38. Balaban, A. T. (1972). *Rev. Roumaine Chim.* 17, 865.
39. van Tamelen, E. E. (1965). *Angew. Chem.* 77, 759; *Also Angew. Chem. Internat. Edit.* 4, 738 (1965).
40. Maier, G. (1972). 'Valenzisomerisierungen', Verlag Chemie, Weinheim.
41. von E. Doering, W. and Rosenthal, J. W. (1966). *J. Amer. Chem. Soc.* 88, 2078.

42. Jones, M. Jr., Reich, S. D. and Scott, L. T. (1970). *J. Amer. Chem. Soc.* **92**, 3118.
43. Zimmerman, H. E., Binkley, R. W., Givens, R. S., Grunewald, G. L. and Sherwin, M. A. (1969). *J. Amer. Chem. Soc.* **91**, 3316.
44. Scott, L. T. and Jones, M. Jr. (1972). *Chem. Rev.* **72**, 181.
45. Masamune, S. and Darby, N. (1972). *Accounts Chem. Res.* **5**, 272.
46. Burkoth, T. L. and van Tamellen, E. E. (1969). *In* 'Nonbenzenoid Aromatics', (J. P. Synder, editor), vol. 1, p. 63, Academic Press, New York, London and San Francisco.
47. van Tamelen, E. E. (1972). *Accounts Chem. Res.* **5**, 186.
48. Bolesov, I. G. (1968). *Uspekhi khim.* **37**, 1567.
49. Schäfer, W. and Hellmann, H. (1967). *Angew Chem.* **79**, 566; *Angew. Chem. Internat. Edit.* **6**, 518 (1967).
50. Cava, M. P. and Mitchell, M. J. (1967). 'Cyclobutadiene and Related Compounds', Academic Press, New York, London and San Francisco.
51. Watts, L., Fitzpatrick, J. D. and Pettit, R. (1965). *J. Amer. Chem. Soc.* **87**, 3253; *Also* **88**, 623 (1966).
52. Shevlin, P. B. and Wolf, A. P. (1970). *J. Amer. Chem. Soc.* **92**, 406, 3253, 5291.
53. Katz, T. J., Wang, E. J. and Acton, N. (1971). *J. Amer. Chem. Soc.* **93**, 3782.
54. Wilzbach, K. E. and Kaplan, L. (1964). *J. Amer. Chem. Soc.* **86**, 2307; *Also* **87**, 4004 (1965).
55. Viehe, H. G. (1965). *Angew. Chem.* **77**, 768; *Angew. Chem. Internat. Edit.* **4**, 746 (1965).
56. De Boer, C. and Breslow, R. (1967). *Tetrahedron Letters*, 1033.
57. Scgröder, G. 'Cyclooctatetraen', Verlag Chemie, Weinheim, 1965.
58. Huisgen, R. and Mietzsch, F. (1964). *Angew. Chem.* **76**, 36; *Also Angew. Chem. Internat. Edit.* **3**, 83 (1964).
59. Avram, M., Dinulescu, I. G., Marica, E., Mateescu, G., Sliam, E. and Nenitzescu, C. D. (1964). *Chem. Ber.* **97**, 382.
60. Criegee, R., Schröder, G., Maier, G. and Fischer, H. G. (1960). *Chem. Ber.* **93**, 1553.
61. Avram, M., Nenitzescu, C. D. and Marica, E. (1957). *Chem. Ber.* **90**, 1857.
62. von E. Doering, W. and Roth, W. R. (1963). *Tetrahedron*, **19**, 715; *Also Angew. Chem.* **75**, 27 (1963).
63. Schröder, G. (1964). *Chem. Ber.* **97**, 3140.
64. Vogel, E. in 'Aromaticity', Chem. Soc. Spec. Publ. 21, p. 113, London, 1967.
65. Balaban, A. T. (1970). *Rev. Roumaine Chim.* **15**, 463. *Also* **19**, 1611 (1974).
66. Lablache-Combier, A. and Remy, M.-A. (1971). *Bull. Soc. Chim. France*, 679.
67. van Tamelen, E. E. and Whitesides, T. H. (1968). *J. Amer. Chem. Soc.* **80**, 3895.
68. Wynberg, H. and van Driel, H. (1965). *J. Amer. Chem. Soc.* **67**, 3998; *Chem. Commun.* 203 (1966).
69. Ullman, E. F. and Singh, B. (1966). *J. Amer. Chem. Soc.* **88**, 1844; *Also* Singh, B. and Ullman, E. F., *ibid.* **89**, 6911 (1967).
70. Kollmar, H., Smith, H. O. and von R. Schleyer, P. (1973). *J. Amer. Chem. Soc.* **95**, 5834.
71. Dewar, M. J. S. and Haddon, R. C. (1973). *J. Amer. Chem. Soc.* **95**, 5836.

72. Stohrer, W. D. and Hoffman, R. (1972). *J. Amer. Chem. Soc.* **94**, 1661.
73. Masamune, S., Sakai, M., Ona, H. and Jones, A. J. (1972). *J. Amer. Chem. Soc.* **94**, 8956.
74. Meyerson, S., Rylander, P. N., Eliel, E. L. and McCollum, J. D. (1959). *J. Amer. Chem. Soc.* **81**, 2606.
75. Hoffmann, M. K. and Wallace, J. C. (1973). *J. Amer. Chem. Soc.* **95**, 5064.
76. von R. Schleyer, P. (1970). *Angew. Chem.* **82**, 889; *Also Angew. Chem. Internat. Edit.* **9**, 860 (1970).
77. Goldstein, M. J. and Kline, S. A. (1973). *J. Amer. Chem. Soc.* **95**, 934.
78. Cava, M. P. and Napier, D. R. (1957). *J. Amer. Chem. Soc.* **79**, 1701.
79. Avram, M., Dinu, D. and Nenitzescu, C. D. (1959). *Chem. and Ind.* 257.
80. Paquette, L. A., Kukla, M. J. and Stowell, J. C. (1972). *J. Amer. Chem. Soc.* **94**, 4920.
81. Elian, M., Banciu, M., Stănescu, L. and Gorănescu, E. (1976) *Rev. Roumaine Chim.* **21**, 90 and further references therein.
82. Balaban, A. T. (1974). *Rev. Roumaine Chim.* **19**, 1185.
83. Clar, E. (1964). 'Polycyclic Hydrocarbons', Academic Press, London, New York and San Francisco; *Also* 'The Aromatic Sextet', Wiley, New York, 1972.
84. Buchanan, B. G., Duffield, A. M. and Robertson, A. V., in 'Mass Spectrometry. Techniques and Applications' (G. W. A. Milne, editor), Wiley–Interscience, New York, 1971, p. 121.
85. Lederberg, J., Sutherland, G. L., Buchanan, B. G., Feigenbaum, E. A., Robertson, A. V., Duffield, A. M. and Djerassi, C. (1969). *J. Amer. Chem. Soc.* **91**, 2973.
86. Duffield, A. M., Robertson, A. V., Djerassi, C., Buchanan, B. G., Sutherland, G. L., Feigenbaum, E. A. and Lederberg, J. (1969). *J. Amer. Chem. Soc.* **91**, 2977.
87. Schroll, G., Duffield, A. M., Djerassi, C., Buchanan, B. G., Sutherland, G. L., Feigenbaum, E. A. and Lederberg, J. (1969). *J. Amer. Chem. Soc.* **91**, 7440.
88. Buchs, A., Duffield, A. M., Schroll, G., Djerassi, C., Delfino, A. B., Buchanan, B. G., Sutherland, G. L., Feigenbaum, E. A. and Lederberg, J. (1970). *J. Amer. Chem. Soc.* **92**, 6831.
89. Rouvray, D. H. (1974). Chem. Soc. Revs., **3**, 355; *Idem* (1971). *Roy. Inst. Chem. Revs.*, **4**, 173.
90. Balaban, A. T. (1975). *Rev. Roumaine Chim.*, **20**, 227.
91. Pilgrim, R. L. C. (1974). *J. Chem. Educ.*, **51**, 316 (cf. also ref. 89).
92. Rouvray, D. H. (1973). *J. South African Chem. Inst.*, **26**, 141; *also* **27**, 20 (1974).
93. Klemperer, W. G. (1972). *Inorg. Chem.*, **11**, 2669.
94. McDaniel, D. H. (1972). *Inorg. Chem.*, **11**, 2678.
95. Riemschneider, R. (1956). *Oesterr. Chem.*, Ztg. 57, 38.
96. Hill, T. L. (1943). *J. Chem. Phys.*, **11**, 294.
97. Taylor, W. J. (1943). *J. Chem. Phys.*, **11**, 532.
98. Leonard, J. E. (1972). *Dissert. Abstr.*, 32B, 5659.
99. Masinter, L. M., Sridharan, N.Z., Carhart, R. E. and Smith, D. H. (1974) *J. Amer. Chem. Soc.* **96**, 7714 (London Math. Soc. Lecture Notes Series).
100. Balaban, A. T. and Schleyer, P. V. R. (to be published).
101. Balaban, A. T. (In press)
102. Balaban, A. T. (1974). *Rev. Roumaine Chim.* **19**, 1323.
103. Masinter, L. M., Sridharan, N. S., Lederberg, J. and Smith, D. H. (1974). *J. Amer. Chem. Soc.*, **96**, 7702.

CHAPTER 6

Metric Spaces and Graphs Representing the Logical Structure of Chemistry

James Dugundji

Department of Mathematics
University of Southern California
Los Angeles, California, U.S.A.

Paul Gillespie, Dieter Marquarding, and Ivar Ugi

Organisch-Chemisches Institut der
Technischen Universität München
München, West Germany

and

Fausto Ramirez

Department of Chemistry
State University of New York
Stony Brook, New York, U.S.A.

1   INTRODUCTION*

Until very recently, little use had been made of the fact that mathematics can be used directly for gaining insight into the logical structure of chemical systems and their interactions.[9,10,25,26,39,40,44,52,63-67,87,93-96,108,110,111,118,121,123]

Among mathematical structures the group[53,60,74,81,83] and topological space[6,24,29,69,70] seemingly have the greatest potential for chemistry. Strikingly successful group theoretical solutions to chemical problems related to symmetry are well documented.[53,60,81,83,106-108] The use of graph theory for the treatment of chemical constitution was introduced by Cayley in 1857,[15] and is almost as old as the concept of constitutional chemistry. The chemical applications of topology have all been intimately associated with graphs[1,2,17,19,23,27,41-44,49,50,62,77,89-91,103] by which certain aspects of chemical constitution or steric features concerning individual molecules are described.

Balaban[1,2] in particular demonstrated the versatility of graph theory[5a,29,54,55,68] in the treatment of chemical problems. It appears certain that the chemical applications of planar graphs will proliferate in the near future, because visually represented statements are in many cases a more effective basis for understanding and communication than their more abstract counterparts.

Since graphs $G(V,L)$, consisting of a set V of v vertices connected by a set L of l lines can serve as models for any system involving l (or v) binary relations between v (or l) objects, their applications to chemistry are not confined solely to the representation of molecules but can include the relations between different chemical entities, e.g. ensembles of molecules (see below) as well.[25,121-123]

In order that the latter type of graph be valid, one must insure that the represented relations between the chemical systems in question are suitably defined to allow the use of graphs for this more general purpose. Very recently, it has been demonstrated that this condition is indeed fulfilled, if the chemical systems under consideration are appropriately defined.[25,121-123] With this, a straightforward foundation for the representation of the mutual relations between chemical systems by graphs is provided.

In this chapter, an attempt is made to present a foundation for the logical structure of chemistry.

---

*The following abbreviations will be used in this chapter: be-matrix = bond and electron matrix; BPR = Berry pseudorotation; cc-matrix = configuration-conformation matrix; EM = ensemble of molecules; FIEM = family of isomeric ensembles of molecules; nc-matrix = nuclear coordinate matrix; TBP = trigonal bipyramid; TR = turnstile rotation; VB = valence bond.

2   THE EXTENDED CONCEPT OF ISOMERISM

2.1   Constitution

Completely interpreting and predicting the behaviour of chemical systems which consist of N quantum mechanical particles would require the knowledge of the 3N position and 3N impulse coordinates of the system over the period of time under investigation.[34]

According to the uncertainty principle, this type of detailed information is not available with the precision required, even if the most perfectly idealized device of observation were available.[104] Thus, the analysis of chemical systems requires the generation of models that represent certain equivalence classes[73,110,130] of states, rather than the states themselves.

We frequently consider two distinguishable objects of our experience or thought as being equivalent if they are equal in some respect (even though they are different in some other). *Equivalence relations* partition sets into equivalence classes, i.e. classes of members of a set possessing certain common features.[8,69,73] The set of all chemical compounds can be divided into equivalence classes whose elements have the same empirical formula,* i.e. the isomers. In this case, isomerism is the defining equivalence relation, and refers to those molecules with like numbers and kinds of atoms.[123]

With good approximation, the atoms that participate in a molecular structure can be treated in terms of a *core* composed of the nucleus and tightly bound inner electrons, together with a *valence shell* of electrons that are less tightly bound.[22]

In the Periodic Table of Elements, the *periods* are equivalence classes of elements whose atoms have the same number of core electrons, and the *groups* are equivalence classes of atoms with the same number of valence electrons. Since the exchange and sharing of valence electrons between atoms accounts for the forces by which aggregates of atoms (the molecules) are primarily held together, the chemistry of elements belonging to the same group is similar.

Molecules are characterized by their set of atoms $A(n) = \{A_1, \ldots, A_n\}$ and are represented by an empirical formula. The *chemical constitution* of the molecule is the precise interrelation between the atoms (i.e., the topological structure of the molecule); this is determined by bonds and bonded neighbors. Since a given set of atoms can be endowed with different topological structures, there can be constitutional isomers.

Since constitution, by definition, is determined by bonds and bonded neighbors, one needs a convention to determine which rearrangements within a molecule may occur without breaking bonds, i.e. one may consider the constitution defined by a set of tension barriers which determine the limits within which the distance between any pair of atoms may vary.[113]

The constitution of most molecules can be adequately described in terms of

---

*In this chapter 'empirical formula' stands for 'molecular formula'.

atomic core pairs, held together by covalent bonds corresponding to valence electron pairs.[22] The important observation is that with suitable reinterpretation, the concept of chemical constitution applies also to *ensembles* of molecules.[25] Here, too, it can be represented by graphs (no longer necessarily connected) $G(V,L)$ where sets of vertices V symbolize the atoms and sets of connecting lines L represent connections given by the bonds. Structural formulas are essentially graphs representing these neighborhood relations of the atoms in a molecule.

If the different kinds of atoms that participate in a given chemical constitution are not distinguished, unlabelled graphs result which represent essentially the skeletons of molecules; a labelled graph[54] permits us to differentiate between distinguishable parts of molecules and to work directly with their complete chemical constitution.[39,123]

2.2.   Double bookkeeping be-matrices

It is precisely the application of the equivalence relation of isomerism to *ensembles* of molecules (rather than only to *individual* molecules) that leads to a unified theory for the relations between chemical systems.[25,120,121,124] Previously, the potential importance of this extended equivalence relation for the interpretation of chemistry and for the systematic planning of chemical experiments seems to have been overlooked.

There are two types of empirical formulas for a given EM, the *brutto empirical formula* which indicates the total number of atoms that are contained in the EM, and the *detailed empirical formula* which represents the set of empirical formulas of the molecules that belong to that EM.

Let A be a given set of atoms. The empirical formula of A is the brutto empirical formula of all the atoms belonging to A. By an EM(A) is meant any compound or collection of chemical species that can be formed from A using each atom belonging to A exactly once. Thus, an EM(A) is a partition $\{\bar{A}_1 \ldots, \bar{A}_s\}$ of A (i.e., the $\bar{A}_i$ are pairwise disjoint and $A = \overset{s}{\underset{1}{\cup}} \bar{A}_i$) such that each $\bar{A}_i$ is some molecule. Letting $\bar{A}_i$ be the empirical formula for the molecule $\bar{A}_i$ we shall call the collection of formulas $\{\bar{A}_1, \ldots, \bar{A}_s\}$ the detailed empirical formula for the EM(A) we are considering; observe that, because isomeric molecules can differ constitutionally, there can be more than one EM(A) with the same detailed empirical formula.

The FIEM(A), the *family of isomeric ensembles of molecules* of A,[25] is the set of all the EM(A). The constitutional formula for the EM(A) is the *be*-matrix for all atoms in A.

The chemistry of a set of atoms A is given by the FIEM(A): the chemical processes are the interconversions of EM(A) by redistributions of the valence electrons between the cores; the left and right sides of chemical equations refer to isomeric EM. Reactions and sequences of reactions correspond to interconversions of EM of the same FIEM. The chemical constitutions of EM, and the constitutional relations between EM, define the fundamental logical structure of chemistry within the FIEM.

For example, the conceivable neutral molecules belong to the following C_2H_6O FIEM:

$$EM_0 = \{2C + 6H + O\}, \; EM_1 = \{CH_3-CH_2-OH\}, \; EM_2 = \{CH_3-O-CH_3\},$$

$$EM_3 = \{CH_3-CHO + H_2\}, \; EM_4 = \{CH_4 + CH_2O\},$$

$$EM_5 = \{CH_3OH + CH_2\},$$

$$EM_6 = \left\{ \begin{array}{c} O \\ \diagup \; \diagdown \\ H_2C-CH_2 \end{array} + H_2 \right\}, \; EM_7 = \{CH_2=CH_2 + H_2O\},$$

$$EM_8 = \{H_2C=CHOH + H_2\}, \; EM_9 = \{CO + CH_4 + H_2\},$$

$$EM_{10} = \{CH_2 + CH_2O + H_2\}, \; EM_{11} = \{H_2C=C=O + 2H_2\},$$

$$EM_{12} = \{HC\equiv C-OH + 2H_2\}, \; EM_{13} = \{HC\equiv CH + H_2O + H_2\},$$

$$EM_{14} = \{CO + CH_2 + 2H_2\}, \; and \; EM_{15} = \{2CH_2 + H_2O\}.$$

It is interesting to note that the EM and the FIEM correspond to a subset of the 'atomic associations' whose adiabatic potential energy surfaces have been calculated by Preuss.[106]

The chemical constitution of an EM which contains a set of n atoms $A(n)$ can be represented by an $n \times n$ *atom connectivity matrix* (ACM). The ACM were introduced in 1963 by Spialter for documentation purposes.[115] In an atom connncectivity matrix, each off-diagonal entry indicates the formal bond order between an indexed pair of atoms, A_i and A_j. The ACM are symmetric matrices because the chemical bond is a symmetric relation.[3,88]

The ACM are not quite adequate for the representation of chemical reactions, since the latter involve not only redistributions of bonds but also unshared valence electrons that do not participate in bonds between atom pairs belonging to the individual molecules of the EM. An adequate representation of interconvertible chemical systems requires a device with *double bookkeeping capability*, i.e. matrices which account for both, *bonds and electrons*. The *be*-matrices E, i.e. *bond and electron matrices*, of EM are such a device. [9,10,25,101,120,121,124] The *be*-matrices are obtained from the ACM by inserting diagonal entries e_{ii} which indicate the number of *free*, unshared valence electrons of the atom A_i.

The set of all EM of a FIEM can be represented by a family $F = \{E_0, \ldots, E_f\}$ of *be*-matrices. Each *be*-matrix contains all the constitutional information of the EM, i.e. all information concerning the bonds and certain aspects of valence electron distributions in that EM, which conventional chemical formulas would contain; moreover, they have the following properties:

M1: The sum e_k over the kth row or column of a be-matrix is the number of valence electrons that belong directly to the atom A_k in that EM. The number e_k lies within an interval $[\bar{e}_{k,min}, \bar{e}_{k,max}]$ which is characteristic of the chemical element to which the atom A_k belongs, and the chemistry under consideration, so that

$$\bar{e}_{k,min} \leq e_k = \sum_{i=1}^{n} e_{ik} = \sum_{i=1}^{n} e_{ki} \leq \bar{e}_{k,max}$$

The formal electrical charge of A_k in the given EM is $e_k^{\oplus} = e_k^{\circ} - e_k$, where e_k° is the number of valence electrons of the free atom A_k. Note that e_k° is equal to e_{kk}°, the kth diagonal entry of the be-matrix of the set of free atoms $A(n)$.

M2: The total number $\hat{e}_k = 2e_k - e_{kk}$ of valence electrons that are associated with A_k lies within an interval $[\hat{e}_{k,min}, \hat{e}_{k,max}]$, which is characteristic of A_k and the chemistry under consideration, so that

$$\hat{e}_{k,min} \leq \hat{e}_k \leq \hat{e}_{k,max}$$

M3: In the case of a closed shell[20] EM the diagonal entries of the be-matrix are even; and each $e_{kk} = 0,2,4,6,8$.

M4: The sum T over all entries is equal to the total number of valence electrons of the EM, and corresponds to the number T_0 of valence electrons in $A(n)$,

$$T = \sum_{k=1}^{n} e_k = \sum_{k=1}^{n} e_{kk}^{\circ} ,$$

M5: Two be-matrices represent the same EM if they are interconvertible by permutations of rows/columns that refer to atoms belonging to the same chemical element.

M6: For an EM containing m distinct molecules, there are $m!$ be-matrices in block form, where each block represents a distinct molecule of the EM.

M7: In the be-matrices E of restricted chemistry (see Section 2.4.2) we have $e_k = e_k^{\circ}$, $\hat{e}_k = 2$ for H, and $\hat{e}_k = 8$ for the other elements.

M8: The strongly contributing valence bond structures of resonance systems with delocalized[22] π-electrons are represented by a class of be-matrices which differ only in bond orders and diagonal entries but not in connectivities, i.e. if one replaces all non-zero e_{ij} in these be-matrices by 'ones' and the diagonal entries by 'zeros' they all yield the same adjacency matrix.

The following representations for the chemical constitution of formaldehyde may illustrate various types of labelled graphs for its constitution and the corresponding matrix representations.

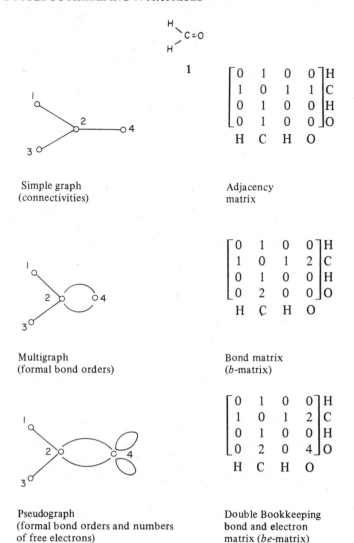

Simple graph
(connectivities)

Adjacency
matrix

Multigraph
(formal bond orders)

Bond matrix
(b-matrix)

Pseudograph
(formal bond orders and numbers
of free electrons)

Double Bookkeeping
bond and electron
matrix (be-matrix)

The be-matrices E can be related to permutation matrices.[25] An $n \times n$ permutation matrix is an $n \times n$ matrix having n entries 1 and all the rest zero, such that no two of the 'ones' are in a common row or common column.

Let e_k be the sum of the kth row, and let

$$e = \max(e_1 - e_{11}, e_2 - e_{22}, \ldots, e_n - e_{nn})$$

that is, the largest row sum when the diagonal term is omitted. If we remove each diagonal term and replace it by $\tilde{e}_i = e - (e_i - e_{ii})$, then clearly all entries in this new matrix $\tilde{E}$ are non-negative, and all row sums and all column sums have the common value $e \neq 0$. Now a theorem of König says that any such matrix must have a system of non-zero entries with no two of them being on the same row or column.[69] Let P_1 be the permutation matrix with the n 'ones' on those rows and columns, and consider $\tilde{E} - P_1$; this still has all entries non-negative, but this time each row/column sum is $e - 1$. Thus, we repeat the argument with this new matrix; after e steps we find $\tilde{E} - P_1 - \ldots - P_e$ is a matrix of non-negative terms, with each row/column sum zero. This matrix is therefore the zero matrix so that $\tilde{E} = P_1 + \ldots + P_e$ and, recalling how $\tilde{E}$ was formed, we find

$$E = \sum_1^e P_i - \mathrm{diag}(e\!-\!e_1, \ldots, e\!-\!e_n).$$

Thus E is, except for a diagonal matrix, a sum of exactly e permutation matrices. We remark that, although E is a symmetric matrix, it is not true that the P_i can be chosen to be symmetric; nor is it true that the representation of E as a sum of permutation matrices is unique.

There is a variety of chemical species whose constitution cannot be adequately described in terms of electron pair bonds between pairs of cores. Examples are the boron hydrides,[80] the 'transition' metal complexes of hydrocarbons and their anions, such as π-allyl nickel[7,51] or ferrocene,[105] or the hypothetical intermediate of the limiting S_N2 mechanism.[45] In these cases the molecular structure can be adequately represented by the model concepts of multicenter bonds and fractional bond orders.

EM with fractional bond orders can be represented by be-matrices N having rational entries which satisfy the properties M1 to M7 and, additionally, the following:

M9: The off-diagonal entries are $e_{ij} = x_{ij}/y_{ij}$; $x_{ij} = 1,2,3, \ldots$ (in most cases $x_{ij} = 2$) is the number of valence electrons that belongs to the multicenter bond in which A_i and A_j participate, and $y_{ij} = 2,3, \ldots$ the number of cores that share the electrons.

Even if only inadequate information on the chemical constitution of a species is available, the double bookkeeping properties of be-matrices avoid serious errors in the algebra of the rational be-matrices and their reaction matrices.

For example, the chemical constitution of diborane is rather described by **2a** than by **2b**.

2a   **2b**

However, both can be used for the rational be-matrices N, but **2b** with ½ bonds is more convenient, than **2a** with $\frac{1}{3}$ bonds and formal $\frac{1}{3}$ electrical charges on two of the hydrogens and the borons. The rational $n \times n$ be-matrices N can be converted into equivalent $n \times n$ matrices with integral entries by multiplication with the smallest common denominator of the fractional bond orders. By this the algebra of the integral be-matrices and of their reaction matrices can also be used for chemical systems with fractional bond orders.

2.3 Chemical metric and the constitutional aspects of chemistry

Every $n \times n$ be-matrix E is symmetric, so that the entries $\{\, e_{ij} \mid 1 \leqslant i,j \leqslant n \,\}$ on the diagonal and the upper triangle determine the matrix E uniquely. We define the *chemical distance*[25] between two matrices E, E' belonging to the same FIEM by

$$D(E, E') = \sum_{1 \leqslant i,j \leqslant n} |\, e_{ij} - e'_{ij} \,|$$

This is a true metric in the set of matrices belonging to the given FIEM; it is simply twice the number of bonds and electrons involved in an interconversion of E to E', and measures how closely E and E' are related with regard to that constitutional aspect.

The *chemical metric* also has a geometric interpretation.* Observe first that any $n \times n$ matrix E corresponds to a point of $\mathbb{R}^{n^2}$ (or to a vector, if $\mathbb{R}^{n^2}$ is regarded as a vector space): specifically, the mapping $E \rightarrow (e_{11}, \ldots, e_{1n}, \ldots, e_{n1}, \ldots, e_{nn})$ is an embedding of the set of all $n \times n$ matrices into $\mathbb{R}^{n^2}$. Since be-matrices are symmetric, we can embed the set of matrices belonging to an n-element FIEM into $\mathbb{R}^{n(n+1)/2}$. Taking the distance between E, E' to be the usual Euclidean distance between their counterparts, we find

$$d(E, E') = \sqrt{\sum_{i,j} (e_{ij} - e'_{ij})^2} \ .$$

Now, because $d(E, E') \leqslant D(E, E') \leqslant d(E, E') \cdot \sqrt{n(n+1)/2}$.

the chemical distance and the Euclidean distance both determine the identical topology in $\mathbb{R}^{n(n+1)/2}$.

Thus, the chemical metric on the EM of an FIEM provides not only a formalism for constitutional chemistry, but also allows us to fully use the *properties of Euclidean spaces in expressing the fundamental logical structure* of the FIEM.

---

*To our knowledge, the term *chemical topology* has been introduced by V. Prelog[99] for the description of those steric features of molecules which are related to their Pólya configurations[39,98,108-110,123] (Section 3.1; see also ref. 18).

2.4   Chemical reactions and reaction matrices

2.4.1   Elementary processes

With the above two center bond concepts of chemical constitution, there are two *elementary processes* by which the redistribution of valence electrons in a FIEM occurs:

R1: *redox* processes, i.e., the exchange of electrons between the pairs of atoms, $A_i \cdot + \cdot A_j \rightleftharpoons A_i + : A_j$;

R2: *bond making* by electron sharing, and its reverse, *homolysis*, $A_i \cdot + \cdot A_j \rightleftharpoons A_i - A_j$.

The following simple reactions may serve as an illustration:

$(R1_\rightarrow)$ $\qquad\qquad\qquad\qquad\qquad$ $Li^{\cdot} + {}^{\cdot}H \rightarrow Li^{\oplus} + :H^{\ominus}$

$(R2_\rightarrow)$ $\qquad\qquad\qquad\qquad\qquad$ $H^{\cdot} + {}^{\cdot}H \rightarrow H - H$

$(R1_\leftarrow + R2_\rightarrow)$ $\qquad\qquad\qquad\quad$ $H^{\oplus} + :\ddot{F}:^{\ominus} \rightarrow H - \ddot{F}:$

$(R2_\leftarrow + 2R2_\rightarrow)$ $\qquad\qquad\quad$ $H - H + \ddot{O}: \rightarrow H - \ddot{O} - H$

$(2R2_\leftarrow + 2R2_\rightarrow)$ $\qquad\qquad$ $H - H + :\ddot{F} - \ddot{F}: \rightarrow 2H - \ddot{F}:$

As we will see in Section 2.4.4, any chemical reaction involving systems with integral bond orders is *uniquely* expressible as a combination of these two elementary reactions.[25]

2.4.2   Closed shell chemistry and restricted chemistry

With few exceptions, in the stable organic compounds all atoms have *closed shells*; precisely:

RC1: The valence shells of all atoms are closed (i.e., singlet states with the same numbers of electrons having α spin as have β spin) and can be approximated by an orbital description in which each orbital is either doubly occupied or empty.[22]

The *be*-matrices of closed shell EM have even diagonal entries ($e_{kk} = 0,2,4,6,8$).

We shall use the term *restricted chemistry* for the chemistry of closed shell molecules which have the following further properties:

RC2: None of the atoms carries net electrical charge, i.e. the number e_k of valence electrons that belong directly to each of the individual atoms A_k is equal to the formal core charge number. That is, it corresponds to the group index of the given element in the Periodic Table of Elements.

RC3: The molecules follow the *Lewis-Langmuir Octet Rule*,[75,79,113] i.e. all atoms, except H, have a valence shell of one s and three p orbitals filled with overall eight electrons, corresponding to a Ne or Ar configuration; H carries two valence electrons in a He configuration.

The majority of stable organic molecules consists of the elements H, C, N, O, F, Si, P, S and Cl and belong to restricted chemistry.

The EM of restricted chemistry are represented by restricted be-matrices $\overset{\circ}{E}$ with $e_k = e_k^{\circ}$, and $\hat{e}_k = 8$, except $\hat{e}_H = 2$.

2.4.3 Double bookkeeping reaction matrices

Representations of EM $1 + 3$ and 4 by be-matrices and their transformations are illustrated by E_{1+3} and E_4 which refer to the EM of Reaction 1.

Reaction 1

$$H^1 - C^2 = O^4 + H^5 - C^6 \equiv N{:}^7 \quad \longrightarrow \quad \begin{matrix} C^6 \equiv N{:}^7 \\ | \\ H^1 - C^2 - O^4 - H^5 \\ | \\ H^3 \end{matrix}$$
$$\underset{\displaystyle H^3}{}$$

$$\quad 1 \qquad\qquad 3 \qquad\qquad\qquad 4$$

$$E_{1+3} = \begin{array}{c} \\ \\ \\ \\ \\ \\ \\ \end{array} \begin{array}{ccccccc} H^1 & C^2 & H^3 & O^4 & H^5 & C^6 & N^7 \\ \left[\begin{array}{ccccccc} 0 & 1 & 0 & 0 & 0 & 0 & 0 \\ 1 & 0 & 1 & 2 & 0 & 0 & 0 \\ 0 & 1 & 0 & 0 & 0 & 0 & 0 \\ 0 & 2 & 0 & 4 & 0 & 0 & 0 \\ 0 & 0 & 0 & 0 & 0 & 1 & 0 \\ 0 & 0 & 0 & 0 & 1 & 0 & 3 \\ 0 & 0 & 0 & 0 & 0 & 3 & 2 \end{array}\right] & \begin{array}{c} H^1 \\ C^2 \\ H^3 \\ O^4 \\ H^5 \\ C^6 \\ N^7 \end{array} \end{array} \qquad (1)$$

$$E_4 = \begin{array}{ccccccc} H^1 & C^2 & H^3 & O^4 & H^5 & C^6 & N^7 \\ \left[\begin{array}{ccccccc} 0 & 1 & 0 & 0 & 0 & 0 & 0 \\ 1 & 0 & 1 & 1 & 0 & 1 & 0 \\ 0 & 1 & 0 & 0 & 0 & 0 & 0 \\ 0 & 1 & 0 & 4 & 1 & 0 & 0 \\ 0 & 0 & 0 & 1 & 0 & 0 & 0 \\ 0 & 1 & 0 & 0 & 0 & 0 & 3 \\ 0 & 0 & 0 & 0 & 0 & 3 & 2 \end{array}\right] & \begin{array}{c} H^1 \\ C^2 \\ H^3 \\ O^4 \\ H^5 \\ C^6 \\ N^7 \end{array} \end{array} \qquad (2)$$

Transformation of E_{1+3} into E_4 can be affected by adding the matrix $R_{1+3\rightarrow4}$ to E_{1+3}. Thus $R_{1+3\rightarrow4} = E_4 - E_{1+3}$ may be called the *reaction matrix* of $1 + 3 \rightarrow 4$.

$$R_{1+3\rightarrow4} = \begin{bmatrix} 0 & 0 & 0 & 0 & 0 & 0 & 0 \\ 0 & 0 & 0 & -1 & 0 & +1 & 0 \\ 0 & 0 & 0 & 0 & 0 & 0 & 0 \\ 0 & -1 & 0 & 0 & +1 & 0 & 0 \\ 0 & 0 & 0 & +1 & 0 & -1 & 0 \\ 0 & +1 & 0 & 0 & -1 & 0 & 0 \\ 0 & 0 & 0 & 0 & 0 & 0 & 0 \end{bmatrix} \qquad (3)$$

Every chemical reaction $B \to E$ is represented by a *reaction matrix* $R = E - B$, the difference between the *be*-matrices of the end products and the starting materials.[9,10,25,101,120,121]

A reaction matrix is essentially a redox and 'make–break' indicator of bonds between pairs of atoms, endowed with *double bookkeeping* capability for bonds and electrons. The off-diagonal entries $r_{ij} = e_{ij} - b_{ij} = 0, \pm1, \pm2, \pm3$ indicate how many bonds between A_i and A_j are made ('+') or broken ('–'). The diagonal entries $r_{ii} = e_{ii} - b_{ii} = 0, \pm1, \dots$ $(- e_{ii} \leqslant r_{ii} \leqslant \bar{e}_{ii,max} - e_{ii})$ indicate how many free electrons A_i gains ('+'), or loses ('–'). A reaction matrix R of a FIEM with n atoms is a symmetric $n \times n$ matrix with entries $r_{ij} = r_{ji}$; for, from $b_{ij} = b_{ji}$ and $e_{ij} = e_{ji}$ follows $r_{ij} = e_{ij} - b_{ij} = e_{ji} - b_{ji} = r_{ji}$.

2.4.4   Elementary reaction matrices

For each pair of integers (i,j) with $1 \leqslant i < j \leqslant n$, let U_{ij} be the $n \times n$ matrix with entries $u_{ii} = +1$, $u_{jj} = -1$, and all others zero. Each U_{ij} represents an elementary redox reaction of type R1$_\leftarrow$ by which one electron is transferred from the atom A_j to the atom A_i in an EM of n atoms; $- U_{ij}$ corresponds to the reverse reaction R1$_\to$. Each U_{ij} is called an elementary reaction matrix and the set $\{U_{ij} \mid 1 \leqslant i < j \leqslant n\}$ is denoted by $U(n)$.

For each pair of integers (i,j) with $1 \leqslant i < j \leqslant n$, let V_{ij} be the $n \times n$ matrix with entries $v_{ii} = v_{jj} = -1$, $v_{ij} = v_{ji} = +1$, and all others zero. Each V_{ij} represents an elementary bond making process R2$_\to$ by which a bond between A_i and A_j is made and A_i and A_j lose one free electron each; $-V_{ij}$ stands for the reverse, the homolytic elementary process R2$_\leftarrow$ of breaking a bond between A_i and A_j. Each V_{ij} is also called an elementary reaction matrix, and the set $\{V_{ij} \mid 1 \leqslant i < j \leqslant n\}$ is denoted by $V(n)$.

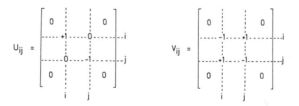

The significance of the elementary reaction matrices is this: Every $n \times n$ reaction matrix can be written as a sum of elementary reaction matrices.[25] In fact, if $R = (r_{ij})$, and if r_i is the sum of the ith row, then

$$R = \sum_{1 \leqslant i < j \leqslant n} r_{ij} V_{ij} + \sum_{i=1}^{n-1} r_i U_{in},$$

and this expression using the V_{ij}, U_{in}, is *unique*; precisely:

There is one, and only one, way to express any reaction matrix as a sum of elementary bond making processes V_{ij} and the particular elementary redox reactions U_{in}.

One could (see Section 2.2) also express each reaction matrix R as a difference of sums of permutation matrices, together with a diagonal matrix; but as remarked before, that representation need not be unique.

2.4.5   Restricted reaction matrices

The reaction matrices $\overset{\circ}{R} \subset R(n)^{2\,5}$ of restricted chemistry have all row/column sums 0, and all diagonal entries 0. In a restricted reaction matrix $\overset{\circ}{R}$, the sum of the terms in the upper triangle (i.e., $\sum_{i<j} r_{ij}$) is zero. Moreover, it is easy to see that any reaction matrix is necessarily singular, i.e. $\det \overset{\circ}{R} = 0$.

There are *no* non-zero restricted reaction matrices of order $\leqslant 3$, but non-zero reaction matrices for each order $\geqslant 4$ exist. In the 4 x 4 case, e.g.

$$
\begin{pmatrix}
0 & 0 & +1 & -1 \\
0 & 0 & -1 & +1 \\
+1 & -1 & 0 & 0 \\
-1 & +1 & 0 & 0
\end{pmatrix}
\quad \text{or also} \quad
\begin{pmatrix}
0 & -1 & 0 & +1 \\
-1 & 0 & +1 & 0 \\
0 & +1 & 0 & -1 \\
+1 & 0 & -1 & 0
\end{pmatrix} ;
$$

and given the existence of such a matrix, a higher-order non-zero one follows by bordering the matrix with 0's in the last row and last column; repeated borderings by

$$
\begin{pmatrix}
0 & -1 \\
-1 & 0
\end{pmatrix}
$$

at beginning and end of the main diagonal,

$$
\begin{pmatrix}
0 & 1 \\
1 & 0
\end{pmatrix}
$$

at the ends of the other diagonal, give such matrices for every odd order; and the difference of two even order symmetric permutation matrices with zeros down the main diagonal provides such matrices for even orders.

No non-trivial restricted reaction matrix of order 4 can have a zero row; for if any 4 x 4 reaction matrix had a zero row, then the corresponding column is also zero; removing that row and column would give a non-zero 3 x 3 reaction matrix, which is impossible.

There are restricted reaction matrices of every order $\geqslant 4$ having no zero row, for example, in the 5 x 5 case,

$$
\begin{pmatrix}
0 & -2 & 0 & 0 & +2 \\
-2 & 0 & +1 & +1 & 0 \\
0 & +1 & 0 & 0 & -1 \\
0 & +1 & 0 & 0 & -1 \\
+2 & 0 & -1 & -1 & 0
\end{pmatrix}
\quad \text{or also} \quad
\begin{pmatrix}
0 & 1 & -1 & +1 & -1 \\
+1 & 0 & +1 & -1 & -1 \\
-1 & +1 & 0 & -1 & +1 \\
+1 & -1 & -1 & 0 & +1 \\
-1 & -1 & +1 & +1 & 0
\end{pmatrix}
$$

There are $\frac{1}{4}\, n(n-1)\,(n-2)\,(n-3)$ restricted $n \times n$ reaction matrices $\overset{\circ}{R}$ with eight non-zero entries ± 1 and all other entries 0. An $n \times n$ restricted reaction matrix is the sum of an even number of elementary reaction matrices $V_{ij} \in V(n)$, $(i \neq j)$ such that the total number of pairs of indices (i,j) is ≥ 4.

Any $n \times n$ restricted reaction matrix $\overset{\circ}{R}$ corresponds to a point in $\mathbb{R}^{n^2}$. Taking the set of all reaction matrices $\overset{\circ}{R}$, we get a set of points $\overset{\circ}{P} \subset \mathbb{R}^{n^2}$.

Let $L(\overset{\circ}{P})$ be the linear subspace spanned by $\overset{\circ}{P}$, i.e. the smallest linear subspace of $\mathbb{R}^{n^2}$ containing $\overset{\circ}{P}$. Then $L(\overset{\circ}{P})$ has dimension $\leq n(n-3)/2$.

The space spanned by the set of reaction matrices having dimension $n(n-3)/2$ means that a maximal linearly independent set of $n \times n$ reaction matrices has no more than $n(n-3)/2$ members. Thus, for each integer n, there is a set

$$\{\overset{\circ}{R}_1, \ldots, \overset{\circ}{R}_s\} \quad s \leq \frac{n(n-3)}{2}$$

of reaction matrices with this property. For any other restricted reaction matrix $\overset{\circ}{R}$, we have

$$\overset{\circ}{R} = \lambda_1 \overset{\circ}{R}_1 + \ldots + \lambda_s \overset{\circ}{R}_s$$

for suitable real $\lambda_1, \ldots, \lambda_s$ (i.e., any restricted reaction matrix is a linear combination of the given ones).

2.4.6   Closed shell reaction matrices

The reactions of closed shell chemistry are described by restricted reaction matrices $\overset{\circ}{R}$ and they are combinations of couples W_{ij} of elementary reaction matrices

$$W_{ij} = (U_{ij} + V_{ij}),$$

which represent *dative bond formation* and *heterolysis* R3 (= R1 + R2):

$$A_i: + A_j \rightleftharpoons A_i - A_j \tag{5}$$

In the couples of elementary reaction matrices $W_{ij} \in W(n)$ all entries are zero except $r_{ij} = r_{ji} = +1$ and $r_{ii} = -2$.

The non-zero diagonal entries of closed shell reaction matrices are even numbers.

2.4.7   Fitting reaction matrices

An $n \times n$ reaction matrix R can be applied to an $n \times n$ *be*-matrix E, only if the result E' is a *be*-matrix; in particular, E' must be a symmetric matrix with non-negative entries.

Since the sum of symmetric matrices is symmetric, the requirement that $E' = E + R$ be a *be*-matrix reduces to: (1) for each negative entry $-\,|\,r_{ij}\,|$ of R

there must be positive entry e_{ij} of E, such that $e'_{ij} = e_{ij} + (- |r_{ij}|) \geqslant 0$; (2) further, since for E and E' there are the restrictions (of M1, M2):

$$\bar{e}_{i,min} \leqslant e_i, e'_i \leqslant \bar{e}_{i,max}$$

$$\hat{e}_{i,min} \leqslant \hat{e}_i, \hat{e}'_i \leqslant \hat{e}_{i,max},$$

we also have the conditions

$$\left. \begin{array}{c} \bar{e}_{k,min} \\ \\ \tfrac{1}{2}(\hat{e}_{k,min} + e_{kk} + r_{kk}) \end{array} \right\} \leqslant \sum_{i=1} (e_{ik} + r_{ik}) \leqslant \left\{ \begin{array}{c} \bar{e}_{k,max} \\ \\ \tfrac{1}{2}(\hat{e}_{k,max} + e_{kk} + r_{kk}) \end{array} \right.$$

We shall say that R *fits* E if $E + R$ is the *be*-matrix of some member of the FIEM of E. Each *be*-matrix $E \in F$ determines a set $R_E \subset R(n)$ of fitting reaction matrices.

Suppose that a reaction matrix R fits a *be*-matrix E, and let $R = (R_1 + \ldots + R_a) + (R_{a+1} + \ldots + R_b) + (R_{b+1} + \ldots + R_c) + (R_{c+1} + \ldots + R_d)$ be the unique expansion of R in terms of the elementary reaction matrices, grouped together so that:

$R_1, \ldots, R_a$ are the elementary homolytic terms with $r_{ij} < 0$,

$R_{a+1}, \ldots, R_b$ are the elementary redox terms with $r_i < 0$,

$R_{b+1}, \ldots, R_c$ are the elementary redox terms with $r_i > 0$,

$R_{c+1}, \ldots, R_d$ are the elementary homoaptic terms with $r_{ij} > 0$.

Then for each $1 \leqslant s \leqslant d$, R_s fits $E + R_1 + \ldots + R_{s-1}$ (that is, R_1 fits E, R_2 fits $E + R_1$, R_3 fits $E + R_1 + R_2$, etc.).

The analogous result is true in closed-shell chemistry.[25]

The resonance structures[84] of an unsaturated EM are a subset of the FIEM whose *be*-matrices contain at least one entry $e_{ij} \geqslant 2(i \neq j$ or $i = j)$ and are interconverted by reaction matrices which fit without breaking connectivities, i.e.

$$\sum_{i \neq j} e_{ij} / |e_{ij}| = \sum_{i \neq j} (e_{ij} - r_{ij}) / |e_{ij} - r_{ij}|;$$

we call this property *resonance fitting*. Resonance fitting can be used to find the complete set of the resonance structures of an EM.

2.4.8   Isoconjunction of cyclic structures and processes

According to the Dewar–Evans Rules[21,31-33] thermal reactions of Hückel systems (anti-Hückel systems)[20,21] with a *cyclic transition state* which is isoconjugate to aromatic hydrocarbons, are preferred (disfavored) over those with anti-aromatic isoconjugate transition states. The opposite is stated for photochemical reactions.

The Dewar–Evans Rules correspond to a simple and comprehensive theory of concerted reactions, and related phenomena; they can be regarded as a counterpart of the Woodward–Hoffmann Rules,[131] as well as corresponding statements.[38,92,133] In this context, the Dewar–Evans Rules, in contrast to their counterparts, can be formalized in a form suited for the selection of preferred concerted reactions and their steric course (see Section 5) on the basis of the algebraic classification of reaction matrices, if the Hückel, or anti-Hückel character of the systems is given due consideration.[20,21]

In fact, the presence of cyclic structures in an EM can be recognized from its be-matrix.

Let E be the given $n \times n$ be-matrix, and let m be the number of molecules in the EM that the matrix E represents (this is the number of blocks down the main diagonal; see M6). Let s be the number of off-diagonal entries in the upper triangle (i.e., the number of non-zero e_{ij} with $i < j$). Then E contains a cyclic structure if and only if $s + m > n$; and $s + m - n$ is the number of distinct (i.e. linearly independent) cyclic structures.

Let E be the be-matrix of an EM whose constitution contains at least one c-membered cyclic moiety such that $4p + 1 \leqslant c \leqslant 4p + 3$, $(p = 0,1,2, \ldots)$. Then a restricted reaction matrix with $4p + 2$ pairs of non-zero entries ± 1 per row/column and all other entries 0 which fits E

$$E + R = E'$$ (6)

corresponds to a reaction whose *transition state* is *isoconjugate*[9,10,21,23,31-33,101] with an *aromatic hydrocarbon* with a c-membered ring and $4p + 2$ π-electrons.

Analogously, a be-matrix E for an EM having a c-membered ring with $4p - 1 \leqslant c \leqslant 4p + 1$, and a restricted reaction matrix $\overset{\circ}{R}$ with $4p$ pairs of *non-zero entires* ± 1 per row/column and all other entries 0, represents a reaction whose transition state is isoconjugate with an *antiaromatic hydrocarbon* with a c-membered ring and $4p$ π-electrons.[20]

By this criterion, the interconversion of the hypothetical Kekulé structures for benzene **5a** and **5b**, the Diels–Alder reaction **6** → **7**,[84] the Cope rearrangement **8** → **9**,[84] the synthesis **10** → **11**,[84] the 1,4-addition of Grignard reagents to α,β-unsaturated carbonyl compounds by a cyclic mechanism **12** → **13**,[84] the McLafferty reaction **14** → **15**,[13,82] all represent reactions whose transition states are isoconjugate to benzene. In fact, all of these reactions are represented by matrices $\overset{\circ}{R}$ whose entries are zero except for the block $\overset{\circ}{R}'$ or its equivalent which can be transformed

$$
\overset{\circ}{R}{}' = \left\{
\begin{array}{cccccc}
0 & -1 & 0 & 0 & 0 & +1 \\
-1 & 0 & +1 & 0 & 0 & 0 \\
0 & +1 & 0 & -1 & 0 & 0 \\
0 & 0 & -1 & 0 & +1 & 0 \\
0 & 0 & 0 & +1 & 0 & -1 \\
+1 & 0 & 0 & 0 & -1 & 0
\end{array}
\right\}
\tag{7}
$$

by permuting the indices of equivalent atoms (see M4, Section 2.2).

Reaction 2

5a   5b

6   7

8   9

10   11

12   13

14   15

2.5   Graphs of FIEM

A labelled graph[48] $T(E,L)$ whose set of points E represents the members of an FIEM and whose connecting lines L correspond to the chemical interconversions of the EM $\in$ FIEM can be used to represent the interconnections of the FIEM, and is called a graph of the FIEM.

The labelled set of points E is in $1-1$ correspondence with the family of *be*-matrices of the FIEM, and the connecting lines correspond to the fitting reaction matrices R_E. Thus the complete graph of the FIEM is obtained in which each point is connected with each other point. The complete graph of an FIEM represents only the stoichiometric interconvertibility of EM, and is therefore of very limited usefulness in the treatment of chemical problems. However, some subgraphs of the complete FIEM graphs, in particular those subgraphs whose connecting lines correspond to certain types of reaction matrices, not only lead to general insights into the fundamental logical structure of chemical systems, but also show great promise to be directly useful as a device for the analysis of reaction mechanisms, mass spectra, as well as non-empirical or semi-empirical design of syntheses, and the elucidation of biosynthetic pathways.

The following types of FIEM subgraphs are based upon the chemical metric, and are of specific interest in this context:

G1: The *limited graphs* $T(E,L_D) \subset T(E,L)$ with the points of E connected by lines of $L_D \subset L$, where L_D is the set of lines which connects points having a chemical distance $D \leqslant D_{lim}$ (see Section 2.3); the upper bound D_{lim} is chosen in accordance with the chemistry which is being represented.

G2: The *elementary graphs* $T(E,L_{el}) \subset T(E,L)$ have connecting lines of $L_{el} \subset L$ which correspond to the elementary reaction matrices of $U(n) \cup V(n)$. (See Section 2.4.4).

G3: The limited *closed shell graphs* $T(E_R^c, L_{RD}^c) \subset T(E,L)$; the points of $E_R^c \subset E$ refer to the closed shell EM_R^c of the FIEM, and the lines of $L_{RD}^c \subset L$ connect the points whose chemical distance $D \leqslant D_{lim}$ (see Section 2.4.2).

G4: The *restricted graphs* $T(E_R^\circ L_{RD}^\circ) \subset T(E,L)$; the points of $E_R^\circ \subset E$ represent the restricted EM_R° of the FIEM, and the lines of $L_{RD}^\circ \subset L$ connect the points whose chemical distance $D \leqslant D_{lim}$.

2.5.1   Multistep reaction mechanisms

The set of conceivable multistep reaction mechanisms[84] for the conversion of an EM into another EM' in terms of sequences of intermediates corresponds to those oriented pathways which connect the points p and p' in the elementary graph G2 of the FIEM to which EM and EM' belong. The points along a pathway from p to p' represent the intermediates characteristic of the reaction mechanism.

The set of reaction mechanisms can be represented by an oriented subgraph of G1, e.g. Graph 1 in which the labels have been omitted, except for p and p'.

Graph 1

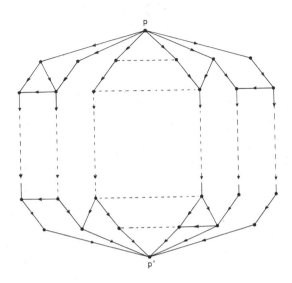

In general it is advisable to consider only those pathways for which the sum over all chemical distances between points along the pathway does not exceed a suitably chosen bound.

The orientation of a reaction mechanisms graph follows from kinetic and thermodynamic considerations on the basis of empirical information. By establishing a reaction mechanisms graph in a systematic manner, it is possible to make sure that complete sets of mechanisms are considered. This is particularly important when an attempt is made to elucidate a reaction mechanism *per exclusionem.*

2.5.2   Mass spectroscopy

The decomposition of parent ions $P^{\oplus}$ by processes which occur in mass spectroscopy[13] proceed within the FIEM defined by the empirical formula of $P^{\oplus}$. These multistep processes proceed stepwise in the direction of increasing numbers of molecular entities in the EM. The observed mass spectrum relates to the molecular species that are contained in some characteristic subset of the FIEM, and are represented by a subgraph of G1 or G2.

The mass spectroscopic processes are described by oriented graphs with the points $p_{m\mu}$ referring to the FIEM that are formed by decomposition of $P^{\oplus}$ (see Graph 2 where the indices $m\mu$ are used as labels of points). This approach differs from that of an existing machine-assisted procedure[12] for the interpretation of mass spectra based upon chemical analogy.

Graph 2

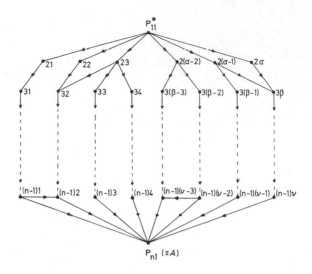

2.5.3   Deductive design of syntheses

In the recent past several computer-assisted procedures for the design of syntheses have been developed.[9,16,57,61,116,120,121,125,126] These methods differ in approach but have in common that they are based upon known chemistry, and use empirical information to solve synthetic problems by analogy.

The combination of *be*-matrices and reaction matrices has great potential as a different type of device for the computer-assisted design of syntheses, on the basis of the logical structure of constitutional chemistry, without direct reliance on chemical analogy.

The chemical graph of an FIEM contains synthetic pathways, e.g. according to Scheme 1, for a target molecule Z, if Z belongs to at least one of the $EM \subset FIEM$, and if at least one available set of starting materials $A + A' + B' + \ldots + Y'$ is contained in an $EM \in FIEM$.

Scheme 1[120,121]

$$EM_A = \{A, A', B', C', \ldots, Y'\} \longrightarrow$$
$$\longrightarrow EM_B = \{B'', B, B', C', \ldots, Y'\} \longrightarrow$$
$$\longrightarrow EM_C = \{B'', C'', C, C', \ldots, Y'\} \longrightarrow$$
$$\longrightarrow \quad ------------- \longrightarrow$$
$$\longrightarrow EM_X = \{B'', C'', \ldots, X'', X, X', Y'\} \longrightarrow$$
$$\longrightarrow EM_Y = \{B'', C'', \ldots, X'', Y'', Y, Y'\} \longrightarrow$$
$$\longrightarrow EM_Z = \{B'', C'', \ldots, X'', Y'', Z'', Z\}$$

Conventionally the synthesis of Z from the starting materials $\{A, A', B', \ldots, Y'\}$ according to a *stepwise strategy** is represented by Scheme 2, an equivalent of Scheme 1, or simply by $A \to B \to C \to \ldots \to X \to Y \to Z$.

Scheme 2

$$A + A' \longrightarrow B + B''$$
$$B + B' \longrightarrow C + C''$$
$$C + C' \longrightarrow D + D''$$
$$--- \longrightarrow ---$$
$$--- \longrightarrow ---$$
$$X + X' \longrightarrow Y + Y''$$
$$Y + Y' \longrightarrow Z + Z''$$

Scheme 2 is obtained from Scheme 1 by omitting at each synthetic step all those compounds which do not participate directly in the reaction of that step. The conversion of a conventional synthetic Scheme 2 into a synthetic pathway according to Scheme 1, is effected by addition of the complementary sets of molecules to the reactants of each step of Scheme 2, obtaining EM which contains all compounds that are present at the considered stage of the synthesis, i.e. those starting materials and intermediates that have not yet been consumed and those intermediates and products which have already been formed.[120,121]

In many cases a target molecule Z can also be synthesized by a *fragment strategy** (see Scheme 3) which often is superior to the stepwise strategy.[11,132]

Scheme 1 also represents Scheme 3, if we consider $\{A, B, \ldots, Y\}$ of Scheme 1 as the precursor sets of Z in Scheme 3, i.e. $A = \{A_1, \ldots, A_a\}$, $B = \{B_1, \ldots, B_b\}$, $\ldots$, $Y = \{Y_1, Y_2\}$.

Therefore, not only the stepwise synthesis of Z from A according to Scheme 1 but also a synthesis by a fragment strategy according to Scheme 3, corresponds to a line in the ensemble space of the FIEM.

Let us assume that this $FIEM_Z$ contains also other synthetic pathways for EM_Z from EM_A and other EM of the starting materials. Then the conceivable syntheses for EM_Z correspond to a labelled oriented Graph 3, or Graph 3a, where the participating EM_P are given by the set of points $E_R^C = \{P_0, P_{11}, P_{12}, \ldots, P_a\}$ and the set of *lines* L_{RD}^C represents their chemical interconversions. In Graph 3 the points $P_{p\pi}$ are represented by their indices; p is the *order* of the point $P_{p\pi}$ and indicates the number of synthetic steps that are at least required to convert $EM_{p\pi}$ $(P_{p\pi})$ into $EM_Z(P_0)$. The Graph 3 contains

---

*The stepwise vs. fragment strategies of peptide syntheses may serve as an illustration.[11,59,97,119,120,127,132]

Scheme 3

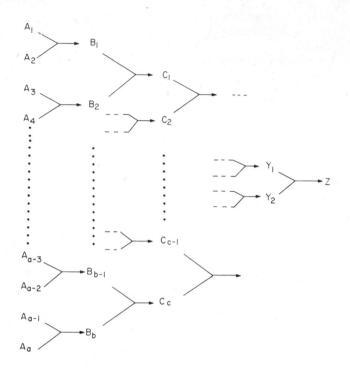

Graph 3

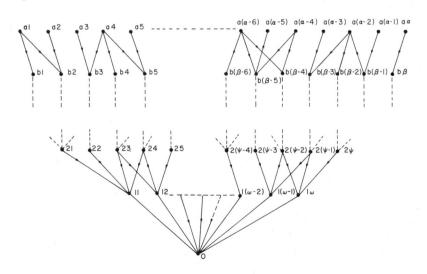

Graph 3a

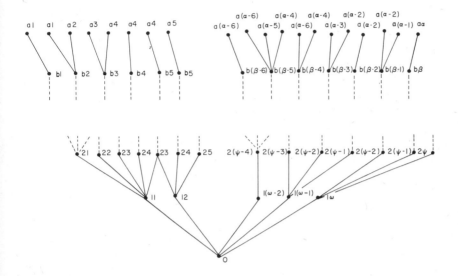

the mappings of sets of 'precursor points' of the order p onto such sets of the order $(p - 1)$.

A graph of synthetic pathways is not necessarily a tree, but a graph of synthetic pathways which contains cycles can be transformed into a tree: the cyclic parts are 'opened' by replacing each labelled point from which ν oriented lines originate ν points.

Thus, the tree of Graph 3a is obtained from Graph 3. The representation of a set of synthetic pathways by a tree rather than by a graph that contains cycles has the advantage that the distinguishable pathways are easier to recognize.

The graph of the synthetic pathways of Z depends upon the FIEM within which the syntheses take place. If syntheses for Z are considered within an $FIEM_i \subset FIEM_{ii}$ then the graph G_i of the synthetic pathways within $FIEM_i$ is a subgraph of G_{ii} which represents the syntheses of Z within $FIEM_{ii}$.

The design of a synthesis of a target molecule Z on the basis of the present concept begins with the choice of the complement EM_Z^c of Z, i.e. the set of by-products $EM_Z^c = \{B'', C'', \ldots, X'', Y'', Z''\}$ (see Scheme 1) that may be expected, such as H_2O, CO_2, NH_3, LiBr, NaCl, CH_3OH, CH_3CO_2H, or $(C_6H_5)_3P = O$. One may also include in EM_Z^c some general formulas, e.g., A − B, X = Y, which may be identified at some later stage of the procedure.[9,120,121] The more potential by-products one includes in EM_Z^c, the more synthetic pathways are conceivable for Z. It is, however, not advantageous to add more conceivable by-products to EM_Z^c than are needed for a small number of conceivable synthetic pathways, because the complexity of the operations that lead to the graph of the synthetic pathways increases considerably with the number of atoms in EM_Z.

A graph of the synthetic pathways is obtained from the $n \times n$ be-matrix E_Z of EM_Z by adding suitable $n \times n$ reaction matrices to E_Z. Generally, the closed shell reaction matrices $\overset{c}{R}$ with $D = 4$, 8 and 12 generate an adequate synthetic pathway Graph 3, a subgraph of G3 from EM_Z.

Even with well-selected EM_Z^c, very large graphs result for complex molecules (i.e. graphs containing a large number of synthetic pathways and intermediates) unless some selective reduction is achieved by eliminating those intermediates and pathways which are not likely to correspond to feasible syntheses.

If the yields of all synthetic steps that Graph 3 contains were known, then the quality of each pathway could be assessed on the basis of the *inverse yields* of the pathways, i.e. the amounts of starting materials that are required for the synthesis of Z by a given set of reactions.

The inverse yields of syntheses[47,119,120] are a better criterion for synthetic pathways than their overall yields, because the overall yield can only be defined in a meaningful manner for stepwise syntheses, but not for syntheses according to a fragment strategy, whereas the inverse yields of all types of syntheses can be determined in a straightforward manner through the inverse yield functions.[47,119,120]

The inverse yields of the pathways of Graph 3 can be computed and compared by using a small amount of computer time,[47,61,119,120] if it is possible to estimate the inverse yields of the synthetic reactions involved. Unfortunately, the latter is confined to syntheses on the basis of a small set of known reactions and their analogues.[47,119,120]

Evans[30] has recently developed a procedure for generating a *charge affinity scheme* for target molecules Z, a refinement of Lapworth's alternating polarity patterns,[76] that shows great potential as a device of synthetic planning. Evans' charge affinity scheme seems to provide selection rules for fitting reaction matrices, and thus for synthetic pathways.

The elimination of reaction matrices which correspond to highly endothermic reactions may also be desirable. This selection can be accomplished by roughly estimating the changes in bond energies that accompany the reactions[121] on the basis of bond energy tables,[84,114] Thus, for instance, one could permit a Beckmann rearrangement[84] but exclude the reverse process.

2.5.4   Elucidation of biosynthetic pathways

In the elucidation of biosynthetic pathways the problem is to establish a metabolic pathway on the basis of the knowledge of the final product Z, the knowledge of some intermediates and processes that may be part of the pathway, and the known set of compounds which are available within a given biochemical system.

The present concept provides a systematic procedure for the elucidation of metabolic pathways, a procedure which is related to both the design of syntheses and the determination of reaction mechanisms.

The metabolic intermediates, with very few exceptions (e.g., if Vitamin B_{12} is involved), belong to closed shell chemistry. Accordingly, a subgraph of G3 is manufactured from the $n \times n$ be-matrix of an EM_{ip} containing the likely initial precursors by repeated fitting of the reaction matrices of $R(n)$, till the be-matrices of all EM_Z are obtained, which contain Z. The resulting graph is then analysed for those pathways which lead from EM_{ip} to EM_Z, and contain the observed intermediates and processes. If only one pathway is found that meets the above requirements, the problem can be considered to be solved; otherwise further experimental evidence is needed, in order to eliminate the alternatives.

3 MONOCENTRIC STEREOCHEMISTRY

3.1 Pólya configurations

If tension barriers were the only barriers that determine the structure of chemical species, there would be no stereoisomers.

Stereoisomers are chemical compounds identical in constitution and differing only in the relative spatial arrangement of their atoms. Directional stability of the chemical bonds is a precondition for the existence of distinguishable stereoisomeric species, i.e. the existence of flexion barriers that keep bond angles constant within certain limits.

The chemical features of a monocentric molecule, i.e. molecules having only one polycoordinated atom, are completely characterized by their constitution and configuration. When one considers the configurational aspect of molecules, one views them as consisting of a skeleton together with a set of ligands attached to the skeleton. The configuration of a monocentric molecule is defined by its skeleton and the distribution of its ligands on the valence skeleton of the central atom. The model concept of a monocentric configuration is also useful for the treatment of many stereochemical problems which involve polycentric molecules, because often the essential stereochemical features are representable by a monocentric configuration model.

The following heuristic criteria are generally useful in the dissection of molecules, and permit an assignment of the various parts of molecules to the skeleton or the ligand set without pretending to define the terms.[26,39,40]

(a) Only one single, or multiple covalent bond is dissected, in separating a ligand from the residual molecule.
(b) What remains after removal of all ligands is the skeleton.

If there are several ways of determining the ligands and the skeleton according to the above criteria, it is advantageous to take that skeleton which is common to as many of the molecules under consideration as possible. With more than one monocentric unit, the overall configuration of molecules with inhibited internal rotations is also determined by the dihedral angles of the bond system between monocentric units. Since the latter may no longer be treated as independent, a representation by polycentric configurations is advisable.

By a monocentric *skeleton*, we shall mean a single atom in its hypothetical valence state; a skeleton is therefore characterized by its group of rotational symmetries and its coordination number.

Two constitutionally equivalent molecules are said to have the same configuration if one can be converted into the other by a permutation of equivalent ligands.

The monocentric skeletons correspond to the most often encountered hypothetical valence states of the chemical elements in their covalent compounds.

The rigid models of monocentric skeletons I–VIII and their combinations in polycentric skeletons[44] suffice to generate the stable configurational features of the majority of stereoisomers.

For most purposes the skeletal sites and ligands are represented by points which in the stereochemically equivalent subsets[39] all have the same distance from the skeletal atom to which they are connected covalently.

The polycentric skeletons, in essence combinations of monocentric skeletons I–VIII, can be formally treated as monocentric skeletons that contain a *superatom* as the central moiety. The concept of superatoms* is most useful in the treatment of the stereochemistry of polycentric molecules. For example, the cyclobutane skeleton, containing four tetrahedral, T_d, monocentric skeletal units can be considered as an octavalent skeleton, or superatom, with a D_{4h} symmetry.

The monocentric molecules with the skeletons I–III can only exist as one isomer. With the skeletons IV and VI all ligands of monocentric molecules must be different in order to have distinguishable isomers, e.g. a pair of chiral antipodes (see below). In the case of skeleton V a ligand set {A,A,B,B} already leads to a pair of diastereomers, and with four distinguishable ligands three isomers result.

We shall use the term permutation isomers[44,123] for those isomers that have the same skeleton and ligand set but differ only by the skeletal distribution of the ligands. With a given skeleton and set of ligands, a permutation isomer is described by stating for each ligand which skeletal site it occupies.

With this terminology, the stereoisomers are exactly those permutation isomers which differ only by ligand permutations at constitutionally equivalent positions.

---

*This concept was introduced by J. Lederberg.[78]

I   sp (C_∞)

II sp^2, sp^3 (C_{2v})

III sp^2 (D_{3h})

IV sp^3 (C_{3v})

V   dsp^2 (D_{4h})

VI sp^3 (T_d)

VII dsp^3 (D_{3h})

VIII d^{2}sp$^3(O_h)$

IX (D_{4h}) $\equiv$

X (D_{4h})

According to a concept which was introduced by Pólya[98] for the enumeration of configurations and was further developed by De Bruijn,[18] a set of permutation isomers has in common and is characterized by its set **L** of n ligands and its molecular skeleton with n ligand sites. The permutational equivalencies of ligands are representable by the ligand equivalence group $H \subset S_n$, and the symmetry equivalencies of skeletal positions by the skeletal rotation group $\overset{\circ}{S} \subset S_n$. The distinct permutation isomers **M** are invariant under ligand permutations which correspond to H or $\overset{\circ}{S}$. They may also be called the Pólya configurations.[18,48,98] These are representatives of the double cosets $Hx\overset{\circ}{S}, x \epsilon S_n$ of the subgroups H and $\overset{\circ}{S}$, as was found by Ruch et al.[108] This permits the enumeration of configurations according to the theorem of Frobenius.[36,117]

3.2   Permutational nomenclature

An analysis of the reasons for the difficulties that one encounters in the nomenclatural treatment of chiral ferrocene derivatives, according to the conventional definitions,[14,85] in particular if they belong to *Ruch's class*

B,[107,109] led us to discover the advantages of a permutational nomenclature of configurations.[123]

The permutational descriptors are defined on the basis of a double numbering system of the skeleton and ligands. This permits us to separately take into account skeletal symmetries and ligand equivalencies. The permutational descriptor of a configuration M_σ is a permutation P_σ of ligand indices which transforms the representation of its reference isomer M_0 by a mapping of ligand indices onto the skeletal indices into a mapping that corresponds to M_σ.

This permutational representation of the mapping of a set of ligand indices, partially symmetrized relative to H, onto the set of skeletal indices, partially symmetrized relative to $\overset{\circ}{S}$ affords not only a universal and consistent nomenclature system for permutation isomers, but also leads to the graphs of permutational isomerization processes.

Let

$$E = \begin{pmatrix} 1, \ldots, n \\ 1, \ldots, n \end{pmatrix}$$

be the reference isomer. This is interpreted to mean that particular geometric configuration in which ligand 1 is attached to site 1, etc.. A permutation λ of the *ligands* changes the reference isomer E to the geometric configuration

$$\begin{pmatrix} \lambda^{-1}(1), \ldots, \lambda^{-1}(n) \\ 1, \ldots, n \end{pmatrix}$$

in which site 1 is equipped with the ligand $\lambda^{-1}(1)$ which replaces ligand 1, etc.; this geometric configuration is exactly the same as

$$\begin{pmatrix} 1, \ldots, n \\ \lambda(1), \ldots, \lambda(n) \end{pmatrix} .$$

A permutation s of the *skeletal* indices changes the reference isomer E to the geometric configuration

$$\begin{pmatrix} 1, \ldots, n \\ s(1), \ldots, s(n) \end{pmatrix}$$

since the ligand 1 will now appear at site $s(1)$, etc. In particular, we note that each configuration obtained from E by permuting skeletal indices can be regarded as obtained by permuting the ligands only, i.e. any skeletally symmetric form of E can be obtained from E by a suitable ligand permutation. Thus, the elements of the symmetric group S_n interpreted as ligand permutations serve to indicate the various permutation isomers obtained from E.

The role of the reference isomer E, and the numbering of the ligands/sites is purely auxiliary:[26] it serves only to identify the physical configuration represented by the given ligand permutation. Any other numbering, and/or any other reference isomer $\hat{E}$ can be used: there is, then, a single inner automorphism of the ligand set which changes the ligand permutation transforming $\hat{E}$ to any given geometric configuration A into that transforming E to A. We remark that, because of this characteristic, any property of a molecule based on some property of its possible ligand permutations is physically meaningful if and only if the ligand permutation property is invariant under inner automorphisms of the ligand permutation group.

3.3 Pentacoordinate molecules

3.3.1 Descriptors

The example of pentacoordinate molecules with a TBP skeleton VII may serve as an illustration of the permutational descriptors of configurations and their interconversions.[26,44,102,103,122,123]

Molecules of this type are conceptually viewed as a set of ligands and a pentacoordinate central atom Z whose valencies form a TBP molecular skeleton VII with D_{3h} symmetry.

Skeleton VII can also be represented by the set of skeletal numbers $s = \{1, \ldots, 5\}$. If we now describe the set of ligands $L = \{L_1, \ldots, L_5\}$ by the set of ligand indices $l = \{1, \ldots 5\}$ we are able to describe the molecule 16 by the formula 17. This is equivalent to

the 2×5 mapping matrix $\left(\begin{smallmatrix} l \\ s \end{smallmatrix}\right)_0$. Thus formula 17 and the matrix $\left(\begin{smallmatrix} l \\ s \end{smallmatrix}\right)_0$ not only correspond to 16 but contain the statement that with skeletal numbering according to VII, the ligand L_1 occupies the skeletal position 1, ligand L_2 occupies skeletal position 2, etc. If we now define the rule by which the ligands are to be indexed, e.g. by employing the *Sequence Rule* of Cahn, Ingold, and

Prelog,[14] then 16, or $\left(\begin{smallmatrix} l \\ s \end{smallmatrix}\right)_0$, corresponds to that isomer which we call for convenience the reference isomer M_0; the reference matrix $\left(\begin{smallmatrix} l \\ s \end{smallmatrix}\right)_0$ is generated when the ligand index set l is mapped onto the skeletal index set s so that equal numbers are 'paired'.

From the reference isomer $17 = M_0$ we obtain its permutation isomers M_σ by applying the permutations P_σ to the ligands $\{L_1, \ldots, L_5\}$, or the permutations Q_σ to the skeletal positions according to eq. (8).

$$\binom{l}{s}_\sigma = P_\sigma \binom{l}{s}_0 = \binom{P_\sigma l}{s}_0 = Q_\sigma \binom{l}{s}_0 = \binom{l}{Q_\sigma s} \tag{8}$$

Representative permutations P_σ or Q_σ respectively can be used as *permutational descriptors* of the isomers of **16**.[125] Two mappings

$$\binom{l}{s}_\sigma \quad \text{and} \quad \binom{l}{s}_{\sigma'}$$

representing the permutational isomers M_σ and $M_{\sigma'}$, belong to the same molecule, i.e.

$$\binom{l}{s}_{\sigma'} = \binom{l}{s}_{\sigma'}$$

and $M_\sigma = M_{\sigma'}$, if they differ only by the order of the connected pairs of the ligand indices l and the skeletal indices s. With no skeletal symmetry, a pentacoordinate molecule with five different ligands could yield $5!$ distinguishable isomers corresponding to the $5!$ different mappings of the set l onto the set s represented by their $\binom{l}{s}$ matrices.

If, however, some elements of skeletal symmetry exist, as in VII, for example, then the number of distinguishable isomers is reduced. Owing to the D_{3h} symmetry of VII, certain permutational isomers resulting from skeletal or ligand permutations on **16** can be interconverted by rotations belonging to the D_{3h} symmetry group and therefore are equivalent. The D_{3h} of VII is represented by the group $S_{VII} = \overset{\circ}{S}_{VII} \cup \overline{S}_{VII}$ where

$$\overset{\circ}{S}_{VII} = \{e,(1\ 2\ 3),(1\ 3\ 2),(1\ 2)(4\ 5),(1\ 3)(4\ 5),(2\ 3)(4\ 5)\} \tag{9}$$

is the skeletal rotation group and

$$\overline{S}_{VII} = \{(1\ 2),(1\ 3),(2\ 3),(4\ 5),(1\ 2\ 3)(4\ 5),(1\ 3\ 2)(4\ 5)\} \tag{10}$$

represents the skeletal reflections.

The C_2 rotations ($= 180°$) of the skeleton VII correspond to the permutations $(1\ 2)(4\ 5)$, $(1\ 3)(4\ 5)$ and $(2\ 3)(4\ 5)$, and the C_3 rotations ($\doteq 120°$) of VII are represented by the permutations $(1\ 2\ 3)$ and $(1\ 3\ 2)$ of the skeletal positions. Here, $(1\ 2\ 3)$ means position 1 replaces position 2, position 2 replaces position 3 and position 3 replaces 1, i.e.,

These five and the identity operation e ('do nothing') form a group of permutations. The rigid skeletal rotation group $\overset{\circ}{S}$ which contains the above permutations and is of order $|\overset{\circ}{S}| = 6$ represents the rotational symmetries of the rigid skeletal model VII. Molecules whose $\begin{pmatrix} l \\ s \end{pmatrix}$ mappings are transformed into each other by permutations belonging to $\overset{\circ}{S}$ are equivalent. With five different ligands on the skeleton VII there are $5!/6 = 20$ distinguishable permutational isomers corresponding to 10 pairs of antipodes. When two of the ligands of **16** are not distinguishable, there are 10 isomers (rather than 20), with 3 pairs of antipodes.

If the ligand set $\mathbf{L} = \{L_1, \ldots, L_5\}$ contains sets of equivalent ligands rather than five different ones, these equivalencies must be accounted for by permutations belonging to the ligand equivalence group H which represents the permutational symmetry of the ligand set.

Formerly, pentacoordinate permutational isomers with a TBP skeleton were nomenclaturally characterized by defining some arbitrary isomer (e.g. the most stable or the first discovered) to be a reference isomer. Ligands are then numbered as a function of their skeletal position, e.g. as in **18**, in an arbitrarily chosen spatial orientation. This results in what is called the 'a' isomer[77] (or the '12'-isomer, or the '21'-isomer,[50] the ligands L_1 and L_2 occupy the apical positions).

$$
\begin{array}{c}
L_1 \\
L_3 \!\!-\!\!\!\!\begin{array}{c}\!\!\!\!\\ \end{array}\!\!\!\!\cdots L_4 \\
\searrow L_5 \\
L_2 \\
\mathbf{18}
\end{array}
$$

These 'binary descriptors', in contrast to our permutational descriptors, do not always represent the ligand or skeletal permutations that result from permutational isomerizations (e.g. of the reference isomer). Within the binary correlation system one can take also into consideration ligand equivalencies and a correspondence of binary and permutational descriptors can be achieved.

However, the binary descriptors have one distinct advantage: in the case of chiral isomers, the descriptors of antipodes can be immediately recognized as they are inverse.

3.3.2   The classification of permutational isomerizations, processes, mechanisms, and cycles

The classification of permutational isomerizations (PI) will be discussed in this Section with particular reference to pentacoordinate molecules with a trigonal-bipyramidal skeleton (TBP). However, some aspects of this classification are better illustrated by examples which involve molecules of other skeletal classes.

Pentacoordinate compounds, **16**, can undergo regular and irregular permutational isomerizations (PI). Regular PI proceed without rupture and reformation of bonds while this is not true for the irregular PI. Intermediates of irregular PI differ in coordination number relative to starting material and products. If bonds are first cleaved and then reformed, the intermediates are lower in coordination number compared to **16**, while in the event that bonds are formed first, then cleaved intermediates possess higher coordination numbers.[44,86]

The interconversion of permutational isomers by regular PI is possible only if the molecular skeleton involved is flexible, i.e. if by deformation of bond angles or by rotations about bonds the molecular skeleton can be represented by some rigid model with indexed positions which have the identical coordination numbers at all skeletal centers.

An instance of monocentric skeletal flexibility is the indexed TBP model VII, which is transformed by certain bond angle deformations (e.g. BPR or TR) with preservation of coordination number 5, into other equivalent reindexed TBP skeletons.

The interconversion $M_\sigma \rightleftharpoons M_{\sigma'}$ of two permutational isomers M_σ and $M_{\sigma'}$ is described by the permutations $I_{\sigma\sigma'} = Q_{\sigma'}Q_\sigma^{-1}$ and $I_{\sigma'\sigma} = I_{\sigma\sigma'}^{-1}$, which are called the *isomerizers*; they are derived from the skeletal descriptors Q_σ and $Q_{\sigma'}$ as shown in eq. (11) and (12).

$$\begin{pmatrix} l \\ s \end{pmatrix}_{\sigma'} = Q_{\sigma'}Q_\sigma^{-1}Q_\sigma \begin{pmatrix} l \\ s \end{pmatrix}_0 = Q_{\sigma'}Q_\sigma^{-1} \begin{pmatrix} l \\ s \end{pmatrix}_\sigma \qquad (11)$$

$$\begin{pmatrix} l \\ s \end{pmatrix}_\sigma = Q_\sigma Q_{\sigma'}^{-1}Q_\sigma \begin{pmatrix} l \\ s \end{pmatrix}_0 = Q_\sigma Q_{\sigma'}^{-1} \begin{pmatrix} l \\ s \end{pmatrix}_{\sigma'} \qquad (12)$$

A *permutational isomerization* has an individual characteristic idealized pathway of motion and a given initial and final situation and is represented by an isomerizer.

Any two permutational isomerizations belong to the same *process*, i.e. are resultwise equivalent, if their results can be interconverted by rotations belonging to the skeletal symmetry. Any two isomerizers I and I' belong to the same process, if they are in the same right coset of $\overset{\circ}{S}$, i.e. $I' = I \cdot g$ with $g \in \overset{\circ}{S}$.

The permutation (2435) transforms **19a** into **19a'**. The same isomer results by applying the permutation (124)(35) as well, shown as **19b** → **19b'**.

Scheme 4

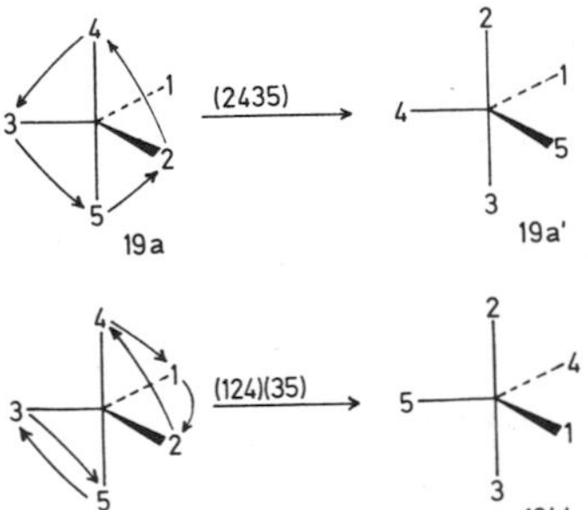

TABLE I

Comparison of the binary and permutational descriptors of pentacoordinate molecules with a TBP skeleton

Five different ligands $L_1,\ldots,L_5$	Binary descriptors			Permutational descriptors									
	(a)	(b)	(c)	$L_1 = L_2$	$L_2 = L_3$	$L_3 = L_4$	$L_4 = L_5$	$L_1 = L_2$, $L_3 = L_4$	$L_2 = L_3$, $L_4 = L_5$	$L_1 = L_2$, $L_4 = L_5$	$L_1 = L_2 = L_3$	$L_2 = L_3 = L_4$	$L_3 = L_4 = L_5$
a	21	$\frac{12}{12}$	(1 4)(2 5)	(1 4)(2 5)	(1 4)(2 5)	(1 4)(2 5)	(1 4)(2 5)	(1 4)(2 5)	(1 4)(2 5)	(1 4)(2 5)	(1 5)	(1 5)	(1 4)(2 5)
ā	12	$\frac{12}{12}$	(1 4 2 5)	(1 4)(2 5)	(1 4)(3 5)	(1 4)(2 5)	(1 4)(2 5)	(1 4)(2 5)	(1 4)(2 5)	(1 4)(2 5)	(1 5)	(1 5)	(1 4)(2 5)
b	31	$\frac{13}{13}$	(1 4 3 5)	(2 4)(3 5)	(1 4)(2 5)	(1 4 5)	(1 4)(3 5)	(1 4)(2 5)	(1 4)(3 5)	(2 5)	(1 4)	(1 4)	(1 4)
b̄	13	$\frac{13}{13}$	(1 4)(3 5)	(1 4)(3 5)	(1 4)(3 5)	(1 5)	(1 4)(3 5)	(1 4)(3 5)	(1 4)(3 5)	(1 5)	(1 4)	(1 4)	(1 4)
c	41	$\frac{14}{14}$	(1 4 5)	(2 5)	(1 5)	(1 4 5)	(1 4)	(2 5)	(1 4)	(2 5)	(1 5)	(1 4)	(1 4)
c̄	14	$\frac{14}{14}$	(1 5)	(1 5)	(1 5)	(1 5)	(1 5)	(1 5)	(1 5)	(2 5)	(1 4)	(1 5)	(1 4)
d	51	$\frac{15}{15}$	(1 4)	(1 4)	(1 4)	(1 4)	(1 4)	(1 4)	(1 4)	(1 4)	(1 4)	(1 4)	(1 4)
d̄	15	$\frac{15}{15}$	(1 4 2)	(2 4)	(1 4)	(1 4)	(1 5)	(1 5)	(1 4)	(1 5)	(1 5)	(1 4)	(1 4)
e	32	$\frac{23}{23}$	(2 4)(3 5)	(2 4)(3 5)	(2 4)(3 5)	(2 5)	(2 4)(3 5)	(2 4)(3 5)	(2 4)(3 5)	(2 5)	(2 5)	(2 5)	(2 4)
ē	23	$\frac{23}{23}$	(2 4 3 5)	(1 4)(3 5)	(2 4)(3 5)	(1 2 5)	(2 4)(3 5)	(2 4)(3 5)	(2 4)(3 5)	(1 5)	(2 4)	(2 5)	(2 4)
f	42	$\frac{24}{24}$	(2 5)	(2 5)	(2 5)	(2 5)	(2 5)	(2 5)	(2 5)	(2 5)	(2 5)	(2 5)	(2 4)
f̄	24	$\frac{24}{24}$	(1 2 5)	(1 5)	(3 5)	(1 2 5)	(2 4)	(2 4)	(2 4)	(1 5)	(2 5)	(2 5)	(2 4)
g	52	$\frac{25}{25}$	(1 2 4)	(2 4)	(3 4)	(2 4)	(2 5)	(2 5)	(1 4)	(1 5)	(1 4)	(2 5)	(2 4)
ḡ	25	$\frac{25}{25}$	(2 4)	(2 4)	(2 4)	(2 4)	(2 4)	(2 4)	(1 4)	(1 4)	(1 4)	(2 5)	(2 4)
h	43	$\frac{34}{34}$	(1 3 5)	(3 5)	(2 5)	(3 5)	(3 4)	(3 5)	(3 4)	(3 5)	(3 5)	E	E
h̄	34	$\frac{34}{34}$	(3 5)	(3 5)	(3 5)	(3 5)	(3 5)	(3 5)	(3 5)	(3 4)	(3 5)	E	E
i	53	$\frac{35}{35}$	(3 4)	(3 4)	(3 4)	E	(3 4)	(3 4)	(3 5)	(3 5)	E	(2 5)	E
ī	35	$\frac{35}{35}$	(1 3 4)	(3 4)	(2 4)	(1 2)	(3 5)	(3 5)	(3 5)	(3 5)	E	(2 5)	E
j	54	$\frac{45}{45}$	E	E	E	E	E	E	E	E	E	E	E
j̄	45	$\frac{45}{45}$	(1 2)	E	E	(1 2)	E	E	(2 5)	(2 5)	E	E	E

(a) According to ref. 77; (b) according to ref. 89; (c) according to ref. 50.

Molecules **19a′** and **19b′** are equivalent since they are interconverted by a $120°$ rotation about the skeletal C_3 axis.

In the matrix representation the equivalence of the permutations $(2\ 4\ 3\ 5)$ and $(1\ 2\ 4)(3\ 5)$ upon **19** as well as the permutations $(1\ 2\ 5)(3\ 4)$, $(1\ 3\ 4)(2\ 5)$, and $(1\ 3\ 5)(2\ 4)$ follows from eq. (13)–(15).

$$\binom{l}{s}_{(19a')} = \binom{(2\ 4\ 3\ 5)l}{s} = \binom{(2\ 4\ 3\ 5)\ 1\ 2\ 3\ 4\ 5}{1\ 2\ 3\ 4\ 5}$$

(13)

$$\binom{1\ 5\ 4\ 2\ 3}{1\ 2\ 3\ 4\ 5} = \binom{1\ 2\ 3\ 4\ 5}{1\ 4\ 5\ 3\ 2}$$

$$\binom{l}{s}_{(19b')} = \binom{(1\ 2\ 4)(3\ 5)l}{s} = \binom{(1\ 2\ 4)\ (3\ 5)\ 1\ 2\ 3\ 4\ 5}{1\ 2\ 3\ 4\ 5}$$

(14)

$$= \binom{4\ 1\ 5\ 2\ 3}{1\ 2\ 3\ 4\ 5} = \binom{1\ 2\ 3\ 4\ 5}{2\ 4\ 5\ 1\ 3}$$

The resulting matrices of eq. (13) and (14) represent the same molecule since they differ only by the skeletal permutations $(1\ 3\ 2)$ representing a C_3 rotation of the skeletal given by eq.(15).

$$\binom{1\ 2\ 3\ 4\ 5}{(1\ 3\ 2)\ 1\ 4\ 5\ 3\ 2} = \binom{1\ 2\ 3\ 4\ 5}{2\ 4\ 5\ 1\ 3}$$

(15)

A permutational isomerization can be described either in terms of ligand permutations, or in terms of the skeletal site permutation. Any two ligand permutations belong to the same mechanism, if they can be represented by the same skeletal index permutation belonging to S. This corresponds to what is called 'symmetry equivalence' by Ruch et al. Mechanistic equivalence involves not only the above type of equivalence, but comprises further the equivalence of pathways of motion for any two PI: two pathways of motion are equivalent if there exists a bijective mapping of the sequences of idealized skeletal geometries such that the idealized skeleton of any situation in one case can be brought into coincidence with the corresponding situation in the other case by a combination of translation and rotation, (and also reflection if 'enantiomeric mechanisms' are not distinguished). Any two isomerizers I and I' belong to the same mechanism, if they are conjugate with regard to S, i.e. $I' = \hat{g} \cdot I \cdot \hat{g}^{-1}$ (with $\hat{g} \in S$).

For example, the permutational isomerizations (14) and (34) belong to the same mechanism, the $(TR)^3$ (see also $\bar{C}_{(14)}$ in Fig. 1).

Scheme 5

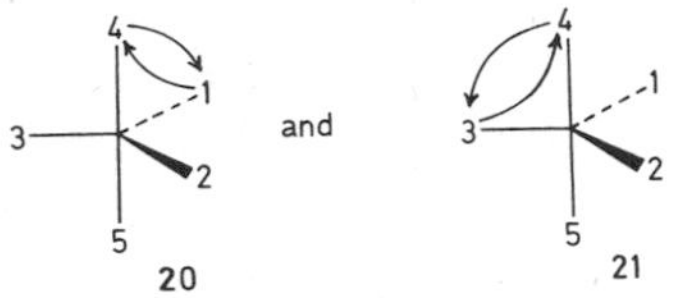

With regard to the pathways of intramolecular motion, the BPR and TR processes are distinguishable mechanisms and they are also the physical interpretations of the two different subclasses $\overline{C}_{(1\,4\,2\,5)}$ and $\overline{C}_{(1\,2\,4)(3\,5)}$ of S_5. They may lead, however, to the same result, because of the skeletal symmetry of VII. Similarly, there are other subclasses of S_5 which also can be interpreted as representatives of distinguishable but resultwise equivalent mechanisms.

Thus the distinguishable PI (124)(35) and (2435) belong to the same *process* (coset) and lead to the same result. The PI (124)(35) and (2435) belong to different *mechanisms* (subclasses).

In a rigorous treatment the index permutations can only then be directly translated into a pathway of motion in the sense of mechanism, if the molecular skeleton of the initial and the final situation is superimposable by a rotation belonging to the skeletal symmetry. Thus, for example, the permutations (1425) and (1524) describe permutational isomerizations according to Scheme 6. The Berry pseudorotation which is also given in Scheme 6 has a result which differs from the results of the latter permutational isomerizations by 90° rotations of the whole molecule. Note that the skeletal symmetry of a TBP does not contain a 90° rotation. Strictly speaking, there exists no index permutation which represents the motion of a BPR faithfully. In particular, the fact that a BPR does not contain a rotational component cannot be described by a permutational representation alone. In the present context permutations of the above-mentioned type such as (1425) are used to represent all permutational isomerizations contained in Scheme 6.

Scheme 6

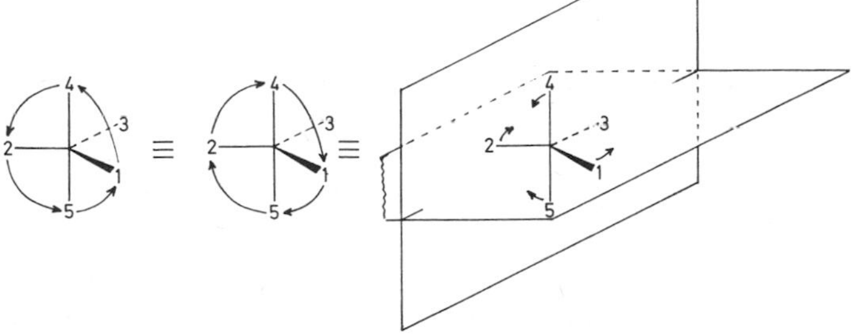

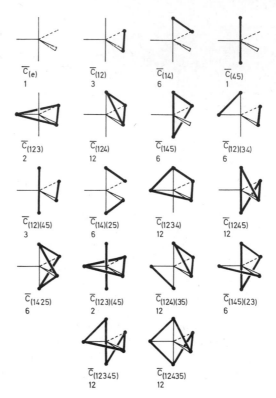

FIG. 1   The PI of VII and the subclasses of S_5 with respect to S_{VII}.

Figure 1 illustrates the PI of VII and corresponding subclasses of S_5 with respect to S_{VII}; the number of elements in each of these subclasses is indicated below the subclass symbol. Subclasses are denoted by their alphanumerically lowest element (permutation).

Those elements $s'_5 \in S_5$ which are conjugate to a given $s'_5 \in S_5$ according to $s'_5 = \hat{g}s_5\hat{g}^{-1}$ with $\hat{g} \in S_{VII}$ form a Wigner subclass of conjugate elements. The subclasses of S_5 with respect to S_{VII} can also be obtained by partitioning S_5 into its classes of conjugate elements by inspection: conjugate permutations have the same cycle structure, and partitioning those classes further into subclasses by inspection according to VII.[44,122]

The *periodicity* of permutational isomerizations is another essential property for their classification. Consider the cyclic group $\{I, I^2, I^3, \ldots, I^q = e\}$ of isomerizers. Any two permutational isomerizations belong to the same cycle, if their isomerizers belong to the same cyclic group.

The *periodicity* number p of a given PI indicates how often the internal motion of the PI must be performed successively in the same direction until the original situation is restored, using the same subsets of ligand repeatedly, i.e. without 'changing the horses'. In other words, the periodicity of a permutational isomerization indicates how often it must be applied to a given isomer in order to recover the latter. The periodicity of the isomerizer I corresponds to the order q of this group unless q has a divisor p such that $I^p \in S$. Then p is the periodicity of I.

A PI which belongs to a cycle of the order q leads from a given permutation isomer back to the latter after it has been repeated q times. However, in this cycle it may pass the original molecule more than once within the cycle, for example, whenever the cycle contains a sequence of PI whose result corresponds to a rotation of the molecule. Then the periodicity of the PI is given by that number of PI whose result corresponds to a rotation.

The lowest power p of a permutation I such that $I^p \in \overset{\circ}{S}$ is the periodicity of the a representing a PI.

For example, the BPR (2435) of **19a** has a periodicity of 2, because $(2435)^2 = (23)(45)$ represents a $180°$ rotation of **19a** (see Scheme 4).

The TR (124)(35) has a periodicity of 6, because $[(124)(35)]^6 = e$. The cyclic group $\{(124)(35), [(124)(35)]^2 = (142), [(124)(35)]^3 = (35), [(124)(35)]^4 = (124), [(124)(35)]^5 = (142)(35), [(124)(35)]^6 = e\}$ contains those simple and multiple TR isomerizations which belong to the same cycle.

The periodicity of the internal rotations of a molecule with a D_{5d} ferrocene skeleton is $p = 5$, (the 'lower' ring is held in a constant orientation, and its indices have been omitted for the sake of simplicity).

Since here the cycle length is q = 5,

Scheme 7

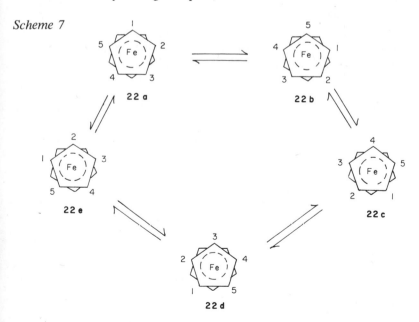

22 a

22 b

22 e

22 c

22 d

a prime number, all powers of the permutation (12345) have the same cycle structure $(12345)^{5n}$, with $n = 1,2,3. \ldots$ A series of five clockwise $72°$ rotations of the type $\mathbf{22a} \rightarrow \mathbf{22b}$, restores the initial molecule. The same holds also for sequences of five internal rotations by $144°$, $216°$, or $288°$, respectively, represented by the cyclic groups of order 5 which are generated from (13524), (14253), or (15432), these are the second, third and fourth power of (12345).

If, however, we considered the conceivable internal rotations of a sandwich complex of cyclobutadiene $\mathbf{23}$, we should observe that the periodicity

Scheme 8

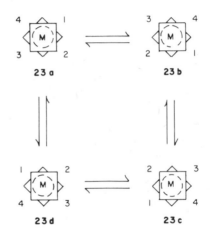

of the $90°$ internal rotation represented by (1234) is $q = 4$, whereas the periodicity of the $180°$ internal rotations corresponding to (13)(24) is $p = 2$.

In the internal rotations of dibenzene-chromium $\mathbf{24}$

we would even observe periodicities of $p = 6$, $p = 3$, or $p = 2$, depending upon whether we use $60°$, $120°$, or $180°$ internal rotations represented by (123456), (135)(246), and (14)(25)(36) and their respective power cycles.

3.3.3   The classification of PI according to Musher's modes and mechanistic types

In many cases the PI have some common features which permit combining them into larger equivalence classes.

The union of processes which contain PI belonging to the same mechanism is simultaneously the union of mechanisms which contain PI belonging to the same process. Such an equivalence class of PI is called a *Musher mode*.[52,64-67,94-96] Any two permutational isomerizations represented by I and I' belong to the same *Musher mode*, if there exists a permutational isomerization I'' such that I' and I'' belong to the same process and simultaneously I and I'' belong to the same mechanism, i.e. $I'' = I \cdot g$ and $I'' = \hat{g} \cdot I' \cdot \hat{g}^{-1}$ with $g \in \overset{\circ}{S}, \hat{g} \in S$.

For example, the BPR (1425) and the TR (15)(234) of **19a** belong to the same mode, because the BPR (1534) belongs to the same process as the TR (15)(234) and to the same mechanism as the BPR (1425). This equivalence of TR and BPR has been discussed in detail in a previous paper.

The permutational isomerization modes of **17** are represented by the following complexes of S_5 which can be simultaneously considered as unions of subclasses representing mechanisms, or as unions of cosets representing processes:

$$K_1 = \{\bar{C}_e, \bar{C}_{(123)}, \bar{C}_{(12)(45)}\} = \{\bar{K}_{11}\}$$

$$K_2 = \{\bar{C}_{(12)}, \bar{C}_{(45)}, \bar{C}_{(123)(45)}\} = \{\bar{K}_{21}\}$$

$$K_3 = \{\bar{C}_{(14)}, \bar{C}_{(1234)}, \bar{C}_{(1245)}, \bar{C}_{(145)(23)}\} = \{\bar{K}_{31}, \bar{K}_{32}, \bar{K}_{33}, \bar{K}_{34}, \bar{K}_{35}, \bar{K}_{36}\}$$

$$K_4 = \{\bar{C}_{(145)}, \bar{C}_{(124)}, \bar{C}_{(12)(34)}, \bar{C}_{(12345)}\} = \{\bar{K}_{41}, \bar{K}_{42}, \bar{K}_{43}, \bar{K}_{44}, \bar{K}_{45}, \bar{K}_{46}\}$$

$$K_5 = \{\bar{C}_{(1425)}, \bar{C}_{(124)(35)}\} = \{\bar{K}_{51}, \bar{K}_{52}, \bar{K}_{53}\} \tag{16}$$

$$K_6 = \{\bar{C}_{(14)(25)}, \bar{C}_{(12435)}\} = \{\bar{K}_{61}, \bar{K}_{62}, \bar{K}_{63}\}$$

Elements of the complexes K correspond to the *modes* of permutational isomerization. The result produced by the various elements of the modes upon **17** is indicated in Scheme 9.

Any two mechanisms $\bar{C}$ and $\bar{C}'$ belong to the same mode if there is a process $\bar{K}$ such that both $\bar{C} \cap \bar{K}$ and $\bar{C}' \cap \bar{K}$ are not empty, and any two processes $\bar{K}$ and $\bar{K}'$ belong to the same mode if there is a mechanism $\bar{C}$ such that $\bar{C} \cap \bar{K}$ and $\bar{C} \cap \bar{K}'$ are not empty.

Note that the permutational isomerizations which belong to the same cycle are only then members of the same mode, if the periodicity is a prime number. Accordingly, the usefulness of the mode concept is limited to permutational isomerizations whose periodicity is a prime.

Hitherto the PI have been classified according to modes and mechanisms only; this classification is based upon well-defined complexes of permutation groups. However, it does not take chemical reality into account in an adequate manner whenever the periodicity is not a prime number.

For example, the 90° and the 180° internal rotations of **23** belong to two different mechanisms or modes respectively, and the internal rotations of **24** to three. Note also that the TR, $(TR)^2$, and $(TR)^3$ of **25** belong to three different mechanisms and modes despite their close similarity.

Scheme 9

K_1 (identity operation)

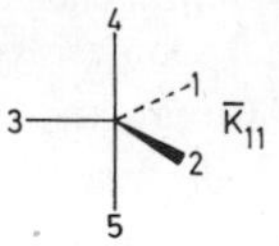

$\overline{K}_{11}$

K_2 (inversion)

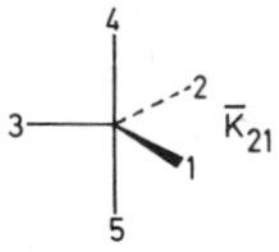

$\overline{K}_{21}$

K_3 (TR)^3 and equiv. PI

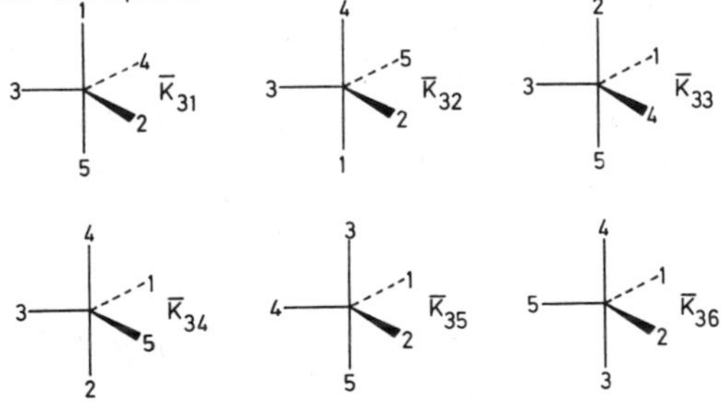

$\overline{K}_{31}$ $\overline{K}_{32}$ $\overline{K}_{33}$

$\overline{K}_{34}$ $\overline{K}_{35}$ $\overline{K}_{36}$

K_4 (TR)^2 and equiv. PI

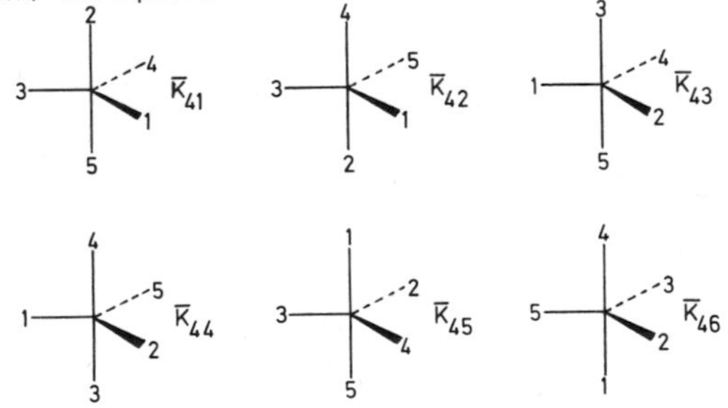

$\overline{K}_{41}$ $\overline{K}_{42}$ $\overline{K}_{43}$

$\overline{K}_{44}$ $\overline{K}_{45}$ $\overline{K}_{46}$

K_5 TR, BPR

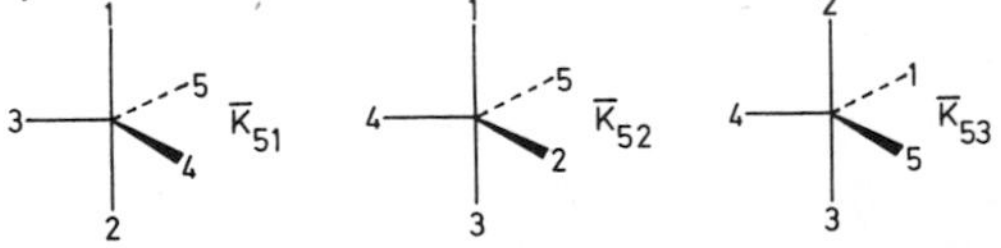

$\overline{K}_{51}$ $\overline{K}_{52}$ $\overline{K}_{53}$

K_6

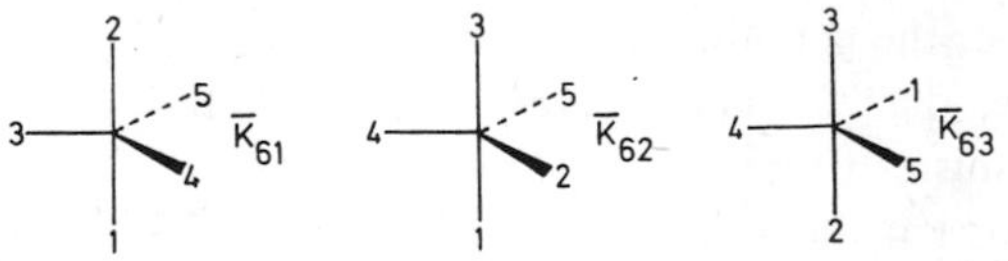

$\overline{K}_{61}$ $\overline{K}_{62}$ $\overline{K}_{63}$

The following concept of mechanistic type affords a classification of permutational isomerizations which is mathematically consistent and takes the common features of PI belonging to the same cycle into account.

A mechanistic type is the union of reaction mechanisms which contain PI belonging to the same cycle and is at the same time the union of cycles which contain PI belonging to the same mechanism.

Any two permutational isomerizations I and I' belong to the same mechanistic type, if there exists a permutational isomerization I'' such that I and I'' belong to the same cycle and simultaneously I' and I'' belong to the same mechanism.

For example, the cycle of the TR (124)(35) consists of permutational isomerizations which are represented by the permutations $\{(124)(35), (142), (35), (124), (142)(35), e\}$. This cycle consists of permutational isomerizations which belong to the TR, $(TR)^2$, and $(TR)^3$ mechanisms. Accordingly, the single and multiple TR mechanisms belong to the same mechanistic type.

This is further illustrated by Scheme 10. In this Scheme the interconversion of 25 $a \rightleftharpoons b$, $b \rightleftharpoons c$, $c \rightleftharpoons d$ etc. belong to the TR mechanism, the direct interconversion (via A and B without intermediacy of b) of $a \rightleftharpoons c$, $c \rightleftharpoons e$ (via C and D without intermediacy of d), etc. belong to the $(TR)^2$ mechanism, and the direct interconversion of $a \rightleftharpoons d$ without participation of b and c belongs to the $(TR)^3$ mechanism.

Scheme 10

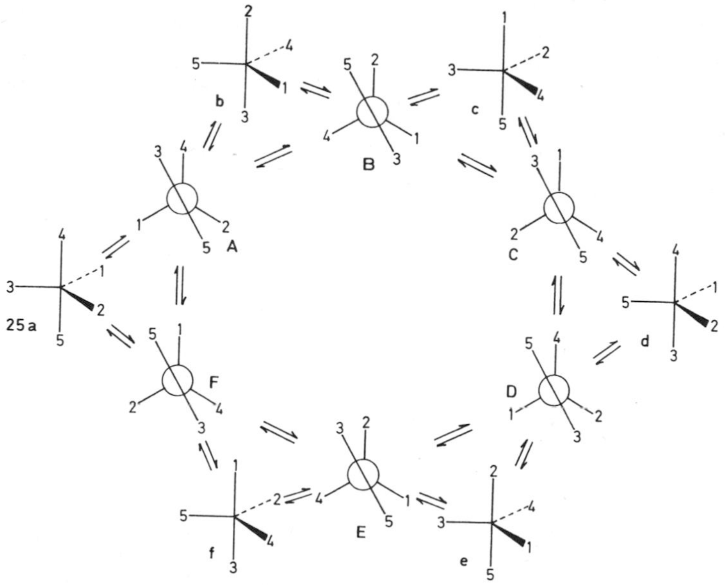

3.3.4   Metric for configurations

The PI, or modes, respectively, provide classes of interconvertible permutation isomers with topological structures.

Let $P_{\overline{C}} = \{M_0, M_1, \ldots, M_f\}$ be a class of interconvertible permutation isomers that are isomerized by the permutational isomerization $I_{\sigma\sigma'}$ belonging to the same mechanism $\overline{C} = \{I_{11}, I_{12}, \ldots, I_{\gamma\psi}\}$, or mode K, respectively. Then the number $D_{\sigma\sigma'}$ of permutational isomerizations $I_{\sigma\sigma'}$ required for the interconversion of a molecule M_σ into its isomer $M_{\sigma'}$ can be defined as the distance between M_σ and $M_{\sigma'}$. The distance $D_{\sigma\sigma'}$ defines a metric on $P_{\overline{C}}$.

The representation of the set $T(P_{\overline{C}}, D)$ by a labelled graph $G(V, L)$ follows from Sections 4.3.1–4.3.3 in a straighforward manner. The vertices V_σ correspond to the individual isomers M_σ, i.e. the Pólya configurations, and are labelled by their descriptors Q_σ. The lines $L_{\sigma\sigma'}$ that connect the vertices V_σ and $V_{\sigma'}$, are given by the isomerizers $I_{\sigma\sigma'}$ that belong to the same subclass $\overline{C}$ of S_n; if a mode is being considered, instead of a mechanism, then the isomerizers belong to a complex K of S_n . Thus the modes define topologies on sets of permutation isomers.

3.3.4.1   *Primary graphs of mechanisms and modes*

The graph $G(V, L)$ of a class of permutation isomers, interconvertible by a mechanism of mode K, with a set of n distinguishable ligands and a skeleton represents a property of it, and may be called a *primary graph*. If all isomers are interconvertible, a connected primary graph results, if not, a graph with disjoint subgraphs, or components, results. With a skeletal rotation group $\overset{\circ}{S}$ of order $|\overset{\circ}{S}|$ the primary graph has $n_1 = n!/|\overset{\circ}{S}|$ vertices, representatives of the cosets of $\overset{\circ}{S} \subset S_n$. If the complex K contains k elements of which κ each represent one process, each vertex is connected by $l = k/\kappa$ lines to other vertices, and the graph contains $n_1 k/2\kappa$ lines. This is illustrated by the BPR and TR mechanisms by which the twenty permutational isomers of **17** are interconverted.

The BPR is represented by the subclass $\overline{C}_{(1425)} \subset S_5$ with $|\overline{C}_{(1425)}| = 6$. Since BPR is defined as a combination of two concerted orthogonal vibrations, it does not involve rotation, and therefore pairs of inverse permutations, e.g. (1425) and (1524) of $\overline{C}_{(1425)}$ describe the same PI (see Scheme 6). Therefore each of the 20 pentacoordinate isomers with five different ligands and the skeleton VII is converted to three other isomers by BPR, and the primary graph of BPR contains overall $20 \cdot 3/2 = 30$ lines.

The TR is represented by $\overline{C}_{(124)(35)}$ with $|\overline{C}_{(124)(35)}| = 12$; since the TR involves internal rotation the pairs of inverse permutations refer to pairs of distinguishable processes with opposite angular momentum, and there are overall 12 different TR processes. However, each four TR lead to the same result, e.g. according to Scheme 11. Therefore a pentacoordinate molecule is converted by TR into $12/4 = 3$ of its isomers.

The BPR of Scheme 6 and the TR of Scheme 11 lead to the same result and can be represented by a *coset* $\bar{K}_{5\,1}$ of the complex K_5. The complex K_5 contains three cosets.

$$K_5 = \{\bar{K}_{5\,1}, \bar{K}_{5\,2}, \bar{K}_{5\,3}\} \tag{17}$$

with

$$\bar{K}_{51} = \{(1\ 4\ 2\ 5),\ (1\ 5\ 2\ 4),\ (1\ 4)(2\ 5\ 3), \tag{18}$$
$$(1\ 5)(2\ 4\ 3),\ (2\ 4)(1\ 5\ 3),\ (2\ 5)(1\ 4\ 3)\}$$
$$\bar{K}_{52} = \{(1\ 4\ 3\ 5),\ (1\ 5\ 3\ 4),\ (1\ 4)(2\ 3\ 5),$$
$$(1\ 5)(2\ 3\ 4),\ (3\ 4)(1\ 5\ 2),\ (3\ 5)(1\ 4\ 2)\}$$
$$\bar{K}_{53} = \{(2\ 4\ 3\ 5),\ (2\ 5\ 3\ 4),\ (2\ 4)(1\ 3\ 5),$$
$$(2\ 5)(1\ 3\ 4),\ (3\ 4)(1\ 2\ 5),\ (3\ 5)(1\ 2\ 4)\}$$

Scheme 11

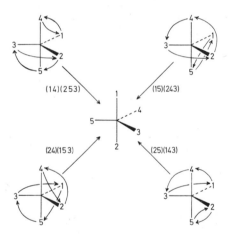

Accordingly, the lines of a primary graph of permutation isomerizations $G(V,L)$ correspond to the cosets of $\overset{\circ}{\mathbf{S}}$, and mechanisms belonging to the same mode are described by the same graph.

A primary graph $G(V,L)$ of a mode K can be constructed in a systematic manner as follows: For a class Z of permutation isomers with different ligands, a skeletal rotation group $\overset{\circ}{\mathbf{S}}_z$ and a subclass $\bar{C}_z$, the cosets $\{\overset{\circ}{\mathbf{S}}_z s_5\}$ and the complex K_Z are determined. A set of $n_1 = n!/\,|\overset{\circ}{\mathbf{S}}_z|$ points is labelled by the descriptors Q_σ of the isomers $\mathbf{M}_\sigma$, representatives of the cosets $\{\overset{\circ}{\mathbf{S}}_z s_5\}$. Those points $\mathbf{M}_\sigma$ and $\mathbf{M}_{\sigma'}$ are connected by lines for which $I_{\sigma\sigma'} = Q_\sigma \cdot Q_{\sigma'}^{-1}$ belongs to K_z. These lines represent the cosets $\{\overset{\circ}{\mathbf{S}}_z s_5\}$.

An alternative procedure for finding $G(V,L)$ is the following:
Two permutation isomers which are interconvertible according to the

mechanism or mode, respectively, under consideration, are both treated as reference isomers. For each one of these the skeletal rotation group is expressed, as $\overset{o}{S}$ and $\overset{o}{S}'$, in terms of ligand index permutations. For each one of the groups $\overset{o}{S}$ and $\overset{o}{S}'$ their cosets in S_5 are established. All those cosets of $\overset{o}{S}$ and $\overset{o}{S}'$ which have in common a permutation are connected by lines. Since these cosets represent the individual permutation isomers the result is a graph of a mechanism or a mode respectively.

If the group generated by $\overset{o}{S}$ and $\overset{o}{S}'$ is not the full group S_5 then the graph contains disjoint parts.

The primary graphs of the BPR and TR (Graph 4)* and of $(TR)^2$ (Graph 5) may serve as an illustration.[44] These isomerizations are properties of the skeleton VII with the skeletal rotation group $\overset{o}{S}_{VII}$ (eq. 10), and the complexes, $K_1 - K_6$ (eq. 16). These PI belong to the modes which are represented by the complexes K_4 and K_5.

By either of the procedures outlined above, Graph 4 is obtained for K_5.

Graph 5 with disjoint components indicates the fact that PI belonging to K_4 do not permit the interconversion of antipodes and thus divide the isomers of 17 into two classes.

Graph 5 can be obtained from Graph 4 by replacing its lines by lines between points that were previously connected by two lines; the results from K_4 correspond to two applications of K_5.

Graph 4

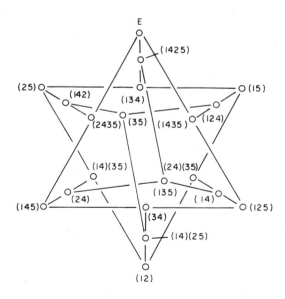

*This corresponds to a graph by Lauterbur and Ramirez.[77]

Graph 5*

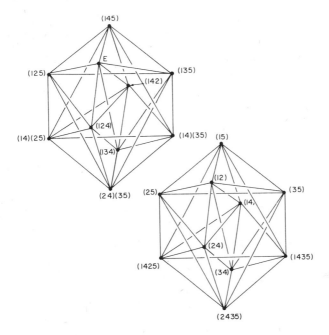

3.3.4.2   Secondary graphs of mechanisms and modes

Ligand equivalencies reduce the number of distinguishable permutation isomers of a skeletal class. With a ligand set that does not contain equivalent ligands all ligand permutations lead to distinguishable isomers unless they represent rotations belonging to the skeletal symmetry. If some ligands are equivalent, their permutations do not interconvert isomers.

This is illustrated by the fact that pentacoordinate molecules form 20 isomers when all ligands are distinguishable, 10 isomers with two equivalent ligands, six isomers with two pairs of equivalent ligands, and 4 isomers with a trio of equivalent ligands (see Table I).

Accordingly, the isomerization graphs of molecules with ligand equivalencies, the *secondary graphs*, are obtained from the primary graphs (see Section 3.3.4.1) by taking into account the equivalencies of certain isomers. The secondary graphs 6–8 of reorganizations according to the mode of K_5 are, for instance, obtained by relabelling Graph 4 according to the equivalencies of descriptors as found in Table I and subsequent removal of the redundant parts of the graphs.

---

*Graph 5 also represents the isomerizations of 17 by the following irregular process which belongs to the subclass $\bar{C}_{(12)(45)} \subset S_5$. The pentacoordinate species is transformed into a tetracoordinate intermediate with a tetrahedral skeleton by removal of an apical ligand (apical departure). The latter ligand is reattached to the central atom by entry in one of the faces of skeletal tetrahedron of the intermediate, thus becoming an apical ligand in a new TBP skeleton (apical entry).[45,86]

Graph 6   *Graph 7*   *Graph 8*

3.3.4.3   Reduced graphs of mechanisms and modes

Permutation isomers which are not chiral enantiomers may differ considerably in energy. For energy reasons, some elements of a set of permutation isomers can often be neglected, since they neither exist at any detectable equilibrium concentration, nor are intermediates of PI that proceed at an observable rate.[44]

In the mechanistic interpretation of the PI of permutation isomers it is sometimes not only advantageous, but necessary for a meaningful treatment of the problem, to omit the *negligible isomers* from the considerations. When some isomers in a class of interconverting permutation isomers are negligible, this class is represented by a *reduced graph* which is obtained from the pertinent primary or secondary graph by omitting the points of the negligible isomers.

The reduced graphs lead often to important mechanistic insights into the observable permutational isomerizations.

Let us assume a class of permutation isomers can conceivably be interconverted by a variety of mechanisms belonging to different modes, and that some of the isomers are negligible. Then one obtains different reduced graphs for the different modes, and only those reduced graphs correspond to acceptable modes which contain connections for isomers that are empirically known to interconvert and contain no connections between isomers that are known not to interconvert. This often enables an assignment of modes, or mechanisms.

Whitesides and Mitchell contributed to the fundamental understanding of PI with an elegant experiment.[129] They demonstrated that at $-100°$ to $-50°C$, **26** undergoes reorganization by pairwise exchange of their fluorine ligands (in **26** the otherwise indistinguishable pairs of atoms have been labelled by asterisks).

$$(CH_3)_2N-P \rightleftharpoons (CH_3)_2N-P$$

26 (14)(25)   **26** (35)

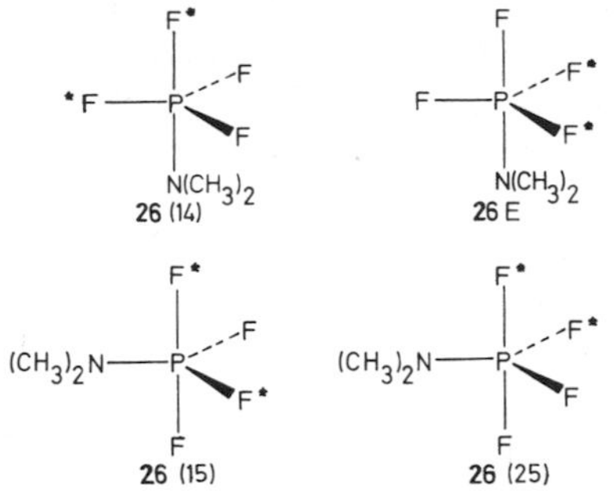

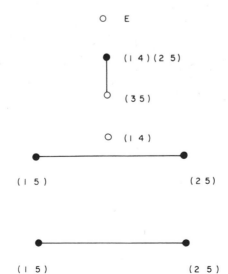

26 (14) 26 E 26 (15) 26 (25)

This means that **26** (3 5) and **26** (1 4)(2 5) interconvert without involvement of the other isomers **26**E, (1 4),(1 5) and (2 5).

The graphs of the various conveivable reorganization modes $(K_3 - K_6)$ are obtained by using **17** and its isomers with $L_1 = L_2 = F^*$, $L_3 = L_4 = F$, and $L_5 = N(CH_3)_2$ as a model (see Table I).

The reference isomer **26**E and also **26**(1 4) are negligible isomers; they violate the *modified polarity rule*[44] and are more than 30 kcal/mole higher in energy than the other permutation isomers **26**(1 5), (2 5), (3 5), and (1 4)(2 5).

For BPR and TR that belong to the mode of K_4, Graph 7 is the secondary graph of this model. The pertinent reduced Graph 9 results from Graph 7 by omission of the points with the labels E and (1 4), as well as the lines that belong to these points.

Graph 9

                  O   E

                ●   (1 4)(2 5)

                ○   (3 5)

                O (1 4)

   ●————————————————●

   (1 5)   (2 5)

Graph 10

   ●————————————————●

   (1 5)   (2 5)

Graph 11

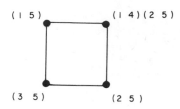

Graph 12 (K_4)

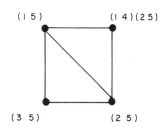

Graph 13 (K_6)

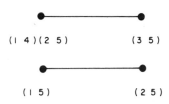

The Graphs 10–13 are reduced graphs of the other modes of PI of **26**.
It is noteworthy that in the primary K_5 graph of **17** the isomers **26** (1 4)(2 5)
and (1 5), (2 5), as well as (3 5), belong to disjoint components of Graph 5, but
are represented by a connected reduced secondary graph, owing to the
equivalence of **26** (1 4)(2 5) and (1 4 2 5). Since only the Graphs 9 and 13 agree
with the experimental evidence the reorganization of **26** proceeds by one of the
mechanisms that belongs to K_5 or K_6.

The only permutation isomers of **27** (**17** with $L_1 + L_5 =$ the ring, $L_2 =$
$L_3 = OCH_3$, $L_4 = C_6H_5$), which can be expected to be stable are **27** (1 4)(2 5)
and (1 4)(3 5).[49,50,100,101,128]

The (1 4) isomer is forbidden, due to diapical placement of the 5-ring. From
quantum mechanical model calculations and from experimental evidence on **27**
it follows that the isomers **27** (2 5), (3 5), and (2 4)(3 5) can occur neither as
equilibrating participants in a measurable concentration nor as intermediates of
PI. The isomers **27E**, (1 5), (2 4), and (3 4) are disfavoured by modified polarity
rule.[44]

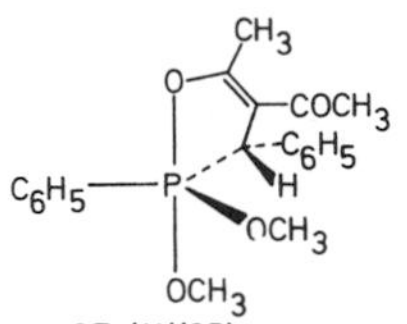

27 (14)(25)

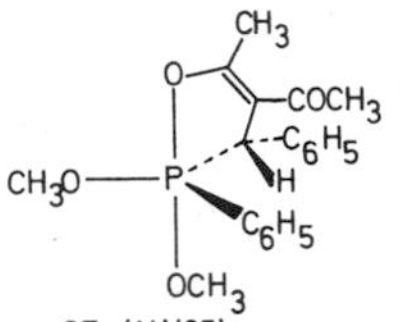

27 (14)(35)

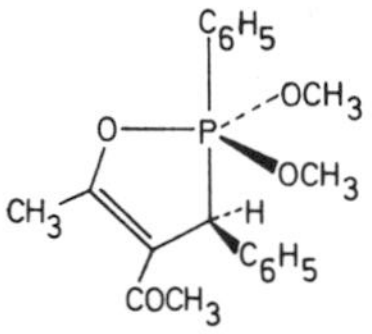

27 E

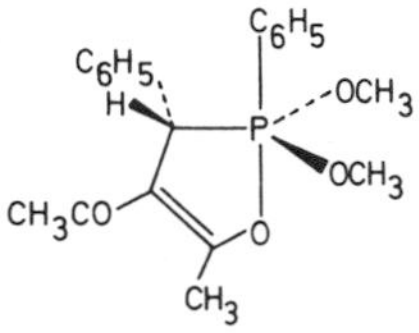

27 (15)

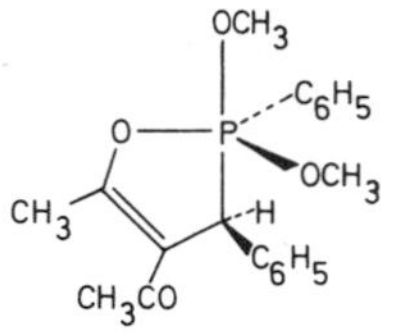

27 (24)

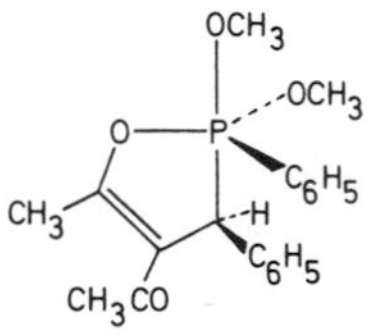

27 (34)

It can be concluded from the NMR data that at $-20°$, **27** exists as a frozen diastereomer mixture of **27** (1 4)(2 5) and (1 4)(3 5) in a ratio of about 1:2. Above $70°$ the interconversion of these isomers is so rapid that it is not observable on the NMR time scale, and yet the methoxyls do not become equivalent.

Below $100°$ the interconversion **27** (1 4)(2 5) ⇌ **27** (1 4)(3 5) takes place as a regular process, since no exchange of C methyl groups is observed. Of all the reduced graphs of the conceivable reorganization modes of **27** only Graph 14, which relates to mechanisms of K_4, is in agreement with the evidence.

Graph 14

(14)(25)   (14)(35)

From this it can be concluded that the isomerization of **27** proceeds by the $(TR)^2$ mechanism, with the 5 ring serving as the 'pair'.

Since the isomers **27** (1 4), (2 5), (3 5), and (2 4)(3 5) are out of the question, the interconversion of **27** (1 4)(2 5) and (1 4)(3 5) by the BPR or a single TR mechanism belonging to K_5 would have Graph 15 as the reduced graph of **27**, a graph that contains four points that correspond to improbable isomers (see above).

Graph 15

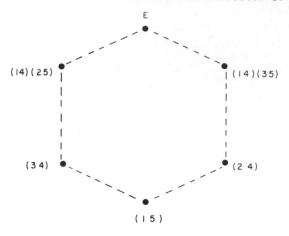

Analogously, it can be shown that for **27** none of the mechanisms belonging to the other modes $(K_2, K_3$ and $K_6)$ is an acceptable alternative to the $(TR)^2$.

The systematic interpretation of configurational interconversions by using reduced graphs that are labelled by permutational descriptors seems to be cumbersome, in comparison with the more intuitive procedures that have previously been used. However, the systematic approach has the advantage that it takes into account the complete set of conceivable interpretations, with a minimum of arbitrariness.

3.3.4.4   Graphs of cycles

The cycle graph of a given permutation isomer, say **17** = E, indicates which other isomers are contained in the cycles involving the afore-mentioned isomer, E, if the PI is accomplished according to a certain mechanism, or mode. The lines in these graphs represent the processes; the cycles correspond to unions of those processes whose sequential application leads from the considered isomer back to it. The nodes in these graphs correspond to isomers: the label is the permutation converting the reference isomer to the isomer in question.

The *cycle graphs* **16**–**18** of the $(TR)^3$, $(TR)^2$ and TR cycles which involve E, the reference isomer, illustrate the cyclic nature of these mechanistic components of the TR mechanistic type.

Graph 16

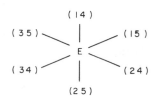

The $(TR)^3$ cycles of E (q = 2).

Graph 17

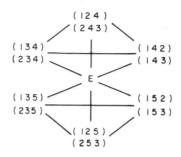

The $(TR)^2$ cycles of E (q = 3).
The double labelling of the nodes indicates the pairs of equivalent $(TR)^2$ by which a given isomer is obtained from E.

Graph 18

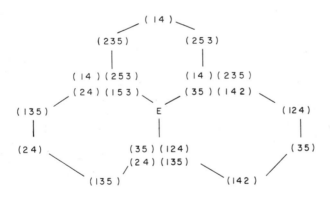

The TR cycles of E (q = 6).
   Any two-dimensional graph of $(TR)^3$ involves so many intersecting lines that it is of little visual help in correlating those interconverting permutation isomers which are interconverted by $(TR)^3$. In such cases an interconverting class of isomers is somewhat better represented by a symmetric PI-matrix M in which the rows and columns refer to the individual isomer and can be labelled by their permutational descriptor. The entries of a PI-matrix are 0, or 1. An entry m_{ij} (= m_{ji}) in the i-th row and j-th column is 1, if the permutation isomers that belong to these rows/columns are interconverted by the considered type of PI, otherwise $m_{ij} = 0$.

The PI-matrices $M(TR)$, $M(TR^2)$ and $M(TR^3)$ correspond to the graphs of $TR, (TR)^2$ and $(TR)^3$, and it is interesting to note that Graph 5 of $(TR)^2$ consists of two disjoint parts, such that the subset of the isomers of a subgraph does not contain any isomers which are mutual antipodes; this reflects the fact that antipodic isomers are not interconverted by $(TR)^2$ which involves only even ligand permutations.

Correspondingly a PI-matrix $M(TR^2)$ can be transformed into a PI–matrix $\overset{\square}{M}(TR^2)$ in block form by row/column relabelling $\overset{\square}{M}(TR^2) = P^* M(TR^2)P$, with P being a 20 x 20 permutation matrix, and P^* its transpose.

The fact that Muetterties' mechanism M_4 which corresponds to what we call $(TR)^2$ can be represented by a matrix of two blocks was already recognized by Muetterties in 1969.[90,91]

Since BPR and TR are resultwise equivalent and are represented by the same graph or matrix respectively, Muetterties' statement: 'A distinction in mechanisms is immediately apparent by process 4 of Figure 7' (which represents the $(TR)^2$ mechanism) indicates that at that time no connection was seen between the resultwise $(TR)^1$-equivalent and the $(TR)^2$-, as well as $(TR)^3$-equivalent permutational isomerization.

The cycle graphs may be used to construct the mode and mechanism graphs of PI by taking each node of a cycle graph as the origin of a new cycle graph, etc., until all isomers which can be connected by processes belonging to the considered mode have been exhausted. If some members of the family of permutation isomers are not reached by this procedure one of these is used as the origin of a new set of cycle graphs, etc., until all isomers of the family have been used.

For example, Graph 4 describes the interconversion of the 20 permutation isomers with a TBP skeleton and five different ligands by the BPR–TR mode. This graph can be obtained from the Cycle Graph 18 by the above procedure. Note that Graph 4 describes a mode as defined previously. Graph 4 contains six membered cycles just like Graph 18, corresponding to the periodicity of the TR mechanism. Since BPR has only a periodicity of 2, it does not provide direct information with regard to the cycle structure of these graphs, in contrast to the TR which belongs to the same mode. This illustrates that the structure of mode and mechanism graphs is determined by all mechanisms which belong to a mode.

It is easy to see that the only significance of the mode is the fact that it is the union of all mechanisms which are represented by the same graph. The mechanisms are of interest, because they are subunits of the mechanistic types.

Let $\emptyset$ be an operator that transforms a matrix $\bar{M}$ with integral entries $\bar{m}_{ij}$ into a matrix M in which the non-zero entries of $\bar{M}$ have been replaced by 1.

Then the PI-matrices $M(TR)$, $M(TR^2)$, and $M(TR^3)$ are related as follows:

$$M(TR^2) = \emptyset \{ [M(TR)]^2 - \tilde{M}(TR \cdot TR') \}$$

$$M(TR^3) = \emptyset \{ [M(TR)]^2 - \tilde{M}(TR \cdot TR'' \cdot TR''') \}$$

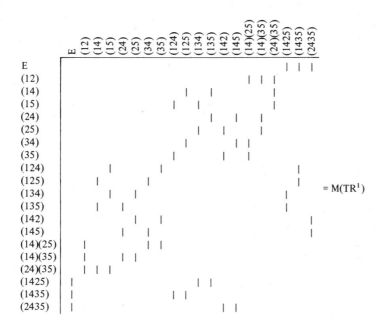

$= M(TR^1)$

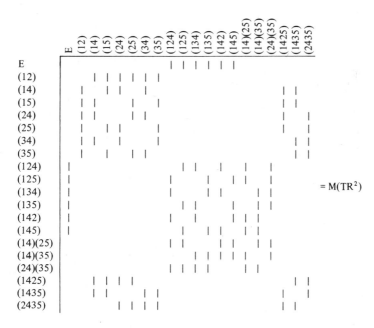

$= M(TR^2)$

Top matrix — $= M(TR^3)$

	E	(12)	(14)	(15)	(24)	(25)	(34)	(35)	(124)	(125)	(134)	(135)	(142)	(145)	(14)(25)	(14)(35)	(24)(35)	(1425)	(1435)	(2435)
E		—	—	—	—	—	—	—												
(12)									—	—	—	—	—	—						
(14)	—									—		—			—		—			
(15)	—									—		—		—		—				
(24)	—									—	—		—		—		—			
(25)	—								—		—		—		—		—			
(34)	—								—		—		—		—		—			
(35)	—								—	—			—			—	—			
(124)		—	—								—	—						—		—
(125)		—		—	—					—			—					—		—
(134)		—	—		—						—		—						—	—
(135)		—		—		—					—		—						—	—
(142)		—		—		—					—	—							—	—
(145)		—	—		—							—	—						—	—
(14)(25)			—	—	—	—												—		—
(14)(35)			—	—					—			—	—					—		—
(24)(35)					—	—	—	—	—									—		—
(1425)									—	—			—	—	—	—				
(1435)											—	—	—	—	—		—			
(2435)									—	—	—	—				—	—			

Bottom matrix — $= \overset{\cup}{M}(TR^2)$

	E	(124)	(125)	(134)	(135)	(142)	(145)	(14)(25)	(14)(35)	(24)(35)	(12)	(14)	(15)	(24)	(25)	(34)	(35)	(1425)	(1435)	(2435)
E		—	—	—	—	—	—													
(124)	—		—	—		—		—												
(125)	—	—			—		—	—												
(134)	—	—			—	—			—											
(135)	—	—		—	—			—		—										
(142)	—	—	—			—	—	—												
(145)	—	—			—	—	—	—												
(14)(25)	—	—			—	—		—	—											
(14)(35)	—			—	—	—	—	—		—										
(24)(35)	—	—	—	—	—				—	—										
(12)											—	—	—	—	—	—	—			
(14)												—	—	—		—		—		—
(15)												—	—		—		—		—	—
(24)												—	—		—	—		—		—
(25)												—		—	—		—		—	—
(34)												—		—	—		—		—	—
(35)												—		—		—	—	—		—
(1425)												—	—	—	—					—
(1435)												—		—		—	—		—	—
(2435)												—	—		—	—	—	—		

The matrices $\tilde{M}$ represent sequences of two, or three TR, respectively, which involve intermediate 'changing the horses'.

The TR mechanistic type is represented by the union of TR, $(TR)^2$, and $(TR)^3$ graphs and, equivalently, by the PI-matrix

$$M(TR)_\Sigma = \emptyset \; [M(TR) + M(TR^2) + M(TR^3)] \,.$$

3.4   Tetra and hexacoordinate molecules

In Section 3.3 the discussion of the graph theoretical treatment of configurational problems was confined to pentacoordinate molecules, because this skeletal class of permutation isomers is particularly illustrative of the configurational aspect of chemistry.

The configurations of central atoms and superatoms with other coordination numbers are, however, equally important.

In this Section, the structures of the permutation group of four and six symbols are discussed, in terms of subgroups, subclasses and complexes which provide the basis of the primary, secondary and reduced graphs for the configuration interconversions of the mechanistically significant tetra- and hexacoordinate skeletal classes.

The permutation group S_4 of four symbols can be partitioned into its classes of conjugate elements according to the cycle structure of its permutations which, in turn, is given by the partitions of the number four.

The skeletons V, VI and XI–XIV yield the indicated subclasses and complexes of S_4.

$S_V = \{e,(13),(24),(12)(34),(13)(24),(14)(23),(1234),(1432)\}$

$K_1 = \{\bar{C}_e,\bar{C}_{(13)},\bar{C}_{(12)(34)},\bar{C}_{(13)(24)},\bar{C}_{(1234)}\}(\text{identity})$

$K_2 = \{\bar{C}_{(12)},\bar{C}_{(123)},\bar{C}_{(1243)}\}$ (isomerization)   (19)

$S_{VI} = \{S_4\}$

$K_1 = \{\bar{C}_e,\bar{C}_{(123)},\bar{C}_{(12)(34)}\}$ (identity)

$K_2 = \{\bar{C}_{(12)},\bar{C}_{(1234)}\}$ (inversion)   (20)

$S_{XI} = \{e,(23),(24),(34),(234),(243)\}$

$K_1 = \{\bar{C}_e,\bar{C}_{(234)}\}$ (identity)

$K_2 = \{\bar{C}_{(23)}\}$ (inversion)   (21)

$K_3 = \{\bar{C}_{(12)},\bar{C}_{(1234)}\}$ (isomerization)

$K_4 = \{\bar{C}_{(12)(34)},\bar{C}_{(123)}\}$ (isomerization)

$$S_{XII} = \{e,(13),(24),(12)(34),(13)(24),(14)(23),(1234),(1432)\}$$

$$K_1 = \{\overline{C}_e, \overline{C}_{(13)(24)}, \overline{C}_{(1234)}\} \quad \text{(identity)}$$

$$K_2 = \{\overline{C}_{(13)}, \overline{C}_{(12)(34)}\} \quad \text{(inversion)} \tag{22}$$

$$K_3 = \{\overline{C}_{(12)}, \overline{C}_{(123)}, \overline{C}_{(1243)}\} \quad \text{(isomerization)}$$

XII (C_{4v})

$$S_{XIII} = \{e,(12)(24),(13)(24),(14)(23)\}$$

$$K_1 = \{\overline{C}_e, \overline{C}_{12(34)}; \overline{C}_{(13)(24)}; \overline{C}_{(14)(23)}\} \quad \text{(identity)}$$

$$K_2 = \{\overline{C}_{(12)}, \overline{C}_{(123)}, \overline{C}_{(1324)}\} \quad \text{(isomerization)}$$

$$K_3 = \{\overline{C}_{(13)}, \overline{C}_{(1234)}\} \quad \text{(isomerization)} \tag{23}$$

$$K_4 = \{\overline{C}_{(14)}, \overline{C}_{(1243)}\} \quad \text{(isomerization)}$$

$$K_5 = \{\overline{C}_{(123)}\} \quad \text{(isomerization)}$$

$$K_6 = \{\overline{C}_{(124)}\} \quad \text{(isomerization)}$$

XIII (D_{2h})

$$S_{XIV} = \{e,(12),(34),(12)(34),(13)(24),(14)(23),(1324),(1423)\}$$

$$K_1 = \{\overline{C}_e, \overline{C}_{(12)(34)}, \overline{C}_{(13)(24)}\} \text{ (identity)}$$

$$K_2 = \{\overline{C}_{(12)}, \overline{C}_{(1324)}\} \quad \text{(inversion)} \tag{24}$$

$$K_3 = \{\overline{C}_{(13)}, \overline{C}_{(123)}, \overline{C}_{(1234)}\} \quad \text{(isomerization)}$$

XIV (D_{2v})

Of the hexacoordinate skeletons the octahedral skeleton VIII is stereochemically by far the most important one.

VIII (O_h)

$$S_{VIII} = \overset{o}{S}_{VIII} \cup \overline{S}_{VIII}$$

$$\overset{o}{S}_{VIII} = \{e,(1\ 4)(2\ 5),(1\ 4)(3\ 6),(2\ 5)(3\ 6),(1\ 2\ 4\ 5),(1\ 5\ 4\ 2),$$

$$(1\ 3\ 4\ 6),(1\ 6\ 4\ 3),(2\ 3\ 5\ 6),(2\ 6\ 5\ 3),(1\ 2)(3\ 6)(4\ 5),$$

$$(1\ 3)(2\ 5)(4\ 6),(1\ 4)(2\ 3)(5\ 6),(1\ 4)(2\ 6)(3\ 5),(1\ 5)(2\ 4)(3\ 6),$$

$$(1\ 6)(2\ 5)(3\ 4),(1\ 2\ 3)(4\ 5\ 6),(1\ 3\ 2)(4\ 6\ 5),(1\ 3\ 5)(2\ 4\ 6),$$

$$(1\ 5\ 3)(2\ 6\ 4),(1\ 2\ 6)(3\ 4\ 5),(1\ 6\ 2)(3\ 5\ 4),(1\ 5\ 6)(2\ 3\ 4),$$

$$(1\ 6\ 5)(2\ 4\ 3)\} \tag{25}$$

$$\overline{S}_{VIII} = \{(14),(25),(36),(12)(45),(13)(46),(15)(24),(16)(34),(23)(56),$$

$$(26)(35),(14)(25)(36),(1245)(36),(1346)(25),(1542)(36),$$

$$(1643)(25),(2356)(14),(2653)(14),(123456),(126453),(132465),$$

$$(135462),(153426),(156423),(162435),(165432)\}$$

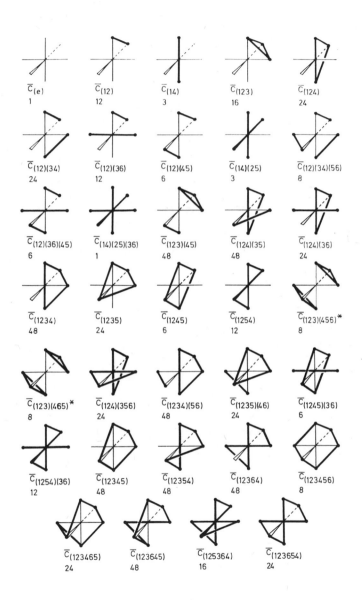

FIG. 2.

*The mechanisms of $\overline{C}_{(123)(456)}$ and $\overline{C}_{(123)(465)}$ differ by the relative directions of the cycles.

In the present context it is interesting to note the following: Let S and S' represent the symmetries of two indexed skeletons. If S_n has a smaller number of subclasses with respect to S than to S', then S represents a higher symmetry than S'.

Figure 2 illustrates the PI of VIII and the corresponding subclasses of S_6 with respect to S_{VIII}. The subclasses are labelled with their alphanumerically lowest element (permutation); the number of elements in each of these subclasses is indicated below the subclass symbol.

Following are the modes for VIII.

$$K_1 = \{\overline{C}_e, \overline{C}_{(14)(25)}, \overline{C}_{(1245)}, \overline{C}_{(12)(36)(45)}, \overline{C}_{(123)(456)}\}$$

$$\text{(identity, } \mathring{S}_{VIII}) \qquad (26)$$

$$K_2 = \{\overline{C}_{(14)}, \overline{C}_{(12)(35)}; \overline{C}_{(14)(25)(36)}; \overline{C}_{(1245)(36)}; \overline{C}_{(123456)}\}$$

$$\text{(inversion, } \overline{S}_{VIII})$$

$$K_3 = \{\overline{C}_{(12)}, \overline{C}_{(124)}, \overline{C}_{(12)(36)}, \overline{C}_{(1254)}, \overline{C}_{(123)(45)}, \overline{C}_{(124)(36)},$$
$$\overline{C}_{(1234)(56)}, \overline{C}_{(1254)(36)}, \overline{C}_{(12345)}, \overline{C}_{(123645)}\}$$

$$K_4 = \{\overline{C}_{(123)}, \overline{C}_{(123)(465)}, \overline{C}_{(124)(35)}, \overline{C}_{(1235)}, \overline{C}_{(1235)(46)}, \overline{C}_{(12364)},$$
$$\overline{C}_{(123654)}\}$$

$$K_5 = \{\overline{C}_{(12)(34)}, \overline{C}_{(1234)}, \overline{C}_{(12)(34)(56)}, \overline{C}_{(124)(356)}, \overline{C}_{(12354)},$$
$$\overline{C}_{(125364)}, \overline{C}_{(123465)}\}$$

A particular permutation isomer with skeleton VIII and six distinguishable ligands e.g. the reference isomer E, is transformed into its antipode by mode K_2, into 12 other isomers by mode K_3 and into eight each by the modes K_4 and K_5.

4   POLYCENTRIC STEREOCHEMISTRY

4.1   Polycentric configurations

Polycentric molecules contain skeletal atoms A_{ni} with coordination numbers $n = 2,3,4,5, \ldots$ and a set of ligands that gives rise to classes of epimers having different Pólya configurations (Section 4.1) at the individual centers A_{ni}. In case $n \geqslant 3$ and independent flexibly connected monocentric configuration units, stereoisomers can be described using cartesian products.

Recall that the cartesian product $\prod_i \{X_i\}$ of an indexed family $\{X_i \mid i \in I\}$ of sets is defined as follows: the members of $\prod_i \{X_i\}$ are all those arrays containing exactly one member from each $\{X_i\}$; for any array the member x_i chosen from the set X_i is called the ith coordinate. The projection $\pi_{i_0}: \prod_i \{X_i\} \to \{x_{i_0}\}$ of the cartesian product onto the i_0th coordinate set is the map sending each array to its i_0th coordinate.

In these terms, a polycentric configuration can be described as the cartesian product of the monocentric configurational units.

Specifically, if each monocentric configuration A_{ni} has the descriptor σ_i then the polycentric molecule has the configuration $\prod_i \sigma$. The following example illustrates this concept.

Let $\{A_R B_R C_R, A_R B_R C_S, A_R B_S C_R, A_R B_S C_S, A_S B_R C_R, A_S B_R C_S, A_S B_S C_R, A_S B_S C_S\}$ be a family of stereoisomers with three tetracoordinate centers A, B, and C that carry such sets of ligands that about each of these centers there can be two configurations, designated by R and S. The configurations A, B and C belong to the sets $\{A_R, A_S\}$, $\{B_R, B_S\}$, $\{C_R, C_S\}$ respectively; the configurations A_R, A_S, B_R, B_S, C_R, C_S are the projections of stereoisomers with regard to these sets.

The classes of epimers, e.g., $A_R B_R C_R$, $A_R B_R C_S$ are the equivalence classes relative to two of these sets.

The interconversion graphs of the epimers are obtained in a slighty different manner. For each center A_{ni}, let G_i be the graph of its interconversions; the possible interconversions of the polycentric molecules are given by the O-join $G_1 x G_2 x \ldots x G_r$ of these graphs.

Recall that the O-join of two graphs G_1 and G_2 can be described as follows: it is the graph formed by the lines in G_1, the lines in G_2 and all lines from each *vertex* of G_1 to each vertex of G_2, no two of these added lines having any points, other than possibly endpoints, in common; the join can always be realized in 3-space, by placing G_1 and G_2 in carefully. The join of three graphs, $G_1 x G_2 x G_3$ is $(G_1 x G_2) x G_3$, etc. this is always realizable in 3-space (Fig. 3).

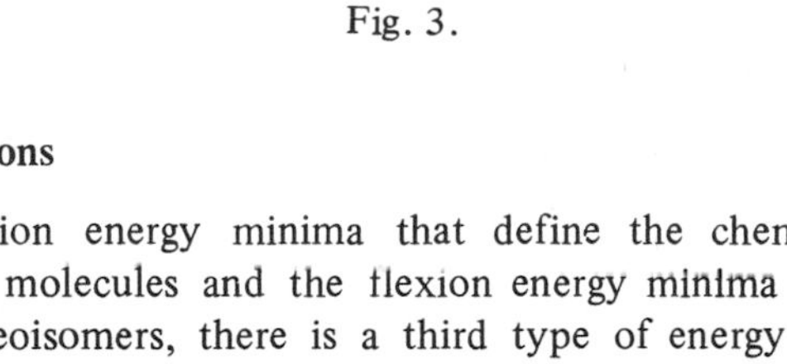

If G_1 = one point, and G_2 = △ , then $G_1 x G_2$ = ◇

If G_1 = two points connected by one line, and G_2 = △ ,

then $G_1 x G_2$ =

Fig. 3.

4.2   Conformations

Besides the tension energy minima that define the chemical constitution (Section 2.1) of molecules and the flexion energy minima which afford the existence of stereoisomers, there is a third type of energy minima which is essential to chemistry. The energy minima that correspond to the *conformers* of polycentric molecules are separated by *torsion barriers* of varying height.

The conformers of a given stereoisomer have in common the same chemical constitution and the same set of monocentric Pólya configurations and differ only by sets of *dihedral angles* between pairs of bonds at adjacent pairs of polycoordinate centers.

Thus **28a–c** are the distinguishable conformers of n-butane which, for example differ by the dihedral angles about the central C–C bond.

only by sets of *dihedral angles* between pairs of bonds at adjacent pairs of

28a   28b   28c

The classification of isomeric molecules by equivalence relations that are based upon chemical constitution (preserved by tension barriers), configurations (preserved by tension and flexion barriers) and conformations (further preserved by torsion barriers) seems to be more advantageous for the present purpose than classifications that involve the heights of interconversion barriers, although also the present classification has some shortcomings.*

For instance, presently available evidence does not permit us to decide whether **29a** and **29b** are interconverted by the flexion barrier of the C–N–O bond, or by overcoming the torsion barrier of the C=N bond. In this context, however, we must define **29a** and **29b** to be conformers, for the sake of uniformity. The conformers are interconverted by rotations about bonds which connect adjacent polycoordinate centers.

29a   29b

The conformers of polycentric molecules with a certain chemical constitution and a set of configurations at the polycoordinate centers are characterized by a set of dihedral angles that pertain to pairs of adjacent centers.

The conformational descriptor κ of a conformation of a molecule $M_{\phi\sigma}$† with distinguishable ligands and r pairs of adjacent polycoordinate centers belongs to a r fold cartesian product $\prod_1^r\{\delta_{ij}\}$. The indices i and j of the dihedral angle descriptor δ refer to the bond A_i–A_j. If a $360°$ rotation about A_i–A_j involves ρ_{ij} potential energy minima, then there are $\prod_1^r\rho_{ij}$ stable conformations and corresponding descriptors κ for $M_{\phi\sigma}$.

The primary graph $G_{\phi\sigma}(V_{\phi\sigma}, L_{\phi\sigma}) = G_{\phi\sigma,1} \times G_{\phi\sigma,2} \times \ldots \times G_{\phi\sigma,r}$ of the conformational interconversions of $M_{\phi\sigma}$ contains a set V_ϕ of $c_{\phi\sigma}$ vertices.

A metric can be defined on a set of conformers by using the number of torsion barriers that separate two conformers as their distance.

---

*The classification of chemical phenomena according to the types of barriers involved originates with V. Prelog and A. Dreiding, to the best knowledge of the present authors.
†Here ϕ refers to the constitution, and σ to the configurations.

The conformers of the propane derivative **30** with the skeleton **XV** are

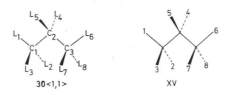

$30 < 1,1 >$ XV

interconverted by $\pm 120°$ rotations about the C_1-C_2 or C_2-C_3 bonds, corresponding to the permutations $(1\,2\,3)$, $(1\,3\,2)$, $(6\,7\,8)$, $(6\,8\,7)$. An abbreviated nomenclature of the conformers of **30** $\langle 1,1 \rangle$ results from assigning the numbers 1, 2, and 3 to each of the three torsion energy minima of the C_1-C_2 and C_2-C_3 bonds. Then, for instance, the $\langle 2,1 \rangle-$ conformer results from **30** $\langle 1,1 \rangle$ by a $+120°$ rotation about the C_1-C_2 bond, i.e., $\langle 2,1 \rangle = (1\,2\,3) \langle 1,1 \rangle$, and the $\langle 1,2 \rangle-$ conformer is generated by a $+120°$ rotation about the C_2-C_3 bond, i.e. $\langle 1,2 \rangle = (6\,7\,8)\langle 1,1 \rangle$. With rotation about only one C–C bond at a time, then Graphs 20 and 21 represent the possible conformers for **XV**. Graph 19 is the primary graph of the conformational interconversions of the skeleton **XV** when rotation about one or both C–C bonds is allowed.

Graph 19

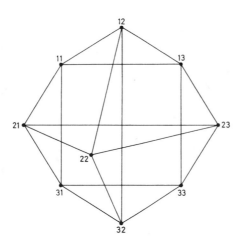

Graphs 20 and 21

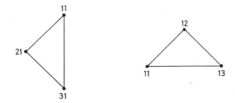

The secondary Graph 22 (see Section 3.3.4.2) of the conformational interconversions of 1,3-dichloropropane $(29, L_2 - L_5, L_7, L_8 = H, L_1, L_6 = Cl)$ is obtained from the primary Graph 19 by taking into account the equivalencies of conformers $\langle 1,2 \rangle = \langle 2,1 \rangle, \langle 1,3 \rangle = \langle 3,1 \rangle$, and $\langle 2,3 \rangle = \langle 3,2 \rangle$ which arise from the ligand equivalencies.

Graph 22

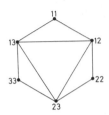

The assumption that the $\langle 2,3 \rangle$ — conformer is negligible because of strong Cl,Cl repulsions the reduced Graph 23 (see Section 3.3.4.3) results from Graph 22, by omitting the 23-point.

Graph 23

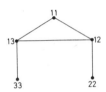

Equivalent results would have been obtained by treating the skeleton XV as a superatom and considering the conformations as Pólya configurations (see Section 3.1 and 3.3.4).

5.   INTEGRATED CHEMICAL METRIC

A polycentric molecule or $EM_{\phi\sigma\kappa}$ is characterized by its chemical constitution (ϕ), and the spatial distribution of its n atoms which is representable by its set of monocentric configurations (σ) and its set of conformations (κ).

The stereoisomers (including the conformers) belonging to a given constitution can be partitioned into equivalence classes either by virtue of the conceivable conformers of a given configurational isomer $(\sigma$ constant) or the configurational isomers for a given conformation $(\kappa$ constant). Thus, in the idealized case, a class of configurational and conformational isomers $EM_{\phi\sigma\kappa}$ belonging to a given chemical constitution can be divided into isomorphic configurational or conformational equivalence classes. This statement forms a part of the stereochemical analogy principle.[10,40,110]

The configurational and conformational features of an EM of $A(n)$ can be represented by an $n \times n$ cc-matrix C which is associated with the be-matrix of the EM. In a cc-matrix C the diagonal entry c_{kk} is the permutational descriptor[123] of the local configuration about the atom A_k, and is zero if no multiplicity of local configurations exists.

The above permutational descriptor of the configuration about A_k is based upon I–VI (or suitably defined superatom skeletons). Depending upon whether we want to distinguish between the conformers of the stereoisomers or only the stereoisomers regardless of their conformation, we use rigid model skeletal symmetries, or the dynamic skeletal equivalencies for the symmetry of the centers. The sequence of the index numbers of the atoms bonded to A_k provide a nomenclatural sequence rule.[9,10,14,123] Off-diagonal entries $c_{ij}(i \neq j)$ of a cc-matrix indicate the dihedral angles of the bonds between the A_i and its bonded neighbor $A_{i'}$ with the lowest index i' in the associated be-matrix and the bond between A_j and $A_{j'}$, its neighbor with the lowest atomic index j'.[39]

The logical fine structure of chemistry of an EM, which corresponds to its configurational and conformational aspects, is represented by a family of cc-matrices. It is possible, in this context, to represent this set of EM, in analogy to the individual molecule, by the cartesian product of two sets, a configuration set and a conformation set. A metric in the configuration space can be based upon the sum of the numbers of transpositions required to transform the configurational descriptors of one stereoisomer of the $EM_{\phi\sigma\kappa}$ into those of another $EM_{\phi\sigma'\kappa}$.[10,44,123] In the conformation space one can use $\Sigma \mid \delta'_{ij} - \delta_{ij} \mid$ as a metric for the cc-matrices.

The interconversion of an $EM_{\phi\sigma\kappa}$ with given configurational (σ) and conformational (κ) features into a constitutionally different isomeric $EM_{\phi'\sigma'\kappa'}$ by some chemical reaction leads, as a rule, to a certain set of configurations (σ') and conformations (κ') in $EM_{\phi'\sigma'\kappa'}$ which is characteristic of the reaction mechanism.

The stereochemical features of an EM can also be represented by a set of $3 \times n$ nc-matrices which are particularly suitable for a computer oriented analysis of reaction mechanisms and synthetic planning with due consideration of the stereochemical aspect of chemical reactions.

In an nc-matrix, the suitably defined cubic lattice coordinates of the atoms A_i, i.e. the suitably approximated nuclear cartesian coordinates, $a_{ti}(a_{1i} = x_i, a_{2i} = y_i, a_{3i} = z_i)$, are entered.[123] Two nc-matrices belong to the same EM and are equivalent if they can be interconverted by transformations corresponding to rotations and/or translations of the molecules. The nc-matrices have a metric defined by $\Sigma_{t,i} \mid a'_{ti} - a_{ti} \mid$, if the atomic indices are assigned such that the chemical distance of the EM is minimal (see Section 2.3).

Since the constitution of the EM requires certain limits in bond lengths and angles between subsets of atoms, the constitutional and stereochemical changes within the EM that occur during chemical reactions are not independent of each other; certain transformations of be-matrices require certain associated transformations of the cc- and/or nc-matrices.

The preferred interconversions of IEM represented by the be-matrix transformation $E \to E'$, are such that there exists at least one nc-matrix associated with both E and E' or with the be-matrix of a feasible intermediate state of EM $\to$ EM'. The be-matrix and at least one nc-matrix of E must be in a suitable metric relation to the be- and at least one nc-matrix of E'.

REFERENCES

1. Balaban, A. T. (1966). *Rev. Roum. Chim.*, 11, 1097, *also* 15, 463, 1243, 1251 (1970), and preceding commun.
2. Balaban, A. T.. Fărcasiu, D. and Harary, F. (1970). *J. Labelled Compounds* 6, 211.
3. Balaban, A. T. and Harary, F. (1971). *J. Chem. Docum.*, 11, 258.
4. Bellman, R. E. (1957). "Dynamic Programming", Princeton Univ. Press, Princeton.
5.a Berge, C. (1962). "The Theory of Graphs and its Applications". Wiley, New York.
   b. Berge, C. (1963). "Topological Spaces", McMillan, New York.
6. Berry, R. S. (1960). *J. Chem. Phys.*, 32, 933.
7. Bird, C. W. (1967), "Transition Metal Intermediates in Organic Synthesis", p. 58, 114, 170. Academic Press, New York and London.
8. Birkoff, G. and McLean, S. (1967). "Algebra", McMillan, New York.
9. Blair, J., Gasteiger, J., Gillespie, C., Gillespie, P., Ugi, I. (1973). in "Proceedings of the NATO/CNA ASI on Computer Representation and Manipulation of Chemical Information", Edit. Wipke, W. T., Heller, S., Feldmann, R. and Hyde, E., John Wiley, New York, p. 129.
10. Blair, J., Gasteiger, J., Gillespie, C., Gillespie, P. D., Ugi, I. (1974). Tetrahedron 30, 1845.
11. Bodanszky M. and Ondetti, M. A. (1966). Peptide Synthesis, Wiley (Interscience), New York.
12. Buchs, A., Duffield, A. M., Schroll, G., Djerassi, C., Delfino, A. B., Buchanan, B. G., Sutherland, G. L., Feigenbaum E. A. and Lederberg, J. (1970). *J. Amer. Chem. Soc.*, 92, 6831.
13. Budzikiewicz, H., Djerassi, C. and Williams, D. H. (1967). Mass Spectrometry of Organic Compounds, Holden-Day, San Francisco.
14. Cahn, R. S., Ingold, C. K. and Prelog, V. (1966). *Angew. Chem.*, 78, 413; *Also Angew. Chem. internat. Edit.*, 5, 385 (1966), and preceding papers cited therein.
15. Cayley, A. (1857). *Phil. Mag*, 13, 19; *Also* 67, 444 (1874); *Mathem. Papers (Cambridge)* 3, 242 (1891).
16. Corey, E. J. and Wipke, W. T. (1969). *Science*, 166, 178.
17. Cram, D. J., Day, J., Rayner, D. R., von Schriltz, D. M., Duchamp, D. J., and Garwood, D. C. (1970). *J. Amer. Chem. Soc.*, 92, 7369.
18. DeBruijn, N. G. *Koninkl. Ned. Akad. Wetenschap. Proc. Ser. A* 62, 59 (1959); *In* (Beckenbach, E. F.) Applied Combinatorial Mathematics. Wiley, New York 1964, p. 144; *Also Nieuw Arch. Wiskunde* (3) 18, 61 (1970).
19. DeBruin, K. E., Naumann, K., Zon, G. and Mislow, K. (1969). *J. Amer. Chem. Soc.*, 91, 7031.
20. Dewar, M. J. S. (1952). *J. Amer. Chem. Soc.*, 74, 3341.

21. Dewar, M. J. S. (1966). *Tetrahedron Suppl.*, **8**, 75; *Angew. Chem.*, **83**, 859 (1971); *Angew. Chem. Internat. Edit.*, **10**, 761 (1971).
22. Dewar, M. J. S. (1969). *Molecular Orbital Theory of Organic Chemistry*, McGraw-Hill, New York.
23. Dubois, J. E. (1969). *Entropie*, **27**, 1.
24. Dugundji, J. (1966). *Topology*, Allyn and Bacon, Boston.
25. Dugundji, J., Ugi, I. (1973). *Topics Curr. Chem.*, **39**, 19.
26. Dugundji, J., Marquarding, D., Ugi, I. (in print). *Chem. Scripta.*
27. Dunitz, J. D. and Prelog, V. (1968). *Angew. Chem.*, **80**, 700, *Angew. Chem. Internat. Edit.*, **7**, 725 (1968).
28. Ege, G. (1971). *Naturwissenschaften*, **58**, 247.
29. Erdös, P. and Katona G. (editors) (1968). "Theory of Graphs" (Proceedings of the Colloquium held at Tihany, Hungary, September 1966), Academic Press, New York and London.
30. Evans, D. (1971). *UCLA Physical Org. Chem.*, Seminar on May 6; *J. Amer. Chem. Soc.*, (submitted).
31. Evans, M. G. (1939). *Trans. Farad. Soc.*, **35**, 824.
32. Evans, M. G. and Polanyi, M. (1938). *ibid.*, **34**, 11.
33. Evans M. G. and Warhurst, E. (1938). *ibid.*, **34**, 614.
34. Eyring, H. (1964). *Statistical Mechanics and Dynamics*, Wiley, New York.
35. Frisch J. L. and Wasserman, E. (1961). *J. Amer. Chem. Soc.*, **83**, 3789.
36. Frobenius, G. (1887). *J. fur reine u angew Math.*, **101**, 273.
37. Fugmann, R., Nickelsen, H., Nickelsen, L. and Winter, J. H. (1970). *Angew. Chem.*, **82**, 611. (1970). *Angew. Chem. Internat. Edit.*, **9**, 589.
38. Fukui, K. (1971). *Acc. Chem. Res.*, **4**, 57.
39. Gasteiger, J., Gillespie, P., Marquarding, D., Ugi, I. (1974). *Topics Curr. Chem.*, **48**, 1.
40. Gasteiger, J., Marquarding, D., Ugi, I. (in print). in "Handbook of Stereochemistry", Edit. Kagan, H. Thieme Verlag, Stuttgart.
41. Gielen, M., (1969), Mededel, *Vlaam. Chem. Ver.*, **31**, 185, 201.
42. Gielen, M., DeClerq, M. and Nasielski, J. (1969). *J. Organometal. Chem.*, **18**, 217.
43. Gielen M. and Vanlautem, N. (1970). *Bull. Soc. Chim. Belges*, **79**, 679.
44. Gillespie, P., Hoffmann, P., Klusacek, H., Marquarding, D., Pfohl, S., Ramirez, F., Tsolis, E. A. and Ugi, I. (1971). *Angew. Chem.*, **83**, 691. (1971). *Angew. Chem. Internat. Edit.*, **10**, 687.
45. Gillespie P. and Ugi, I. (1971). *Angew. Chem.*, **83**, 493. (1971). *Angew. Chem. Internat. Edit.*, **10**, 503.
46. Glasstone, S., Laidler, K. J. and Eyring, H. (1942). "The Theory of Rate Processes," McGraw-Hill, New York.
47. Gokel, G., Hoffmann, P., Kleimann, H., Klusacek, H., Lüdke, G., Marquarding D. and Ugi, I. (1971). "Isonitrile Chemistry," Academic Press, New York and London. p. 211.
48. Golomb, S. W. (1961). "Information Theory," The Universities Press, Belfast.
49. Gorenstein, D. (1970). *J. Amer. Chem. Soc.*, **92**, 644.
50. Gorenstein D. and Westheimer, F. (1970). *J. Amer. Chem. Soc.*, **92**, 634.
51. Green, M. L. H. (1968). "Organometallic Compound," Vol. II, Methuen, London, p. 39.
52. Hässelbarth, W., Ruch, E. (1973). *Theor. Chim. Acta*, **29**, 259.
53. Hamermesh, M. (1962). "Group Theory and its Applications to Physical Problems". Addison-Wesley, Reading, Mass.

54. Harary, F. (1969). "Graph Theory." Addison-Wesley, Reading, Mass.
55. Harris, B. (1970). "Graph Theory and its Applications," Academic Press, New York and London.
56. Heilbronner, E. (1962). *Helv. Chim. Acta,* **45**, 1722.
57. Hendrickson, J. B. (1971). *J. Amer. Chem. Soc.,* **93**, 6847, 6854.
58. Hoerner, S. and Schaifers, K. (1967). *Meyers Handbuch über das Welt-All,* Bibliographisches Institut, Mannheim.
59. Ireland, R. E. (1969). "Organic Synthesis" Prentice-Hall, Englewood Cliffs, N.J.
60. Jaffé, H. H. and Orchin, M. (1967). 'Symmetrie in der Chemie". "Anwendungen der Gruppentheorie auf chemische Probleme". Alfred Hüthig-Verlag, Heidelberg.
61. Kaufhold, G. and Ugi, I. (1967). unpubl. results.
62. King, R. B. (1970). *J. Amer. Chem. Soc.,* **92**, 6455, and preceding papers.
63. Klein, D. J. and Cowley, A. H. (1975). *J. Amer. Chem. Soc.,* **97**, 1633.
64. Klemperer, W. G. (1972). *J. Chem. Phys.,* **56**, 5478.
65. Klemperer, W. G. (1972). *J. Amer. Chem. Soc.,* **94**, 5662.
66. Klemperer, W. G. (1972). *J. Amer. Chem. Soc.,* **95**, 380.
67. Klemperer, W. G. (1972). *Inorg. Chem.,* **11**, 2668.
68. Knödel, W. (1969). *Graphentheoretische Methoden und ihre Anwendung,* Springer, Heidelberg, 1969.
69. König, D. (1936). "Theorie der endlichen und unendlichen Graphen," Leipzig, (reprinted Chelsea, New York, 1950).
70. Kowalski, J. (1965). "Topological Spaces," Academic Press, New York and London.
71. Kresze, G. (1970). *Angew. Chem.,* **82**, 563: (1970), *Also Angew. Chem. Internat. Edit.* **9**, 545.
72. Kuratowski, K., Topologie, *P.A.N. Monog. Math.,* Polish Scientific Publishers, Warszawa, Vol. I, 1958; Vol. II, 1961.
73. Kuratowski, K. and Mostowski, A. (1968). "Set Theory," North Holland Publ. Comp., Amsterdam.
74. Kurosch, A. (1955). "The Theory of Groups". Chelsea Publ. Co., New York.
75. Langmuir, I. (1919). *J. Amer. Chem. Soc.,* **41**, 868, 1543.
76. Lapworth, A. (1922). *J. Chem. Soc.,* **121**, 416.
77. Lauterbur, P. C. and Ramirez, F. (1968). *J. Amer. Chem. Soc.,* **90**, 6722.
78. Lederberg, J. (1969). "Topology of Molecules, in the Mathematical Sciences". The MIT Press, Cambridge, Mass.;(1965), *Also Proc. Nat. Acad. Sci.,* **53**, 134.
79. Lewis, G. N. (1916). *J. Amer. Chem. Soc.,* **38**, 762.
80. Lipscomb, W. N. (1963)' "Boron Hydrides," Benjamin, New York.
81. Loebl, E. (1968). "Group Theory and its Applications," Academic Press, New York and London.
82. McLafferty, F. W. (1959). *Anal. Chem.,* **31**, 82.
83. McWeeny, R. (1963) "Symmetry," Pergamon, New York.
84. March J. (1968, and references cited therein). (see e.g.), "Advanced Organic Chemistry, Reactions, Mechanisms and Structure," McGraw-Hill, New York.
85. Marquarding, D., Klusacek, H., Gokel, G., Hoffmann P. und Ugi, I. (1970). *Angew. Chem.,* **82**, 360; (1970), *Also Angew. Chem. Internat. Edit.,* **9**, 371; (1970). *J. Amer. Chem. Soc.,* **92**, 5389.

86. Marquarding, D., Ramirez, F., Ugi, I., Gillespie, P. (1973). *Angew. Chem.*, **85**, 99; (1973), *Also Angew. Chem. Internat. Edit.*, **12**, 91.
87. Mead, C. A. (1974). *Topics Curr. Chem.*, **49**, 1.
88. Meyer, E. (1970). *Angew. Chem.*, **82**, 605; (1970), *Also Angew. Chem. Intl. Ed.*, **9**, 583.
89. Mislow, K. (1970). *Accounts Chem. Res.*, **3**, 321.
90. Muetterties, E. L. (1969). *J. Amer. Chem. Soc.*, **91**, 1636.
91. Muetterties, E. L. (1969). *J. Amer. Chem. Soc.*, **91**, 4115.
92. Mulder, J. J. C. and Oosterhoff, L. J. (1970). *Chem. Comm.*, **305**, 307.
93. Musher, J. I. (1972). *J. Amer. Chem. Soc.*, **94**, 5662.
94. Musher, J. I. (1972). *Inorg. Chem.*, **11**, 2335.
95. Musher, J. I., Agosta, W. C. (1974). *J. Amer. Chem. Soc.*, **96**, 1320.
96. Musher, J. I. (1974). *J. Chem. Educ.*, **51**, 94.
97. Norman, R. O. C. (1968). "Principles of Organic Syntheses," Methuen, London.
98. Pólya, G. (1937). *Acta Math.*, **68**, 145.
99. Prelog, V. (1969). Abstract of the Roger Adams Award Lecture on June 17, at the A.C.S. Meeting, Salt Lake City, Utah.
100. Ramirez, F. (1968). *Accounts Chem. Res.*, **1**, 168.
101. Ramirez, F. (1970). *Bull. Soc. Chim. Fr.*, 3491.
102. Ramirez F. und Ugi, I. (1971). *Progr. Phys. Org. Chem.*, Gold, V. (ed.), Academic Press, New York and London, p. 25.
103. Ramirez, F., Pfohl, S., Tsolis, E. A., Pilot, J. F., Smith, C. P. Ugi, I., Marquarding, D., Gillespie, P. and Hoffmann, P. (1971). *Phosphorus*, **1**, 1.
104. Rice, O. K. (1961). *J. Phys. Chem.*, **65**, 1588.
105. Rosenblum, . (1965). "The Iron Group Metallocenes, Wiley New York.
106. Preuss, H. (1969). *Int. J. Quant. Chem.*, **3**, 123, 131; and references contained therein.
107. Ruch, E. (1968). *Theor. Chim. Acta*, **11**, 183.
108. Ruch, E., Hässelbarth, W. and Richter, B. (1970). *Theor. Chim. Acta*, **19**, 288.
109. Ruch, E. and Schönhofer, A. (1968). *Theor. Chim. Acta*, **10**, 91.
110. Ruch, E. and Ugi, I. (1966). *Theor. Chim. Acta*, **4**, 287; (1969), *Also Top. in Stereochem.*, **4**, 99.
111. Ruch, E. (1972). *Acc. Chem. Res.*, **5**, 49.
112. Ryser, H. H. (1963). *Combinatorial Mathematics, Carus Mathematical Monographs, Math. Assoc. of America*, p. 57.
113. Sidgwick, N. V. (1933). "The Covalent Link in Chemistry," Cornell Press, Ithaca, New York.
114. Skinner, J. A. and Pilcher, G. (1963). *Quart. Rev.*, (London) **17**, 264.
115. Spialter, L. (1963). *J. Amer. Chem. Soc.*, **85**, 2012; (1964), *Also J. Chem. Document.*, **4**, 261, 269. See also Hiz, H. J. (1964). *J. Chem. Document.*, **4**, 173.
116. Stevens R. (priv. commun. 1971).
117. Stone, H. S. (1973). "Discrete Mathematical Structures and their Applications". Science Research Associates, Inc., Chicago, Palo Alto, Toronto, Henley-on-Thames, Sydney, p. 113.
118. Ugi, I. (1965). *Jahrb. 1964 Akad. Wiss.*, Vandenhoeck and Ruprecht, Goettingen 1965, p. 21; S. Naturforsch. **20B**, 405.
119. Ugi, I. (1969). *Rec. Chem. Progr.*, **30**, 289.
120. Ugi, I. (1971). *Intra-Science Chem. Rep.*, **5**, 229.

121. Ugi, I. and Gillespie, P. (1971). *Angew. Chem.*, **83**, 980, 982; (1971), *Angew. Chem. Internat. Edit.*, **10**, 914, 915.
122. Ugi, I., Marquarding, D., Klusacek, H., Gillespie P. and Ramirez, F. (1971). *Accounts Chem. Res.*, **4**, 288.
123. Ugi, I., Marquarding, D., Klusacek, H., Gokel, G. and Gillespie, P. (1970). *Angew. Chem.*, **82**, 741; (1970), *Also Angew. Chem. Internat. Edit.*, **9**, 703.
124. Ugi, I. Gillespie, P., Gillespie, C. (1972). *Trans. N.Y. Acad. Sci.*, **34**, 416.
125. Ugi, I. (1974). *IBM Nachrichten*, **24**, 180.
126. Ugi, I., Brandt, J., Brunnert, J., Gasteiger, J., Schubert, W. (1974). *IBM Nachrichten*, **24**, 185.
127. Ugi, I. (1975). et al. in "Peptides 1974, Proceedings of the 13th European Peptide Symposium", Edit Wolman, Y., John Wiley, New York; Israel Universities Press, Jerusalem p. 71.
128. Westheimer, F. H. (1968). *Accounts Chem. Res.*, **1**, 70.
129. Whitesides G. M. and Mitchell, H. L. (1969). *J. Amer. Chem. Soc.*, **91**, 5384.
130. Wigner, E. P. (1968). "Spectroscopic and Group Theoretical Methods in Physics" (Racah Mem. Vol.), North Holland Publ. Co., Amsterdam, p. 131; (1971), *Also Proc. Roy. Soc.* (London), **A322**, 181.
131. Woodward R. B. and Hoffmann, R. (1969). *Angew. Chem.*, **81**, 797; (1969), *Also Angew. Chem. Internat. Edit.*, **8**, 781.
132. Wünsch, E. (1971). *Angew. Chem.*, **83**, 773; (1971), *Also Angew. Chem. Internat. Edit.*, **10**, 786.
133. Zimmerman, H. E. (1966). *J. Amer. Chem. Soc.*, **88**, 1564.

CHAPTER 7

The Topological Matrix in Quantum Chemistry

D. H. Rouvray*

Chemistry Department,
University of the Witwatersrand,
Johannesburg, South Africa

Present address: Max-Planck-Institut für Kohlenforschung, D-4330 Mülheim/Ruhr, German Federal Republic.

1   GENERAL INTRODUCTION

The extensive development of quantum chemistry over the last half century is due in no small measure to the widespread application of graph-theoretical concepts in this field. Since its inception, quantum chemistry has been founded upon the graphical representation of molecules and, throughout its considerable evolution, has relied on many combinatorial and graph-theoretical techniques in its description and characterization of molecular systems. In Table I we present a listing of some of the more commonly encountered graph-theoretical concepts in

TABLE I

List of equivalent mathematical and chemical terms used to describe graph-theoretical concepts arising in quantum chemistry.

Mathematical term	Chemical term
Vertex	Atom
Edge	Chemical bond
Chemical graph	Structural formula
Tree graph	Acyclic molecule
Bipartite graph	Alternant molecule
Degree of a vertex	Valency of an atom
Chain on n vertices	n-Polyene
Cycle on n vertices	n-Annulene
Cyclomatic number	Number of rings
Characteristic polynomial	Secular polynomial
Adjacency matrix	Topological matrix
Eigenvalue	Energy level
Zero eigenvalue	Non-bonding level
Positive eigenvalue	Bonding level
Negative eigenvalue	Anti-bonding level
Spectral theory	Hückel theory

quantum chemistry together with their equivalent chemical terms. From this table it is evident that much of the modern discipline of quantum chemistry is infused with the concepts and methods of graph theory. In fact, had not frequent appeal been made to such methods, the remarkable growth of the discipline would not have been possible.

The earliest use of graphs in a purely chemical context appears to have been made by Crum Brown,[28,29] who used them for the depiction of the structural formulae of covalently-bonded molecules. Since then, graphs have also been used to depict canonical structures, such as the Kekulé structures for benzene. In more recent years the graphical representation of bonds and molecules has become an integral part of the valence-bond approach to chemical bonding; Hückel theory also relies heavily upon this type of representation.

A number of other applications of graph theory have frequently been of a less explicit nature. Examples of these include the derivation of rules for giving the allowed spectral terms of equivalent electrons under Russell-Saunders coupling, where recourse is made to a combinatorial method of generating the appropriate canonical structures for the species in question.[115] A graph-theoretical method for enumerating all the canonical structures for a given species was presented by Gordon and Davison.[50] In a similar way, implicit use was made of a graph-theoretical colouring process in establishing the well-known Pairing Theorem.[24] In the last two decades, authors in this field have become increasingly aware of their indebtedness to graph and combinatorial theory, and several have made explicit acknowledgement of this.[54,55,122]

One prevalent technique in quantum chemistry, especially during the last decade, has been the direct representation of molecules by matrices rather than by their graphs. (The use of matrices in chemistry has a time-honoured history which we cannot discuss here.) Many differing types of matrix have been employed in the representation of chemical systems, some of which we describe in the next section. On occasion, matrices have been used in the form of codes,[98] in the expanded form of their characteristic polynomial,[128] or as topological indices.[111]

Although basically the matrix provides no more than a convenient device for representing molecules, its use in quantum chemistry has had a number of far-reaching implications. Thus, the use of matrices has considerably facilitated the calculation of a large number of physico-chemical parameters of molecules. The most common calculation to date has been that to determine the electronic energy levels in molecules; these are obtained from the eigenvalues of certain matrices. Parameters, such as bond orders, deriving from the eigenvectors of these same matrices have also been very frequently calculated. In this article the rôle of matrices in these and other calculations is examined. The discussion will be confined within the framework of elementary molecular orbital theory, and, wherever possible, use will be made of a simple Hückel-type formalism.

Until very recent times the use of matrices had been restricted mainly to the representation of organic species. This was because elementary quantum-chemical calculations were particularly well suited to the treatment of conjugated systems, which abound in organic chemistry. A number of extensions to inorganic systems, and, in general, to systems containing no π-electrons have now been made. This sequence of development is not surprising since in many ways organic molecules were the natural starting point for such calculations. Structurally, many of them could be regarded as consisting of two virtually independent components: a σ-electron framework, which provided the backbone of the molecule, and a π-electron cloud, which was deemed to be mobile and susceptible to delocalization. The σ-electrons could usually be neglected, except insofar as they constituted a background 'effective field' in which the π-electrons moved. Although difficult to justify in any rigorous sense, this model has nevertheless provided an extraordinarily successful working description of such systems and has provided a basis for much of the development of quantum chemistry.

This model has also been responsible for colouring our approach to the study of bonding in organic molecules. Thus, when organic species containing both σ and π-electrons are represented graphically, it is usual to depict them as though they contained only σ-electrons. They are depicted by their σ-framework, where one covalently bonded pair of electrons is represented by a single bond. Examples of this very commonly encountered mode of representation are shown in Fig. 1 for the cases of butadiene, benzene and naphthalene. Multiple bonds have not been drawn in, i.e. all bonds are drawn using modulo one arithmetic. Bent or zig-zag shaped molecules are normally represented as straight chains, non-planar species are similarly represented as planar species. Parameters, such as charge density, which are evaluated for the π-electrons, are frequently inscribed on the graph of the σ-framework at points close to the vertices to which they refer.

In representing molecules by their graphs, one drawback is evident, namely that all the stereochemical features of the molecule, such as its chirality, are lost. This loss is not always as serious as it may seem, however, since the graph does provide us with the full topology of the molecule, and many important parameters of molecules, such as their energy, their bond orders and their charge densities are now known to be essentially topology-dependent. This is especially so in the case of alternant hydrocarbons, and other systems, where each atom makes a net contribution of one electron to the π-cloud.

In addition to providing a very useful facilitation of the calculation of the physico-chemical parameters for a molecule, conversion of a graph into matrix format also makes possible the establishment of a criterion for deciding which of the properties of an individual molecule are topology-dependent. We shall return to this theme after discussing first some of the matrices which have proved particularly valuable in the representation of chemical species.

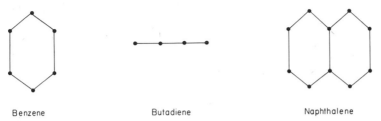

Benzene   Butadiene   Naphthalene

FIG. 1   The graphical representation of some organic molecules. Only the σ-electron framework is depicted in each case.

2   THE MATRIX REPRESENTATION OF MOLECULES

As indicated above, the chief reason for using matrix notation when calculating molecular properties is the very considerable facilitation this formalism provides in such calculations. In many cases some kind of matrix notation is indispensible in quantum-chemical calculations, although there are a few notable exceptions to this generalization. Thus, a formulation of perturbation theory due to Dewar[38] enabled Flurry[44] to calculate parameters, such as charge densities, for carcinogenic hydrocarbons without resort to matrix notation. Similarly Urch[135] has shown how the secular equations for a molecule may be set up without reference to a secular determinant. These cases are comparatively rare, however, and the great majority of calculations in this field do require matrix formalism at some stage.

We describe now several of the more important matrices which have been used to represent molecular species.

2.1   The Hückel matrix

This matrix is the most widely used and well-known matrix for quantum-chemical calculations and, still today, provides the starting point for elementary work in this field. As a typical example, we show in Fig. 2 the *Hückel matrix* **H** for the carbon skeleton of benzene. The term α represents the Coulomb integral for two adjacent atoms in the ring, β the corresponding resonance integral between them, and ϵ_k' is an energy eigenvalue for this system. In simple Hückel theory all of the α terms, and all of the β terms, are taken to be equal, even when bonds are inequivalent.

To determine the eigenvalues of this matrix, the determinant of this matrix, known as the *secular determinant*, is evaluated. Secular determinants have been known for well over a century, although their application to quantum chemistry was first described only some forty years ago.[78] Because all the bonds are treated as equivalent, the matrix will be essentially topology-dependent. Unfortunately, this dependence is somewhat obscured by the terms used in the

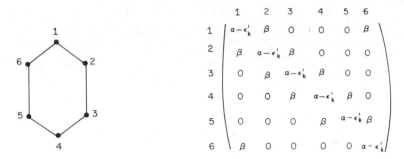

$$
\begin{array}{c c}
 & \begin{array}{cccccc} 1 & 2 & 3 & 4 & 5 & 6 \end{array} \\
\begin{array}{c} 1 \\ 2 \\ 3 \\ 4 \\ 5 \\ 6 \end{array} &
\left(\begin{array}{cccccc}
a-\epsilon'_k & \beta & 0 & 0 & 0 & \beta \\
\beta & a-\epsilon'_k & \beta & 0 & 0 & 0 \\
0 & \beta & a-\epsilon'_k & \beta & 0 & 0 \\
0 & 0 & \beta & a-\epsilon'_k & \beta & 0 \\
0 & 0 & 0 & \beta & a-\epsilon'_k & \beta \\
\beta & 0 & 0 & 0 & 0 & a-\epsilon'_k
\end{array}\right)
\end{array}
$$

FIG. 2   The Hückel matrix for the π-electrons in benzene.

matrix. This difficulty is largely overcome, however, by use of the adjacency matrix, which reflects far more clearly the topology of the species it represents.

2.2   The adjacency matrix

Following the appearance of a monograph by König (1936), chemists in general began to realise that molecules and other chemical species could be conveniently represented by a variety of different matrices.[3] A fuller account of such matrices is to be found in the texts of Busacker and Saaty[13] and Harary.[65] Probably the first of these matrices to be investigated in a chemical context was the *adjacency matrix* **A**. Along with the Hückel matrix, this matrix has proved of the greatest importance in quantum chemistry. One reason for this is that the topology of a chemical species is reflected in the structure of the matrix in a way that is particularly easy to visualize.

In Figs. 3 and 4 two differing types of adjacency matrix are represented. Figure 3 shows an *edge adjacency matrix* for the carbon skeleton of benzene.

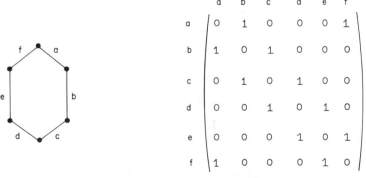

$$
\begin{array}{c c}
 & \begin{array}{cccccc} a & b & c & d & e & f \end{array} \\
\begin{array}{c} a \\ b \\ c \\ d \\ e \\ f \end{array} &
\left(\begin{array}{cccccc}
0 & 1 & 0 & 0 & 0 & 1 \\
1 & 0 & 1 & 0 & 0 & 0 \\
0 & 1 & 0 & 1 & 0 & 0 \\
0 & 0 & 1 & 0 & 1 & 0 \\
0 & 0 & 0 & 1 & 0 & 1 \\
1 & 0 & 0 & 0 & 1 & 0
\end{array}\right)
\end{array}
$$

FIG. 3   The edge adjacency matrix for a benzene molecule.

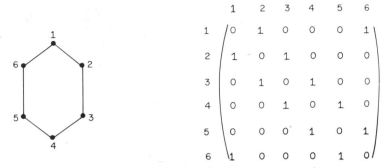

$$\begin{array}{c|cccccc} & 1 & 2 & 3 & 4 & 5 & 6 \\ \hline 1 & 0 & 1 & 0 & 0 & 0 & 1 \\ 2 & 1 & 0 & 1 & 0 & 0 & 0 \\ 3 & 0 & 1 & 0 & 1 & 0 & 0 \\ 4 & 0 & 0 & 1 & 0 & 1 & 0 \\ 5 & 0 & 0 & 0 & 1 & 0 & 1 \\ 6 & 1 & 0 & 0 & 0 & 1 & 0 \end{array}$$

FIG. 4   The vertex adjacency matrix for a benzene molecule.

This matrix is constructed for any graph having m edges by setting up an $m \times m$ array. The number of edges in the graph thus determines the order of this matrix. Whenever two edges r and s in the graph are adjacent, an entry of one is inserted into both the (r,s) and (s,r) positions of the matix array; zeros are inserted into all the other positions. In Fig. 4 the *vertex adjacency matrix* for the carbon skeleton of benzene is given. This matrix is constructed for any graph on p vertices by forming a $p \times p$ array. Whenever two vertices are joined by an edge a one is inserted at the corresponding positions in the array; zeros are inserted elsewhere. Because of their mode of construction, it is evident that both types of adjacency matrix will be symmetrical about the principal diagonal.

Although these two types of adjacency matrix reflect the topology of any species represented by them equally well, they have not been utilized equally in

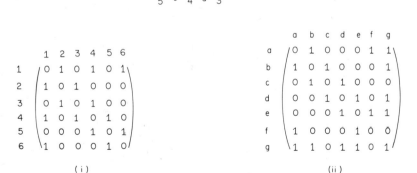

(i)

$$\begin{array}{c|cccccc} & 1 & 2 & 3 & 4 & 5 & 6 \\ \hline 1 & 0 & 1 & 0 & 1 & 0 & 1 \\ 2 & 1 & 0 & 1 & 0 & 0 & 0 \\ 3 & 0 & 1 & 0 & 1 & 0 & 0 \\ 4 & 1 & 0 & 1 & 0 & 1 & 0 \\ 5 & 0 & 0 & 0 & 1 & 0 & 1 \\ 6 & 1 & 0 & 0 & 0 & 1 & 0 \end{array}$$

(ii)

$$\begin{array}{c|ccccccc} & a & b & c & d & e & f & g \\ \hline a & 0 & 1 & 0 & 0 & 0 & 1 & 1 \\ b & 1 & 0 & 1 & 0 & 0 & 0 & 1 \\ c & 0 & 1 & 0 & 1 & 0 & 0 & 0 \\ d & 0 & 0 & 1 & 0 & 1 & 0 & 1 \\ e & 0 & 0 & 0 & 1 & 0 & 1 & 1 \\ f & 1 & 0 & 0 & 0 & 1 & 0 & 0 \\ g & 1 & 1 & 0 & 1 & 1 & 0 & 1 \end{array}$$

FIG. 5   For the bicyclo[2,2,0]hexa2,5 diene molecule are shown (i) the vertex adjacency matrix and (ii) the edge adjacency matrix.

quantum chemistry. The essential difference between these two matrices is illustrated in Fig. 5, which shows the two adjacency matrices for the bicyclo [2,2,0] hexa-2,5-diene molecule. Other types of adjacency matrix which have been described include the *variable adjacency matrix* of Rescigno and Segre,[107] and the *symmetry adjacency matrix* of Billes[5] which was used in the algebraic characterization of molecular structure. However, because the vertex adjacency matrix may be used as a substitute for the Hückel matrix, this particular matrix has been employed in quantum-chemical studies to the virtual exclusion of the other types of adjacency matrix. For this reason, whenever the adjacency matrix is referred to in this article without further qualification the vertex adjacency matrix will be meant.

2.3   The incidence matrix

The *incidence matrix* **B** has found little application in chemistry to date. It was employed by Balandin[3] in his study of the physico-chemical parameters of molecular species, and later by De Chiossone[35] who used it in isomer enumeration work. The incidence matrix for the carbon skeleton of benzene is illustrated in Fig. 6. The matrix may be constructed for any graph having n vertices and m edges by setting up $n \times m$ matrix array; the rows and columns of the matrix then correspond respectively to the vertices and the edges of the graph. A one is inserted in the (i,j)th position in the array if the jth edge is incident with the ith vertex; all the other entries in the array are zeros. A number of differing types of incidence matrix, such as the *cycle, tree* and *co-tree incidence matrices*, have also been defined. These latter matrices are discussed comprehensively by Ponstein.[104]

A graph may be completely determined by either its adjacency or its incidence neighbourhood relations, and thus two graphs possessed of the same

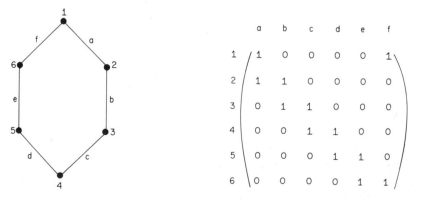

FIG. 6   The incidence matrix for a benzene molecule.

adjacency or incidence matrix may be said to be *isomorphic.* A general formula relating the vertex adjacency matrix to the incidence matrix, originally due to Kirchhoff,[84] has been given by Harary.[65] For any (p,q) graph G we have

$$A(L(G)) = \mathbf{B} \cdot \mathbf{B}^t - 2.1 \tag{1}$$

where $\mathbf{B}$ is the incidence matrix of G, and $\mathbf{B}^t$ its transpose, 1 is the $q \times q$ unit matrix, and $A(L(G))$ represents the adjacency matrix for the line graph of G. The line graph $L(G)$ of G is formed by replacing the edges of G by vertices in such a way that vertices in $L(G)$ are connected whenever the corresponding edges in G are adjacent.

2.4   The cycle matrix

The *cycle matrix* $\mathbf{C}$ has no useful application to chemistry though it is conceivable that it might play a rôle in future in the study of diamagnetic anisotropy in large aromatic systems.[91,96] The matrix is defined in terms of the independent cycles contained within a graph, as illustrated in Fig. 7 for the graph of an hypothetical molecule containing four independent cycles. The construction of the matrix for a graph having c independent cycles and m edges involves the setting up of a $c \times m$ array such that the rows correspond to the cycles and the columns to the edges. When an edge forms part of a given cycle in the graph a one in inserted at the appropriate position in the matrix; in all other cases zeros are inserted.

Unlike the adjacency matrix or the incidence matrix, the cycle matrix does not determine a graph up to isomorphism, since edges which do not form part of a cycle are not indicated in this matrix. For any given graph, the cycle matrix is related to the incidence matrix by the orthogonal relationship:

$$\mathbf{C} \cdot \mathbf{B}^t \equiv 0 \quad (\text{mod } 2) \tag{2}$$

where $\mathbf{B}$ is the incidence matrix of the graph, and $\mathbf{C}^t$ the transpose of its cycle matrix $\mathbf{C}$. A *cocycle matrix*, analogous with the cycle matrix, has also been defined,[65] though it too has no chemical applications at present.

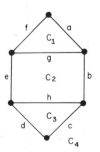

$$
\begin{array}{c c c c c c c c c}
 & a & b & c & d & e & f & g & h \\
C_1 & 1 & 0 & 0 & 0 & 0 & 1 & 1 & 0 \\
C_2 & 0 & 1 & 0 & 0 & 1 & 0 & 1 & 1 \\
C_3 & 0 & 0 & 1 & 1 & 0 & 0 & 0 & 1 \\
C_4 & 1 & 1 & 1 & 1 & 1 & 1 & 0 & 0 \\
\end{array}
$$

FIG. 7   The cycle matrix for a hypothetical molecule.

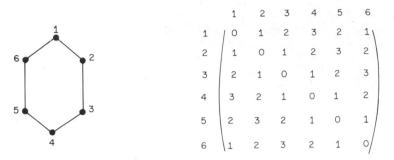

FIG. 8   The distance matrix for a benzene molecule.

2.5   The distance matrix

This less commonly encountered matrix has found only very limited application to chemistry. It was employed in implicit form by Wiener[136] in his study of additive physico-chemical parameters of molecules, and later adapted to further studies on the same topic.[76,103] The *distance matrix* **D** is constructed for any graph on n vertices by setting up an $n \times n$ array. Each entry d_{ij} is equal to the minimum number of edges in the graph connecting vertex i to vertex j. When there is no connection at all the distance d_{ij} is taken to be infinite. The distance matrix for the carbon skeleton of benzene is illustrated in Fig. 8.

The distance matrix has been investigated in its own right by Hakimi and Yau;[59] it has also been discussed by Harary.[65] The entries in the distance matrix are related to the entries in powers of the adjacency matrix **A**. All entries of the type d_{ii} are by definition zero, whereas d_{ij} entries $(i \neq j)$ are given by the least value of ξ such that the a_{ij}^{ξ} entry in **A** is greater than zero. Other similar matrices, also discussed by Harary,[65] are the *reachability matrix* and the *detour matrix*, but to date these latter two have found no application to chemistry.

2.6   A nomenclature proposal

A certain amount of confusion has arisen in the chemical literature over the correct terminology to be used when describing the adjacency matrix **A**. The matrix has been variously labelled by different authors as the topological matrix,[114] the structural matrix,[5,66] the simplexial matrix,[83] and even the incidence matrix.[54,137] This latter usage is particularly unfortunate since, as described above, the term incidence matrix has already been reserved for another type of matrix.

To avoid such complications in future we suggest that the term incidence matrix be confined only to the type of matrix defined in Section 2.3 above. This leaves then the choice of a suitable term for the adjacency matrix. As the term adjacency matrix is now well established, and particularly favoured by mathematicians, we recommend its continued use in this context. In a more chemical context, however, we feel that the term topological matrix should also be retained, especially as it is coming increasingly into favour. In our following discussion we shall treat these two designations as interchangeable. We recommend a standardization of terminology along these lines in the chemical literature.

3.   THE TOPOLOGICAL MATRIX IN QUANTUM CHEMISTRY

Although several types of matrix have been characterized above, our further discussion will relate to only two of them, namely the Hückel and topological matrices. The use of the topological matrix is very old and goes back at least to the work of Kirchhoff[84] on electrical circuits. In a more chemical context an early application involved its use in the form of an invariant to represent chemical structural formulae.[12,132] Much later such matrices were used not only to represent formulae but also to interpret the behaviour of molecules.[3] The first use of the Hückel matrix came with the early development of quantum chemistry, in the molecular orbital approach due to Hückel.[78] In recent years the topological matrix has frequently replaced the Hückel matrix in quantum-chemical calculations.

The principal reason for the prominence of the topological matrix in quantum-chemical studies has been the demonstration by Ruedenberg[112] that for a given molecule the matrix has closely related eigenvalues and the same eigenvectors as those for the Hückel matrix. These two matrices may thus be regarded as equivalent in quantum chemistry. For many purposes the topological matrix may therefore be substituted for the Hückel matrix. The greater simplicity of the former matrix also serves to facilitate calculation of the energy eigenvalues.

It is evident that, because the topological matrix so clearly reflects the topology of a molecule, all the properties of molecules which may be derived from this matrix by any mathematical manipulation must be topology-dependent. These include not only the eigenvalues and eigenvectors of the matrix but also a large number of derived parameters such as the bond orders and the charge densities of molecules so represented. In fact, use of this matrix provides a criterion for deciding which of the many physico-chemical parameters of a molecule are topology-dependent. All those parameters which may be derived from the topological matrix by any set of mathematical operations are to be regarded as topology-dependent.

It is hardly surprising that a great deal of attention has focused on the evaluation of the eigenvalues of topological matrices, since the eigenvalues give directly the Hückel energy levels in the molecules so represented. A proof of the equivalence of these two matrices is given in Appendix I. Except in certain special cases considered below, this evaluation is not usually a simple procedure. In general, the topological matrix cannot even be partitioned into blocked canonical form by interchange of its rows and columns. However, the existence of comprehensive tabulations of eigenvalues for Hückel matrices[26,71,130] implies the availability of solutions for the equivalent topological matrix.

Molecules which are identically connected topologically will, of course, have the same topological matrix and will therefore possess identical sets of eigenvalues. Thus, molecules differing from one another only in their stereochemistry may have the same set of Hückel energy levels. It is known that, on certain occasions, it is even possible for molecules having non-isomorphic graphs to be possessed of identical sets of eigenvalues.[2,134,142]

Several of the more general methods of evaluating the eigenvalues of the topological matrix are discussed in Section V. Before considering these methods, however, we look first at a few well-known cases where exact, analytic solution of the matrix is possible.

4   ANALYTICAL METHODS FOR EVALUATING EIGENVALUES OF THE TOPOLOGICAL MATRIX

Sometimes the structure of a topological matrix suggests a method for the analytic evaluation of its eigenvalues. We consider a few examples of such matrices below. Analytic solutions for many different topological matrices, all representing chemical species, have been presented in tabular format by Gouarné.[52]

4.1   Chains of atoms

A conjugated, unbranched linear chain of carbon atoms, which need not necessarily be straight, is referred to as a *polyene* if it has the general formula $CH_2(CH)_{n-2}CH_2$. It will be represented by an unbranched tree graph on n vertices, if hydrogen atoms are not considered as is usually the case. The polyenes have been studied in great detail by theoretical chemists.[20,36]

The topological matrix for any polyene molecule always has the general form illustrated in Fig. 9. The matrix provides a clear reflection of the topology of such a species in that non-zero terms appear only alongside the principal diagonal. The eigenvalues are readily obtainable in closed form[28,116] from the equation:

$$\epsilon_k = 2 \cos\left[\frac{\pi k}{n+1}\right] \tag{3}$$

where k is an integer which may assume only the values

$$k = 1,2,3, \ldots, n \tag{4}$$

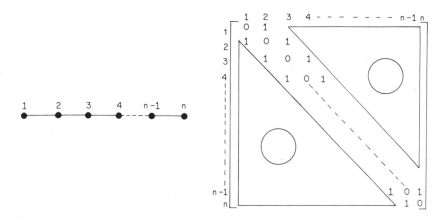

FIG. 9   The topological matrix for a linear polyene molecule.

and n is the number of carbon atoms in the polyene chain. Use of the equivalent Hückel matrix yields the eigenvalues in the following form:

$$\epsilon'_k = \alpha + 2\beta \cos\left[\frac{\pi k}{n + 1}\right]$$ (5)

where α and β represent respectively the Coulomb and resonance integrals for any adjacent pair of carbon atoms in the chain.

A convenient mnemonic device for determining the energy levels in chain molecules has been devised by Frost and Musulin.[47] For any chain consisting of n carbon atoms a fictitious $n+2$ carbon atoms are added in order that a polyene having $2n + 2$ carbon atoms may be constructed. This polygon is inscribed as a regular $(2n + 2)$-gon with a circumscribing circle such that one of its vertices

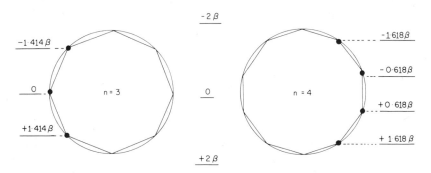

FIG. 10   The mnemonic device of Frost and Musulin depicting the eigenvalues for polyene chains containing three and four atoms in the chain.

touches the circle at its lowest point. If the radius of the circle is taken to be 2 units, the horizontal projection corresponding to the centre of the circle will then represent the zero energy level, i.e. in the Hückel approach this will be the α level of a non-bonding $2p\pi$-electron. Similar projections of all the other points of contact on to one side of the polygon yield the exact eigenvalues for the chain molecule. The device is illustrated in Fig. 10 for polyenes having 3 and 4 atoms in the chain. It is noteworthy that for all chain molecules the energy eigenvalues always occur in pairs, i.e. if λ be one eigenvalue for the chain then so is $-\lambda$. This is a particular case of the Pairing Theorem we discuss later.

4.2   Rings of atoms

When the two ends of a linear chain molecule are linked up to form a ring species, an *annulene* is said to be formed if the ring has an even number of carbon atoms. Those rings having an odd number of atoms produce radical or ionic species. The general formula for an unsubstituted annulene is $CH(CH)_{n-2}CH$, written thus to differentiate it from the corresponding polyene. The topological matrix for an annulene is similar to that for a polyene, but, in addition to the ones alongside the principal diagonal, two extra entries of one now appear in the top right hand and bottom left hand corners of the matrix. This is illustrated for the general n carbon atom annulene in Fig. 11.

The eigenvalues for an annulene system are given in closed form by the relation:[78]

$$\epsilon_k = 2 \cos \left[\frac{2\pi k}{n} \right] \tag{6}$$

where n is the number of carbon atoms in the ring, and k is a running index having the values

$$k = 0, \pm 1, \pm 2, \pm 3, \ldots, \begin{cases} \pm \dfrac{n-1}{2} \ (n \text{ odd}) \\[2mm] \pm \dfrac{n}{2} \ (n \text{ even}) \end{cases} \tag{7}$$

Expressed in Hückel notation, equation (6) may be rewritten as:

$$\epsilon'_k = \alpha + 2\beta \cos \left[\frac{2\pi k}{n} \right] \tag{8}$$

where symbols have their previous significance.

A mnemonic device, again due to Frost and Musulin,[47] is also applicable to ring systems. For a ring containing n carbon atoms a regular n-gon is inscribed within a circle of radius 2 units, such that one vertex touches the circle at its lowest point. In this case n may be either odd or even. As before, projections of

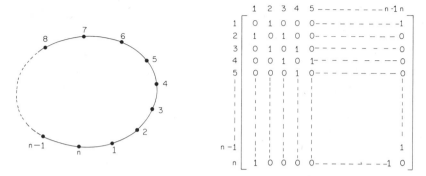

FIG. 11   The topological matrix for a cyclic annulene molecule.

the contact points between the circle and the polygon yield the eigenvalues for the species in question; the centre of the circle also corresponds to the α non-bonding level for carbon. Again, the contact points give eigenvalues which are exact on an absolute basis. This remains true even if the chemical species is non-planar, as, for example, in the case of cycloöctatetraene.

The device is illustrated in Fig. 12 for the ring species cyclopentadiene and cycloheptatriene. It may readily be seen that the eigenvalues obtained from it are the exact eigenvalues for the species under study. Thus, from equation (6) we may write, using a well-known property of cosines,

$$\epsilon_k = 2 \cos\left[\frac{2\pi k}{n}\right] = 2 \cos\left[\frac{2\pi(n-k)}{n}\right] \tag{9}$$

When n is odd, there will always be a unique orbital of energy 2 units for $n = k$, since $\cos 2\pi = \cos 0 = 1$. All the remaining energy levels must occur in degenerate pairs, as is apparent from Fig. 12. When n is even, there will be an orbital having energy -2 units, in addition to one of energy $+2$ units, since the value of the cosine function is -1 for $k = n/2$. All the other orbitals will again

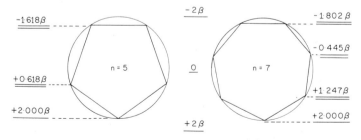

FIG. 12   The mnemonic device of Frost and Musulin depicting the eigenvalues for cyclic species containing five and seven atoms in the ring.

occur in degenerate pairs; this is another instance of the applicability of the Pairing Theorem.

4.3   The polyacenes

When benzene rings are fused together to form a straight chain, as illustrated in Fig. 13, a *polyacene* species is said to be formed. The general formula for members of this series is $C_{4n+2}H_{2n+4}$. By use of a method described by

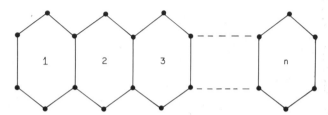

FIG. 13   The graphical representation of a linear polyacene chain.

Coulson,[18] the determinant of the topological matrix for such species may be separated into the product of two determinants. The eigenvalues then appear as two separate sets as indicated:

$$\epsilon_k^{(1)} = 1, \quad \frac{1}{2}\left\{ 1 \pm \left(9 + 8 \cos\left[\frac{k\pi}{n+1}\right]\right)^{\frac{1}{2}} \right\} \tag{10}$$

and

$$\epsilon_k^{(2)} = -1, \quad \frac{1}{2}\left\{ -1 \pm \left(9 + 8 \cos\left[\frac{k\pi}{n+1}\right]\right)^{\frac{1}{2}} \right\} \tag{11}$$

where k is a running integer which may take the values

$$k = 1,2,3, \ldots , 2n + 1 \tag{12}$$

The set of eigenvalues given in equation (10) represents the energies of molecular orbitals of type SZ, whereas those given in equation (11) represent the energies of molecular orbitals of type AZ.

4.4   The radialenes

A *radialene* molecule is formed when a carbon atom is attached by a double bond to each of the carbon atoms forming a vertex of a regular polygon of carbon atoms. Examples of radialene species are illustrated in Fig. 14. All are

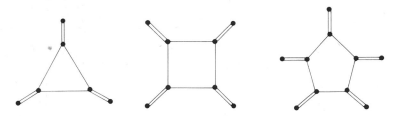

FIG. 14 The graphical representation of some radialene species.

represented by the general formula $C_{2n}H_{2n}$, where $n \geqslant 3$. The eigenvalues for such species were obtained in closed form by Hess and Schaad[72] and are as follows:

$$\epsilon_k = \cos\left[\frac{2k\pi}{n}\right] \pm \left(\cos^2\left[\frac{2k\pi}{n}\right] + 1\right)^{\frac{1}{2}} \tag{13}$$

Their method rests essentially upon a finite difference technique,[106] as do the other closed solutions presented in this section.

4.5 Aromatic species

We demonstrate now a general rule for aromatic species using as our example an annulene molecule. In Hückel notation, it is evident that, whether n is odd or even, the orbital of lowest energy in any annulene will have energy $\alpha + 2\beta$. In the Aufbau process for the π-electrons in such systems the first two electrons are, of course, assigned to this orbital. All the remaining π-electrons will occupy orbitals above this level in degenerate pairs. Because the lowest orbital holds two electrons and each pair of orbitals above this can accommodate a total of four electrons, the filling pattern for these orbitals will be of the general type $4m + 2$, where m is some positive integer. Accordingly, when π-electrons occupy the higher orbitals of a chemical species in two-fold degenerate pairs, a $4m + 2$ system is said to result.

Since $4m + 2$ systems possess a closed electron shell structure, they should display aromatic stability. This prediction is known as the $4m + 2$ or *Hückel rule*. Present evidence, based largely on diamagnetic ring current measurements, indicates that the rule is valid for values of m up to at least 6,[39,97] although it may break down for higher values of m, as such large molecules cannot attain even an approximately planar configuration.

In $4m + 2$ systems where m is very large it has been shown[90] that the occupied π-orbitals do not constitute a single half-filled band but rather give rise to two distinct half-filled bands separated by a finite gap. The energy levels are given by the general formula[16]

$$\epsilon'_k = \alpha \pm \beta \left(1 + 2 \cos\left[\frac{2\pi k}{2m+1}\right]\right)^{\frac{1}{2}} \qquad (14)$$

where the running index k can take on the values

$$k = 0, \pm 1, \pm 2, \pm 3, \ldots, \pm m \qquad (15)$$

Clearly each half-band will contain $2m + 1$ levels and these will lie within the ranges

$$\alpha + 2\beta \leqslant \epsilon'_k < 0 \qquad (16)$$

for the bonding levels and

$$0 < \epsilon'_k \leqslant \alpha - 2\beta \qquad (17)$$

for the non-bonding levels. Thus, both very large annulene and polyene systems should behave as insulators having low excitation energy rather than as one-dimensional strips of metal.[105]

Since its formulation, many extensions have been made to the $4m + 2$ rule and a number of similar rules have also been propounded. As it stands, the rule is applicable only to systems possessing at least a three-fold axis of symmetry, but was extended by Craig[27] to systems having site symmetry of A_2 or B_2 types, and also to polycyclic molecules. The rule has been considered in detail for the case of bicyclic hydrocarbons by Bochvar and Stankevich,[9] who made explicit use of the topological matrix. Several analogous rules have been advanced for a

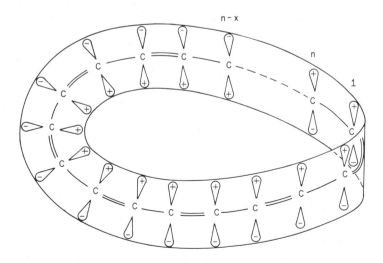

FIG. 15   A representation of a cyclic annulene molecule twisted into the form of a Möbius strip.

variety of other systems. These include the $6m + 2$ and $8m - 2$ rules for the polycyclic polyenes;[60] the $4m + 3$ rule for the spirarenes;[75] the $4m$ rule for bicycloaromatic hydrocarbons;[48] and the $4m - 1$ and $4m + 1$ rules for non-alternant hydrocarbons.[58]

The case of annulenes twisted into a Möbius strip configuration has been studied by Heilbronner.[69] The rule analogous to the $4m + 2$ rule in these systems is the $4m$ rule. A convenient mnemonic device giving the energy levels in such systems has also been devised.[140] An illustration of an annulene ring having all n carbon atoms twisted into the form of a Möbius strip is given in Fig. 15.

To conclude this section, we note that all the above rules, and several others which have now been adduced for many different systems, stem basically from the structure of the topological matrix for the type of species considered. As a necessary consequence of this, it is evident that all such rules are topology-dependent. Thus, from the topology of a chemical species it is possible to explain not only the stability of the aromatic sextet of electrons, but also to explain and interpret a host of other empirical observations.

5   GENERAL METHODS FOR EVALUATING EIGENVALUES OF THE TOPOLOGICAL MATRIX

Having now indicated the wide significance of eigenvalues to quantum chemistry we consider some of the more common methods used to evaluate them in the case when an analytic, closed solution is not possible. As analytic solutions are possible only in a few very special cases, resort is often made to very sophisticated methods and much ingenuity has been evidenced in this work. We shall not discuss the more elementary and standard methods of solving for eigenvalues, such as those described by Margenau and Murphy.[92] Neither shall we treat computer methods such as the frequently adopted Jacobi diagonalization.[41] We consider instead some of the more prominent manipulative techniques developed in this field. These are not normally found in elementary texts, though Daudel, et al.[144] describe the method of polygons, and Streitwieser[129] discusses the recursive approach of Heilbronner.

5.1   Reduction to blocked form

The earliest method of evaluating eigenvalues from secular determinants often involved an attempt to reduce the topological matrix to some standard form, to which analytical solution could then be applied. We discuss briefly a few examples of this approach.

In determining the eigenvalues for molecules having a skew strip graph, illustrated in Fig. 16, Coulson and Rushbrooke[25] showed that, provided two

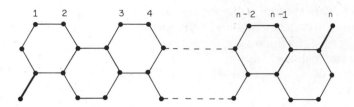

FIG. 16   The graphical representation of a skew strip polyacene chain with two extra terminal bonds attached.

extra terminal bonds were added to the strip, as indicated by the heavy lines in the figure, the matrix for this system could then be partitioned as follows:

$$A = \begin{pmatrix} P & 1 & \cdot & \cdot \\ 1 & Q & 1 & \cdot \\ \cdot & 1 & P & 1 \\ \cdot & \cdot & 1 & Q \end{pmatrix} \tag{18}$$

Here **A** is the topological matrix for a skew strip having n carbon atoms in each of its rows, **1** is the unit matrix of order n, and P and Q also represent $n \times n$ matrices thus:

$$P = \begin{bmatrix} 01 & & & \\ 10 & & & \\ & 01 & & \\ & 10 & & \\ & & \cdot & \\ & & & \cdot \\ & & & \cdot \\ & 01 & \\ & 10 & \\ & 0 & \end{bmatrix} ; \quad Q = \begin{bmatrix} 0 & & & \\ 01 & & & \\ 10 & & & \\ & 01 & & \\ & 10 & & \\ & & \cdot & \\ & & \cdot & \\ & & \cdot & \\ & & 01 \\ & & 10 \end{bmatrix} \tag{19}$$

After rearrangement of row and columns, the determinant of **A** may be brought to the standard form

$$A = \prod_{k=1}^{2} \begin{pmatrix} P & z_k 1 \\ z_k \bar{1} & Q \end{pmatrix} \tag{20}$$

where the parameters z_k are given by the equation

$$z_k = 2 \cos\left[\frac{k\pi}{5}\right] \tag{21}$$

The eigenvalues of **A**, which are not much changed by the two extra terminal bonds provided the strip is very long, are now readily obtainable by the method of Rutherford.[116]

A considerable simplification of secular determinants, and, in certain cases, reduction to blocked form, was effected by Dewar.[37]

Any complex, conjugated molecule is imagined to be split into symmetrical parts such that the component parts are not only conjugated themselves but also mutually conjugated upon reconstitution of the original molecule. Although we give no details of the method here, we may note some of its advantages over the usual linear combination of atomic orbitals (LCAO) approach. In Dewar's method, known as the linear combination of molecular orbitals (LCMO) method, the matrices for the component parts of a molecule are necessarily smaller than that for the molecule itself, and therefore more readily evaluated. Furthermore, since the same components will recur in many of the calculations, a library of standard results may be built up, which is available for repeated usage. Dewar obtained eigenvalues for several polyphenylmethyl molecules and demonstrated that in all cases the secular equation he obtained was identical to that given by the LCAO method. This method of evaluating eigenvalues has probably not received the attention it deserves.

A more recent method of partitioning topological matrices has been propounded by Bochvar and Stankevich.[8] They investigated several large macromolecular ring systems, constituted from identical molecular fragments of the type illustrated in Fig. 17. Each atom is considered to contribute one electron to the π-cloud of the ring. If each fragment consists of n atoms, the Hückel matrix **H** for such a system has the general form:

$$\mathbf{H} = \begin{pmatrix} H & W & \cdot & \cdot & 0 & W^t \\ W^t & H & \cdot & \cdot & 0 & 0 \\ \cdot & \cdot & & & \cdot & \cdot \\ \cdot & \cdot & & & \cdot & \cdot \\ 0 & 0 & \cdot & \cdot & H & W \\ W & 0 & \cdot & \cdot & W^t & H \end{pmatrix} \qquad (22)$$

where H represents the Hückel matrix for one fragment of the ring, W is an

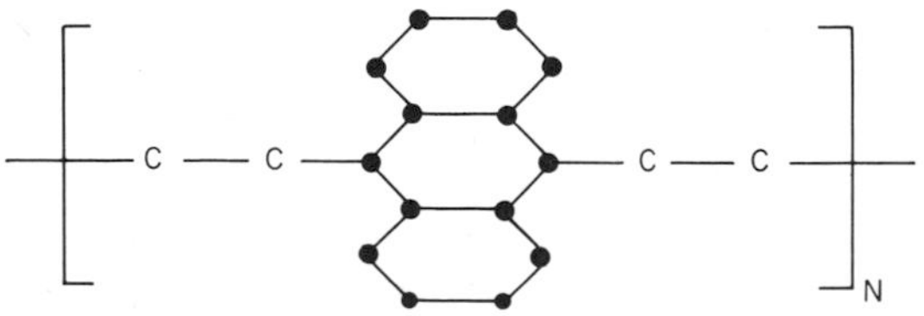

FIG. 17   A representation of a molecular fragment of a large macromolecular ring system.

interaction matrix having only one entry of β to represent the interaction between adjacent fragments, and W^t represents the transpose of W. All the three submatrices are square and of order n.

At this stage a new matrix U is introduced as follows:

$$U = \begin{pmatrix} 0 & 1 & \cdot & \cdot & 0 & 1 \\ 1 & 0 & \cdot & \cdot & 0 & 0 \\ \cdot & \cdot & & & \cdot & \cdot \\ \cdot & \cdot & & & \cdot & \cdot \\ 0 & 0 & \cdot & \cdot & 0 & 1 \\ 1 & 0 & \cdot & \cdot & 1 & 0 \end{pmatrix} \tag{23}$$

where 1 represents the unit matrix of order n, and 0 an $n \times n$ array of zeros. The matrix U is of the same order as H, i.e. of order $N \times n$, where N is the number of fragments in the ring. The eigenvalues of U are readily obtainable as this matrix is in a well-known standard form [cf. equation (6)] :

$$\epsilon_k = -2 \cos\left[\frac{2\pi}{nk}\right] \tag{24}$$

Each of the n eigenvalues will be of multiplicity N, since k is a running index ranging from 1 to N. Moreover, because H and U commute, it may be shown that the eigenvalues of H are given by the eigenvalues of matrix T, where

$$T = H + W \exp\left[\frac{2\pi i k}{N}\right] + W^t \exp\left[\frac{-2\pi i k}{N}\right] \tag{25}$$

As for matrix U, this matrix will have n eigenvalues for each value of k.

5.2   The method of polygons

This method, originally put forward by Samuel[119,120] and later extended by Gouarné,[51] has been adumbrated by several workers over the last thirty years.[4,19,68] Although less fashionable today, it remains a particularly valuable method when the topological matrix contains a large number of zeros, as is normally the case in simple molecular orbital theory.

The graph of the cyclobutadiene molecule, shown in Fig. 18a, is used to illustrate the method. When expanded as a determinant, the topological matrix for cyclobutadiene will yield several terms, each consisting of the product of four elements of the matrix. Terms such as $a_{12}\,a_{23}\,a_{34}\,a_{41}$ and $a_{14}\,a_{43}\,a_{32}\,a_{21}$ will correspond to a full polygon, i.e. in this case a square as indicated in Fig. 18a, whereas other terms will represent only fragments of a polygon. Thus, the term $a_{11}\,a_{22}\,a_{33}\,a_{44}$ will represent the set of vertices shown in Fig. 18b and the term $a_{12}\,a_{21}\,a_{34}\,a_{43}$ will correspond to the graph in Fig. 18c.

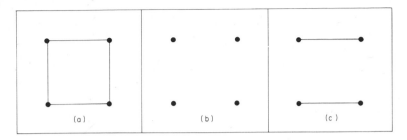

FIG. 18   Some of the component parts of the graph of the cyclobutadiene molecule used in the method of polygons.

Since all terms in the expansion will be represented by either a partial or a complete polygon, the total determinant may be evaluated by enumerating the polygons in each case. An appropriate sign convention must be adopted for each term. Terms of the type $a_{11} \, a_{22} r, \, a_{33} \, a_{44}$ are represented as ω^4, terms such as $a_{12} \, a_{23} \, a_{34} \, a_{41}$ as unity, and so on. The *characteristic polynomial* for butadiene thus becomes

$$P_n(\omega) = \omega^4 - 4\omega^2 = 0, \tag{26}$$

where ω represents the quantity $(\alpha - \epsilon_k)/\beta$. For large molecules the method generally becomes onerous, but for smaller species it is one of the simplest available.

5.3   Use of recursion formulae

Recursion formulae have been used extensively by Heilbronner[67] and a number of other workers[94,95] to factorize secular determinants into smaller, more manageable determinants. We illustrate the approach of Heilbronner by applying it in the case of the polyenes and the annulenes.

The Hückel matrix for a polyene molecule containing n carbon atoms may be expressed in the form of a characteristic polynomial thus:

$$P_n(\omega) = \sum_{i=0}^{n} \kappa_i \omega^i, \tag{27}$$

where P_n is a polynomial of degree n expanded in terms of ω, and ω, as above, is $(\alpha - \epsilon_k/\beta$. It is not difficult to show that the following recursion formula applies to $P_n(\omega)$:

$$P_n(\omega) = P_{n-1}(\omega) - P_{n-2}(\omega) \tag{28}$$

This formula thus allows the expansion of $P_n(\omega)$ in terms of polynomials of lower degree. Similarly, for an annulene molecule, a corresponding expansion is possible using the recursion formula

$$R_n(\omega) = P_n(\omega) - P_{n-2}(\omega) - 2(-1)^n \tag{29}$$

where R_n is the polynomial of degree n for an annulene containing n carbon atoms. Heilbronner[67] has given tables of values for the coefficients κ_i of the polynomials P_n and R_n for small values of n. These tables enable the rapid evaluation of eigenvalues for a large number of chemical species, including polycyclic molecules and molecules having heteroatoms in their structure. Tables of κ_i coefficients for many different types of graph have also been given by Collatz and Sinogowitz[14] and Mowshowitz.[100]

5.4   Use of Sachs' theorem

In recent years a theorem due to Sachs[118] has found several applications in quantum chemistry. We discuss first the use of this theorem in determining the eigenvalues of a graph by providing a convenient method of setting up its characteristic polynomial. This approach was first presented in the work of Coulson[19] but only in the last few years has it been extensively employed.[53,56]

The characteristic polynomial for the graph of the molecule methylene-cyclopropene, illustrated in Fig. 19a, may be set up as follows. All the subgraphs containing vertices of degree one and those containing cycles are constructed for the species, as illustrated in Fig. 19b. The coefficients κ_i for the polynomial are now evaluated from the general formulae:

$$\begin{aligned} \kappa_0 &= 1 \\ \kappa_n &= \sum_{S_n} (-1)^c \, 2^r \end{aligned} \tag{30}$$

where c is the number of components of a subgraph S_n on n vertices, and r is the number of rings in S_n. For the graph in Fig. 19a the coefficients assume the values

$$\begin{aligned} \kappa_0 &= 1 \\ \kappa_1 &= 0 \\ \kappa_2 &= (-1)^1 2^0 + (-1)^1 2^0 + (-1)^1 2^0 + (-1)^1 2^0 = -4 \\ \kappa_3 &= (-1)^1 2^1 = -2 \\ \kappa_4 &= (-1)^2 2^0 = 1 \end{aligned} \tag{31}$$

and the corresponding characteristic polynomial has the form

$$P_n(\omega) = \omega^4 - 4\omega^2 - 2\omega + 1 \tag{32}$$

As before, by setting $P_n(\omega)$ to zero and solving for ω, the energy eigenvalues for the chemical species may be determined.

In addition to yielding the eigenvalues for chemical graphs, Sachs' theorem has also proved invaluable in analysing the eigenvalue spectra of chemical graphs.

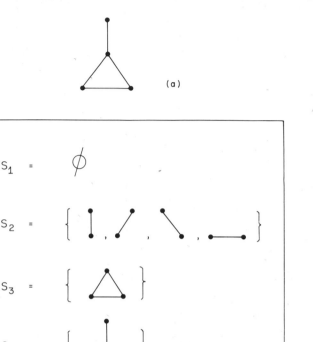

FIG.   19   Some   of   the   component   parts   of   the   graph   of   the methylene-cyclopropene molecule (a) used in Sachs' method.

Of the important results obtained to date we mention four. As no Sachs graph can have only one vertex, it follows that κ_1 must always be zero. This in turn implies that the sum of the whole eigenvalue spectrum is always zero i.e. that

$$\sum_{k=1}^{n} \epsilon_k = 0 \tag{33}$$

It may easily be shown that the number of Sachs' graphs with two vertices is equal to the number of edges n_e in the graph, i.e.

$$\kappa_2 = -n_e \tag{34}$$

Similarly, it can be shown that

$$\kappa_3 = -2n_3 \tag{35}$$

where n_3 is the number of three-membered cycles in the graph. For graphs which are *bipartite*, i.e. graphs whose vertices may be coloured with only two colours, we have for all odd n

$$\kappa_n = 0 \qquad\qquad\qquad\qquad\qquad\qquad\qquad\qquad (36)$$

Further applications of the theorem have lead to graph-theoretical proofs of the Hückel $4m + 2$ rule, the Pairing Theorem, and the Longuet-Higgins and Dewar rule.[53] This work underlines our earlier assertion that such rules and theorems have a topological basis.

5.5   The Hosoya topological index

In 1971 Hosoya[76] first put forward a new topological index for characterizing the physico-chemical properties of hydrocarbon molecules. Such indices have been known for over three decades in chemistry.[111] The Hosoya index Z has the general form

$$Z = \sum_{k=0}^{m} p(G,k) \qquad\qquad\qquad\qquad\qquad (37)$$

where p(G,k) represents the number of Kekulé structures for some unsaturated hydrocarbon molecule possessed of a graph G, and the summation extends over all m edges of G. It was shown recently[77] that this index is related to the characteristic polynomial of a molecule by the relation

$$P_n(\omega) = \sum_{k=0}^{m} (-1)^k\, p(G,k)\, \omega^{n-2k}, \qquad\qquad\qquad (38)$$

and this opened up the possibility of a new method of determining $P_n(\omega)$. The method has proved both simple and effective in its application and further results are to be expected, especially for molecules containing several rings in their structure, for which the method appears to be particularly well adapted.

6   THE STRUCTURE OF THE TOPOLOGICAL MATRIX IN CERTAIN SPECIAL CASES

In certain instances partitioning of the topological matrix becomes possible by direct use of either the topology or the symmetry of a chemical species. We discuss first molecules which possess a bipartite graph, and this leads us to the Pairing Theorem. Finally, in this section, we give a brief outline of some of the applications of group theory in the simplification of secular determinants.

6.1   Alternant molecules

The vertices of any bipartite graph may be divided into two mutually exclusive sets, not necessarily of the same cardinality, such that a given vertex will always

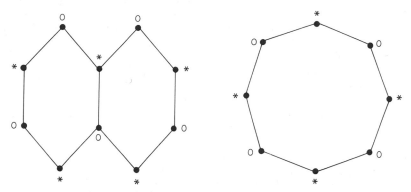

FIG. 20   Two examples of graphs which may be starred and used to represent alternant molecules.

be found in a set different from that to which all its immediately neighbouring vertices belong. This process of assigning vertices into sets is known as a graph-theoretical *colouring process*. When applied to the graphs of molecules the process is usually referred to as *starring*. If the atoms in a molecule can be starred in such a way that no two adjacent atoms are starred the molecule is said to be *alternant*. Examples of alternant molecules are shown in Fig. 20.

Alternant molecules have a number of interesting properties. It is evident that any polyene molecule must be alternant, as must any annulene molecule having an even number of carbon atoms. In fact, a very large number of hydrocarbon molecules can be starred, and thus possess bipartite graphs. The colouring process for the vertices of a bipartite graph results in a partitioning of its topological matrix as follows:

$$\begin{pmatrix} 0 & \vdots & R \\ ---&+&--- \\ R^t & \vdots & 0 \end{pmatrix}$$

where the submatrices R and R^t are in general rectangular, and R^t is the transpose of R.

Molecules possessing graphs which require three colours in the colouring process are said to possess *tripartite* graphs. It has been shown[109] that for such species a partitioning of the topological matrix of the following form results:

$$\begin{pmatrix} 0 & \vdots & Q & \vdots & R \\ --&+&--&+&-- \\ Q^t & \vdots & 0 & \vdots & V \\ --&+&--&+&-- \\ R^t & \vdots & V^t & \vdots & 0 \end{pmatrix}$$

where Q, R and V and their transposes are in general rectangular submatrices.

6.2   The Pairing Theorem

Although the Pairing Theorem was first put forward in 1940,[24] earlier work had laid the foundations for the theorem. The mathematicians Perron[101] and Frobenius[45,46] had shown that if the highest eigenvalue of an irreducible matrix with non-negative elements was paired, than all the other eigenvalues would also be paired. Pairing means that an eigenvalue ϵ_k will always be matched by another eigenvalue $-\epsilon_k$. Moreover, Hückel[79] had demonstrated that in straight chain hydrocarbons of the type C_nH_{2n} and in C_nH_n rings with n even, the energy eigenvalues would be paired.

The modern formulation of the Pairing Theorem states that any species possessed of a bipartite graph will have complementary pairs of eigenvalues, i.e. the eigenvalues will be symmetrical about the zero eigenvalue and matched in plus and minus pairs. Thus, expressed in Hückel terminology, it follows that if $\alpha + \lambda\beta$ is an eigenvalue, where λ is a variable, then $\alpha - \lambda\beta$ must also be an eigenvalue for the species. As mentioned earlier, both the polyenes and the annulenes have bipartite graphs, and will thus obey the theorem.

Since the enunciation of the theorem many workers have either rediscovered or reproved the theorem. Among the mathematicians may be mentioned the work of Collatz and Sinogowitz,[14] Sachs,[117] Hoffman,[74] Marimont[93] and Cvetković.[30] In the chemical world the theorem has been investigated by Ham and Ruedenberg,[63] Bochvar and Stankevich,[7] Koutecký[87] and Graovac, et al.[53] The first graph-theoretical proofs of the theorem were given by Cvetković[30] and Rouvray.[108] An attempt to extend the theorem to molecules possessed of a tripartite graph by Rouvray[109] was only partially successful. A graph-theoretical proof of the theorem is given in Appendix II.

6.3   Symmetric molecules

The utilization of the symmetry of a chemical species for the factorization of secular determinants by group-theoretical methods is generally well known. Numerous expositions, explaining at an elementary level how the determinants may be simplified by such methods are now available,[102,131] including standard texts such as those of Cotton[15] and Hall.[61] We therefore omit the details of these methods here.

In general, use of group theory permits a considerable simplification of the secular determinant through the possibility of forming irreducible represent-ations from the reducible representations for the point group of the molecule being studied. As an example, we mention the reduction of the determinant for the acenaphth-(1,2−a)-acenaphthylene molecule, illustrated in Fig. 21. By using the irreducible representations of the point group C_{2v} for this molecule, it may be shown that

$$\Gamma = 7A_1 + 4A_2 + 6B_1 + 5B_2 \tag{39}$$

FIG. 21 The graphical representation of the acenaphth—(1,2—a)—acenaphthylene molecule.

where symbols have their usual significance. Bose[10] was thereby able to factorize this determinant of order 22 into determinants of seventh, sixth, fifth and fourth orders only. Even in the case when a molecule as a whole possesses little symmetry, provided that component parts of the molecule possess a higher symmetry than the overall symmetry of the molecule, factorization of the secular determinant is still possible.[126]

Group-theoretical methods are, however, not without certain drawbacks. Thus, whereas group theory is able to give the number as well as the degeneracies of the orbital energy terms in a symmetric molecule, the order of these terms is determined rather by the topology of the species in question. Furthermore, the application of group theory can be irksome owing to the complexity of the various mathematical manipulations involved and because each particular case requires an individual solution. Such difficulties have promoted the development of simplified group-theoretical procedures for factorizing secular determinants.

A modified group-theoretical method requiring no knowledge of the theory of group representations was put forward by Davtyan,[33,34] and based on the use of permutation matrices. A *permutation matrix* is a matrix having only one entry of unity per row (or column), all the other entries being zero. Such a matrix Ξ will be in the automorphism group of the graph G representing a chemical species if and only if

$$\Xi A = A\Xi \qquad (40)$$

where A is the adjacency matrix of G. Since the fundamental feature of any symmetry transformation is the permutation of equivalent atoms, permutation matrices provide a means of characterizing these transformations. In fact, to compute the automorphism group of any graph one need only determine all of the matrices Ξ which satisfy equation (40). This method results in a considerable saving of the tedium involved in normal group-theoretical methods.

A similar approach, adopted by Bochvar and Chistyakov[6] proved especially suitable in the study of molecules which could be regarded as built up from several fragments, each of high symmetry. Permutation matrices, as well as the adjacency matrix, were also employed by Wild, *et al.*[137] in an examination of the symmetry properties of the Hückel matrix. This work provided valuable insight into the phenomenon of excessive degeneracy displayed by the eigenvalues calculated from the Hückel matrix.

7   CALCULATION OF QUANTUM PARAMETERS FROM THE   TOPOLOGICAL MATRIX

Having discussed at some length the various methods of evaluating eigenvalues of the topological matrix, we now turn our attention to the calculation of several other quantum-chemical parameters from this matrix. In general, the task of eliciting eigenvalues from the matrix is an exacting one, whereas the determination of the parameters we discuss below is relatively straightforward once the eigenvalues are known.

7.1   The eigenvectors

The molecular orbital corresponding to a given eigenvalue for a molecule is constructed in the linear combination of atomic orbitals (LCAO) approached from a summation of the type

$$\Phi_k = \sum_{m=1}^{n} c_{km} \phi_m \tag{41}$$

where Φ_k is the molecular orbital corresponding to the eigenvalue ϵ'_k, c_{km} is a coefficient giving the contribution of the mth atomic orbital ϕ_m to this molecular orbital, and the summation extends over all n contributing atomic orbitals (atoms). The coefficients c_{km} are obtainable by substitution of the eigenvalue ϵ'_k into the secular equation, and use of the supplementary normalization condition

$$\sum_{m=1}^{n} c_{km}^2 = 1 \tag{42}$$

The set of n coefficients corresponding to ϵ'_k are usually written as a column matrix thus:

$$C_k = \begin{pmatrix} c_{k1} \\ c_{k2} \\ \cdot \\ \cdot \\ \cdot \\ c_{kn} \end{pmatrix} \tag{43}$$

Although each C_k is an eigenvector of the Hückel matrix of a chemical species, because of the equivalence of the Hückel and topological matrices (see Appendix I), this latter matrix will also have the C_k as its set of eigenvectors. Thus, for a topological matrix **A** we may write

$$AC_k = \epsilon_k C_k \tag{44}$$

This equation may be used in determining the eigenvectors of **A**, though an equivalent formulation of the type

$$(A - \epsilon_k 1)C_k = 0, \tag{45}$$

where **1** is a unit $n \times n$ matrix, offers a more expedient means. Furthermore, if the eigenvectors C_k form the elements of a new matrix **Y** as follows:

$$Y = (C_1 \; C_2 \ldots C_n) \tag{46}$$

then clearly **Y** will satisfy the matrix equation

$$AY = YX, \tag{47}$$

where **X** is a diagonal matrix having as its general element $\epsilon_k \delta_{km}$, and δ_{km} is the Kronecker delta.

In certain cases solutions for eigenvectors are available in closed, analytical form as was the case for some eigenvalues (see Section 4). Thus, the coefficients c_{km} for a linear polyene are given by the equation:[22]

$$c_{km} = \left[\frac{2}{n+1} \right]^{1/2} \sin \left[\frac{mk\pi}{n+1} \right] \tag{48}$$

where n is the number of carbon atoms in the polyene and other symbols have their previous significance. Similarly, the coefficients for an annulene molecule are of the form:

$$c_{km} = \left[\frac{2}{n} \right]^{1/2} \cos \left[\frac{2mk\pi}{n} \right] \tag{49}$$

where n is the number of carbon atoms in the annulene. The corresponding molecular orbitals for the annulenes will thus be given as

$$\Phi_k = \left[\frac{2}{n} \right]^{1/2} \sum_{m=0}^{k-1} \cos \left[\frac{2mk\pi}{n} \right] \phi_m \tag{50}$$

7.2   Charge densities

The physical significance of the coefficients c_{km} obtained from the topological matrix is readily apparent from molecular orbital theory. A simple, anti-symmetrized product of orthonormal molecular orbitals $\Phi_k^* \Phi_k$ yields the electron

probability density at any point in space around an atomic nucleus. For real
orbitals Φ_k we may write this density $\mathbf{p}$ as

$$\mathbf{p} = \sum_{k=1}^{\mu} a_k \Phi_k^2 \tag{51}$$

where a_k is the number of electrons occupying orbital Φ_k, and the summation
extends over all μ occupied orbitals. By use of equation (41) an expansion of
equation (51) is possible as follows:

$$\mathbf{p} = \sum_{k=1}^{\mu} a_k (c_{km}^2 \phi_m^2 + 2c_{kl}c_{km}\phi_l\phi_m) \tag{52}$$

The two types of coefficient represented in this equation are given special
designations. The so called π-charge density $\mathbf{p}_{mm}$ on the mth atom in a molecule
is defined as

$$\mathbf{p}_{mm} = \sum_{k=1}^{\mu} a_k c_{km}{}^2 \tag{53}$$

whereas the mobile bond order $\mathbf{p}_{lm}$ of a bond $l-m$ in the molecule has the form:

$$\mathbf{p}_{lm} = \sum_{k=1}^{\mu} a_k c_{kl}c_{km} \tag{54}$$

We consider first the $\mathbf{p}_{mm}$ parameters.

Evaluation of the summation (53) is easily achieved, once the eigenvectors of
the topological matrix are known. It is evident that the $\mathbf{p}_{mm}$ values for each
atom are dependent on the molecular topology of the species under study. For
species such as the alternant hydrocarbons having a bipartite graph, the $\mathbf{p}_{mm}$ are
unity for all atoms. This implies that all species having as many π-electrons as
carbon atoms will display a full unit π-charge on each carbon nucleus. For ionic
and radical species, however, this is usually not the case. A general formula
giving the π-charge densities for ions having an even number of carbon atoms n
is[70]

$$\mathbf{p}_{mm} = 1 \pm c_{k(n/2)}^2 \tag{55}$$

where the plus sign refers to negatively charged ions and the minus sign to
positively charged ions. A similar result obtains for ions having an odd number
of carbon atoms, when m is odd:

$$\mathbf{p}_{mm} = 1 \pm \frac{2}{n+1} \tag{56}$$

where n is the number of carbon atoms, and the sign convention is as above.
For an aromatic cyclic $4m + 2$ system the charge density on each centre is given

by the expression

$$p_{mm} = \frac{4m + 2}{n} \tag{57}$$

where n is the number of carbon atoms in the species.

7.3   Bond orders

The p_{lm} of equation (54) are frequently described as the Coulson bond orders, in contradistinction to several other types of bond order which have also been defined. A discussion of the various bond orders viewed in terms of their relation to the topological matrix has been given by Ruedenberg.[114] A number of workers have considered bond orders in relation to the topology of the species characterized.[43,55,62,64,113] Ruedenberg[114] has demonstrated that all mobile bond orders are determined by the topology of a molecule and are thus all interrelated. Because of this he defined a 'bond order-like quantity' F to characterize all mobile bond orders:

$$F_{lm} = \sum_{k=1}^{\mu} a_k c_{kl} c_{km} f(\epsilon_k) \tag{58}$$

where a_k is the orbital occupation number, $f(\epsilon_k)$ is some function of the eigenvalue ϵ_k, and other symbols have their previous significance. The terms in the summation may be regarded as elements in a new matrix $\mathbf{F}$.

Because the topological matrix $\mathbf{A}$ of a species is a symmetric matrix, any function of $\mathbf{A}$ will satisfy the identity:[114]

$$f(\mathbf{A}) = \sum_{k=1}^{n} c_{kl} c_{km} f(\epsilon_k) \tag{59}$$

where $f(\mathbf{A})$ is some mathematical function of $\mathbf{A}$, n the number of atoms, and the other symbols have their earlier significance. The summation thus now extends over all orbitals, whether occupied or not. Accordingly, the matrices F_{lm} and $f(\mathbf{A})$ are not identical: $f(\mathbf{A})$ may be directly computed from $\mathbf{A}$ whereas F_{lm} cannot. Several useful relationships emerge if F_{lm} is related to $f(\mathbf{A})$.

As an example, if one sets

$$F_{lm} = F(\mathbf{A}) \tag{60}$$

in conjunction with the new definition

$$F(\epsilon_k') = a_k f(\epsilon_k') \tag{61}$$

it is easy to establish that

$$F_{lm} = \mathbf{p} f(\mathbf{A}) = f(\mathbf{A}) \mathbf{p} \tag{62}$$

where $\mathbf{p}$ is the Coulson bond order matrix, the elements of which are given by the summations in equation (54). Thus, any bond order-like matrix F_{lm} is calculable from p and $f(A)$ without reference to the eigenvalues or eigenvectors of A.

A number of other identities may be similarly established. For systems having all their eigenvalues non-zero, i.e. when A has an inverse, p may be computed from the relation

$$p = 1 + A^+ \cdot A^{-1}$$ (63)

where 1 is a unit matrix of the same order as A, and A^+ is defined as follows:

$$A_+ = \sum_{k=1}^{n} c_{kl} c_{km} \mid \epsilon_k \mid = \sum_{k=1}^{n} c_{kl} c_{km} (+ [\epsilon_k^2]^{1/2}) = + [A^2]^{1/2}$$ (64)

For several further relations the reader is referred to the original paper.[114]

In the case of polyene molecules Lennard-Jones and Coulson[88] have shown that the elements of p are given by the analytic formula

$$p_{lm} = \frac{1}{n+1} \ \mathrm{cosec} \left[\frac{\pi}{2k+2} \right] + (-1)^{l-1} \ \mathrm{cosec} \left[\frac{(2l+1)\pi}{2n+2} \right]$$ (65)

when n is even, and by the same expression with the cotangent function replacing the cosecant function when n is odd. The corresponding formulae for the annulenes take the form

$$p_{lm} = \frac{2}{n} \cot \frac{\pi}{n} \quad (n \text{ even})$$ (66)

$$p_{lm} = \frac{2}{n} \mathrm{cosec} \frac{\pi}{n} \quad (n \text{ odd})$$ (67)

where n is the number of carbon atoms in each species. Further similar results have been collected together in tabular form by Gutman, *et al.*[57]

7.4   Other parameters

Because of limitations of space it is not possible here to discuss all of the other uses to which the topological matrix can be put in determining quantum-chemical parameters. We shall therefore make only a brief mention of some of these in this paragraph. Thus, the topological matrix may be employed in the investigation of resonance and stabilization energy in species;[73,85,99,121,133] the study and enumeration of the Kekulé structures of species;[50,53,55,62,64,77,138] the prediction of non-bonding molecular orbitals;[1,31,32,40,89,110,141,] and even in the study of bond length.[11,17,21,55,143]

8   THE STUDY OF INORGANIC SYSTEMS USING THE TOPOLOGICAL MATRIX

In concluding our survey of the major application of the topological matrix in quantum chemisty we take a brief look at some of the inorganic systems to which the matrix has been applied. Such systems are of especial interest as they frequently contain no mobile π-electrons. The methods of study, however, usually parallel those outlined in previous sections.

8.1   The inorganic cumulenes

Over the last decade much attention has been directed toward the study of the quantum chemistry of inorganic systems. Typical of such studies has been the investigation of the inorganic cumulenes by Shustorovich.[127] On the basis of chemical and spectroscopic data it is known that the linear phosphonitrilic chloride polymer behaves similarly to the organic cumulenes $CH_2(=C=)_{n-2}CH_2$, and may therefore by represented as illustrated in Fig. 22. The polymer chain arises basically from the $p\pi - d\pi$ interaction of the nitrogen and phosphorus atoms. Free rotation is possible within the chain, although some slight inhibition may result from the alternately attached chlorine atoms. We consider the eigenvalue spectrum of a system consisting of n pairs of nitrogen and phosphorus atoms, assuming that all internal bonds are equivalent to one another. Siloxane polymers could be described by an analogous procedure.

The π-interactions may be treated independently of one another as the two π-systems in the cumulenes are mutually perpendicular. In setting up the secular determinant the chlorine atoms are neglected, as were the hydrogen atoms in the polyenes. The topological matrix is thus identical to that for polyenes, illustrated in Fig. 9. The determinant, however, differs in that the entries along the diagonal now alternate, and some of the off-diagonal entries take a minus

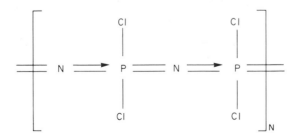

FIG. 22   A representation of two units of the phosphonitrilic chloride polymer chain.

sign:

$$
\begin{vmatrix}
\omega_1 & 1 & 0 & & & \\
1 & \omega_2 & -1 & & & \\
& -1 & \omega_1 & 1 & & \\
& & & \cdot & & \\
& & & & \cdot & \\
& & 0 & 1 & \omega_2 &
\end{vmatrix}
$$

where ω_1 refers to nitrogen atoms and represents the quantity $(\alpha_1 - \epsilon_k')/\beta$, and ω_2 refers to phosphorus atoms and equals $(\alpha_2 - \epsilon_k')/\beta$. The sign alternation of the unit entries allows for the symmetry of the $p\pi - d\pi$ interaction.

If the product $\omega_1 \cdot \omega_2$ is set equal to Ω, the determinant in equation (68) may be readily transformed into a determinant all of whose diagonal entries are Ω. Analytical solution for this transformed determinant is now possible, the eigenvalues being given by the equation:

$$
\Omega = \pm 2 \cos\left[\frac{\pi k}{2n + 1}\right] \tag{69}
$$

where k may take values from 1 to n. (cf. equation (3)). From this equation it is evident that the determinant may be factorized into n determinants of second order thus:

$$
\begin{vmatrix}
\alpha_1 - \epsilon_k' & 2\cos\left[\dfrac{\pi k}{2n + 1}\right] \\
2\cos\left[\dfrac{\pi k}{2n + 1}\right] & \alpha_2 - \epsilon_k'
\end{vmatrix} = 0 \tag{70}
$$

By setting α_1 equal to $\alpha_2 + x\beta$, the eigenvalues of determinant (70) appear now as

$$
\epsilon_k' = \frac{\alpha_1 + \alpha_2}{2} \pm \frac{\beta}{2}\left[x^2 + 16 \cos^2\left[\frac{\pi k}{2n + 1}\right]\right]^{\frac{1}{2}}. \tag{71}
$$

The parameter x allows for the difference in electronegativity between the nitrogen and phosphorus atoms. The total energy for the system E when all n orbitals are fully occupied thus becomes

$$
E = n(\alpha_1 + \alpha_2) + \beta\left[\sum_{k=1}^{n} x^2 + 16 \cos^2\left[\frac{\pi k}{2n + 1}\right]\right]^{\frac{1}{2}}. \tag{72}
$$

It is of interest to note that the delocalization energy for this species is of the form

$$
\Delta E = \beta \sum_{k=1}^{n}\left[x^2 + 16 \cos^2\left[\frac{\pi k}{2n + 1}\right]\right]^{\frac{1}{2}} \tag{73}
$$

and that this reduces to the result obtained for polyenes by Coulson[16] in the special case when $x = 0$, i.e. when the atoms in the chain are identical.

8.2   Symmetric inorganic species

A systematic investigation of heteroatomic molecules which display a high degree of symmetry was undertaken by Schmidtke.[122-125] The symmetry of a molecule is important because highly symmetric molecules will possess a large number of topologically equivalent atoms. This can mean a considerable simplification of the procedure for determining their energy eigenvalues. For asymmetric species resolution of the Hückel matrix, as given in Appendix I, will in general be very approximate, since not all the Coulomb integrals will equal α, but will differ for different pairs of atoms. For a tetrahedral molecule such as P_4, however, where four identical atoms occupy the corners of a regular tetrahedron, all atoms are topologically equivalent and resolution of the matrix is exact.

One difficulty in this approach arises from the fact that molecules which differ chemically may have the same topology, and thus be possessed of identical topological matrices. For instance, a square planar and a tetrahedral AB_4 species have equivalent graphs, as illustrated in Fig. 23. This particular case was investigated by Schmidtke, who used the matrix

$$\begin{pmatrix} 0 & 1 & \theta & 1 \\ 1 & 0 & 1 & \theta \\ \theta & 1 & 0 & 1 \\ 1 & \theta & 1 & 0 \end{pmatrix}$$

to represent the four B atoms in the species. This matrix he called the *structural matrix* for this array of atoms. It clearly reduces to the topological matrix when

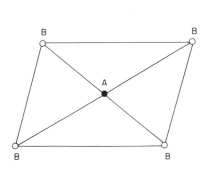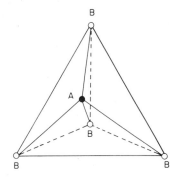

FIG. 23   Two equivalent graphs for the square planar and tetrahedral configuration of an AB_4 species.

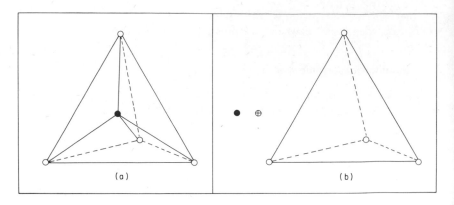

FIG. 24   The resolution of the graph of a tetrahedral AB_4 species (a) into two subgraphs.

$\theta = 1$. The parameter θ, which is never greater than 1, enables the characterization of all B_4 species having configuration between square planar (D_{4h}) and tetrahedral (S_4). When $\theta < 1$ the matrix describes the square planar configuration, whereas when $\theta = 1$ the matrix describes the tetrahedral arrangement.

The eigenvalues of the above structural matrix are, using the notation of Eisenstein:[42]

$$\epsilon_{1,2} = -\theta(e_u); \epsilon_3 = 2 + \theta(a_{1g}); \epsilon_4 = -2 + \theta(b_{2g}) \tag{74}$$

Upon substitution of appropriate θ values, exact determination of the eigenvalues becomes possible. Thus, for the case of tetrahedral symmetry where $\theta = 1$, the eigenvalues reduce to

$$\epsilon_1 = 3(a_1); \epsilon_{2,3,4} = -1(t_2) \tag{75}$$

These results may be used in the study of AB_4 species, as the latter can be represented as the sum of two subgraphs, as illustrated in Fig. 24. Each subgraph depicts atoms of one type only and may be studied separately. However, the sum of the two subgraphs will not represent the AB_4 species, as the connections to the A atom will be missing. Full representation can be achieved only by a supergraph of the two subgraphs. The components are thus added together in a special way indicated by use of the special sign $\oplus$. Quantum parameters may be evaluated for each of the subgraphs, as described above for the eigenvalues. The interaction of the two subgraphs to produce the supergraph of AB_4 is then treated as a quantum perturbational problem. Details of the calculations involved are to be found in Schmidtke.[122]

8.3   The boron hydrides

The boron hydrides, noted for their apparent violations of the usual rules of chemical valency, have been studied in some detail by Kettle and co-workers.[80-83] In these studies the topological matrix appears in the guise of a *simplicial matrix*. A *simplex* is the n-dimensional analogue of an equilateral triangle, where n is an integer which may range from zero upwards. A zero-simplex is thus represented by a point, a one-simplex by a line, a two-simplex by a triangle and a 3-simplex by a tetrahedron.

The boron hydrides generally exist either in the form of complete polyhedra, the so-called 'cage molecules', or as incomplete polyhedra, the so-called 'basket molecules'. Both types can be rationalized in terms of two-centre and three-centre bonds. Such bonds may be seen to correspond to the edges or faces of a triangular polyhedron, i.e. a simplex. This has lead to the use of topological basis sets based on the simplexes in the molecular orbital treatment of the boron hydrides. Use of these sets rests upon the isomorphism existing between the Hückel matrix of a species and the various topological, i.e. simplicial matrices for the triangular polyhedra. In fact, the method may be seen as a topologically correct extension of Hückel theory to three-dimensional molecules. In addition, the interrelations between the various matrices discussed by these workers shed a good deal of light on the topological origins of the Hückel matrix.

9.   CONCLUSION

We have discussed the principal uses to date of the topological matrix in the field of quantum chemistry. The uses to which the matrix have been put may be summarized into three main categories. First, and most important at present, has been the calculation of quantum-chemical parameters. We have considered in some detail the determination of energy eigenvalues, and a major part of the work has been devoted to this task. In lesser detail we considered the evaluation of several other parameters such as bond orders. Some parameters have not been discussed at all, e.g. atom polarizabilities, although these are readily calculable from the parameters which have been considered.

The basic reason for the topological dependence of the above parameters is the fact that, in all the species we have considered, short range forces are dominant as the long range forces between electrons and nuclei tend largely to cancel out one another. The parameters discussed must as a consequence be topology, rather than geometry, dependent. Thus, even a parameter such as bond length may be seen to be determined largely by the topology of the species.

A second important use of the topological matrix is in establishing quantum-chemical theorems. Although only a limited amount of work has been done in this area, the matrix has been employed in establishing a few theorems, such as the Pairing Theorem, discussed in Appendix II. It has been used by Ruedenberg to establish several theorems based on a so-called bond order-like parameter. It has also been demonstrated by Gutman and Trinajstić that the Hückel rule is a specific instance of the more general Loop Rule.

The third application of the matrix, which at present remains largely unexplored, lies in studies which throw light on origins of quantum-chemical phenomena or parameters. Thus, Wild, *et al.*, were able to give an explanation of the excessive degeneracy of eigenvalues calculated using the Hückel procedure by reference to the matrix. Similarly, Kettle, *et al.* have shown how the topological matrix features in the extension of Hückel theory to three-dimensional species. In so doing they elucidated the relationship of this matrix to topological n-simplexes. It should be evident that use of the matrix to shed new light on the origins of the parameters and phenomena of quantum chemistry can lead to many novel and fruitful lines of inquiry, and therefore deserves greater attention than it has hitherto received.

ACKNOWLEDGEMENTS

I should like to thank Dr. Robin J. Wilson and Professor F. Harary for reading the original version of this manuscript and for making helpful comments.

REFERENCES

1. Balaban, A. T. (1972). *Rev. Roum. Chim.* **17**, 1531.
2. Balaban, A. T. and Harary, F. (1971) *J. Chem. Docum.* **11**, 258.
3. Balandin, A. A. (1940). *Acta Physicochim. U.R.S.S.* **12**, 447.
4. Balandin, A. A. (1943). *Izvest. Akad. Nauk S.S.S.R., Otdel. Khim. Nauk.* **35**.
5. Billes, F. (1969). *Acta Chim. Akad. Sci. Hung.* **62**, 7.
6. Bochvar, D. A. and Chistyakov, A. L. (1967). *Zhur. Fiz. Khim.* **41**, 2731.
7. Bochvar, D. A. and Stankevich, I. V. (1965). *Zhur. Fiz. Khim.* **39**, 2028.
8. Bochvar, D. A. and Stankevich, I. V. (1967). *Zhur. Strukt. Khim.* **8**, 943.
9. Bochvar, D. A. and Stankevich, I. V. (1969). *Zhur. Strukt. Khim.* **10**, 680.
10. Bose, S. K. (1966). *Theor. Chim. Acta* **5**, 84.
11. Boyd, G. V. and Singer, N. (1966). *Tetrahedron* **22**, 3383.
12. Buchheim, M. A. (1885). *Proc. London Math. Soc.* **17**, 80.
13. Busacker, R. B. and Saaty, T. L. (1965). "Finite Graphs and Networks", McGraw-Hill, New York.
14. Collatz, L. and Sinogowitz, U. (1957). *Abh. Math. Sem. Univ. Hamburg* **21**, 63.
15. Cotton, F. A. (1963) "Chemical Applications of Group Theory", Wiley, New York.

16. Coulson, C. A. (1938). *Proc. Roy. Soc.* **A164**, 383.
17. Coulson, C. A. (1939). *Proc. Roy. Soc.* **A169**, 413.
18. Coulson, C. A. (1948). *Proc. Phys. Soc.* **60**, 257.
19. Coulson, C. A. (1950). *Proc. Cambridge Phil. Soc.* **46**, 202.
20. Coulson, C. A. and Dixon W. T. (1962). *Tetrahedron* **17**, 215.
21. Coulson, C. A. and Golebiewski, A. (1961). *Proc. Phys. Soc.* **78**, 1310.
22. Coulson, C. A. and Longuet-Higgins, H. C. (1947a). *Proc. Roy. Soc.* **A192**, 16.
23. Coulson, C. A. and Longuet-Higgins, H. C. (1947b). *Proc. Roy. Soc.* **A191**, 39.
24. Coulson, C. A. and Rushbrooke, G. S. (1940). *Proc. Cambridge Phil. Soc.* **36**, 193.
25. Coulson, C. A. and Rushbrooke, G. S. (1947). *Proc. Roy. Soc. Edinburgh* **A62**, 350.
26. Coulson, C. A. and Streitwieser, A. (1965). "Dictionary of π-electron Calculations", Pergamon, Oxford.
27. Craig, D. P. (1959). "Theoretical Organic Chemistry, Chemical Society Kekulé Symposium", Butterworths, London.
28. Crum Brown, A. (1861). "On the Theory of Chemical Combination", Thesis, University of Edinburgh.
29. Crum Brown, A. (1864). *Trans. Roy. Soc. Edinburgh* **23**, 707.
30. Cvetković, D. (1969). *Mat. Biblio. Belgrade* **41**, 193.
31. Cvetković, D. and Gutman, I. (1972). *Mat. Vesnik Belgrade* **9**, 141.
32. Cvetković, D., Gutman, I. and Trinajstić, N. (1972). *Croat. Chem. Acta* **44**, 365.
33. Davtyan, O. K. (1960a). *Zhur. Fiz. Khim.* **34**, 108.
34. Davtyan, O. K. (1960b). *Zhur. Fiz. Khim.* **34**, 295.
35. De Chiossone, E. N. A. (1962). "Numeros Fundamentales de Grafes y Matrices de Incidencia de los Compuestos Quimicos Organicos", Thesis, University of Buenos Aires.
36. De Llano, C. R. (1968). "SCF MO Treatment of Conjugated Hydrocarbons", Thesis, University of Texas.
37. Dewar, M. J. S. (1949). *Proc. Cambridge Phil. Soc.* **45**, 638.
38. Dewar, M. J. S. (1952). *J. Amer. Chem. Soc.* **74**, 3341.
39. Dewar, M. J. S. (1969). "The Molecular Orbital Theory of Organic Chemistry", McGraw-Hill, New York.
40. Dewar, M. J. S. and Longuet-Higgins, H. C. (1952). *Proc. Roy. Soc.* **A214**, 482.
41. Dickson, T. R. (1968). "The Computer and Chemistry", Freeman, San Francisco.
42. Eisenstein, J. C. (1956). *J. Chem. Phys.* **25**, 142.
43. England, W. and Ruedenberg, K. (1971). *Theor. Chim. Acta* **22**, 196.
44. Flurry, R. L. (1964). *J. Med. Chem.* **7**, 668.
45. Frobenius, G. (1909). *Sitzungsber. Deutsch. Akad. Wiss. Berlin, Math. Nat. Kl,* 514.
46. Frobenius, G. (1912). *Sitzungsber. Deutsch. Akad. Wiss. Berlin, Math. Nat. Kl,* 456.
47. Frost, A. A. and Musulin, B. (1953). *J. Chem. Phys.* **21**, 572.
48. Goldstein, M. J. (1967). *J. Amer. Chem. Soc.* **89**, 6357.
49. Goldstein, M. J. and Hoffmann, R. (1971). *J. Amer. Chem. Soc.* **93**, 6193.
50. Gordon, M. and Davison, W. H. T. (1952). *J. Chem. Phys.* **20**, 428.

51. Gouarné, R. (1954). *Comptes Rend. Acad. Sci. Paris* **239**, 383.
52. Gouarné, R. (1956). *Thesis, University of Paris.*
53. Graovac, A., Gutman, I., Trinajstić, N. and Živković, T. (1972). *Theor. Chim. Acta* **26**, 67.
54. Günthard, H. H. and Primas, H. (1956). *Helv. Chim. Acta* **39**, 1645.
55. Gutman, I. and Trinajstić, N. (1973a). *Topics Curr. Chem.* **42**, 49.
56. Gutman, I. and Trinajstić, N. (1973b). *Croat. Chem. Acta* **45**, 423.
57. Gutman, I., Trinajstić, N. and Živković, T. (1972). *Chem. Phys. Lett.* **14**, 342.
58. Gutman, I., Trinajstić, N. and Živković, T. (1973). *Tetrahedron* **29**, 3449.
59. Hakimi, S. L. and Yau, S. S. (1965). *Quart. Appl. Math.* **22**, 305.
60. Halford, J. O. (1967). *J. Amer. Chem. Soc.* **89**, 5338.
61. Hall, L. H. (1969). "Group Theory and Symmetry in Chemistry", McGraw-Hill, New York.
62. Ham, N. S. (1958). *J. Chem. Phys.* **29**, 1229.
63. Ham, N. S. and Ruedenberg, K. (1958a). *J. Chem. Phys.* **29**, 1191.
64. Ham, N. S. and Ruedenberg, K. (1958b). *J. Chem. Phys.* **29**, 1215.
65. Harary, F. (1969). "Graph Theory", Addison-Wesley, Reading, Massachusetts.
66. Hartmann, H. (1963). *Adv. Chem. Phys.* **5**, 1.
67. Heilbronner, E. (1953). *Helv. Chim. Acta* **36**, 170.
68. Heilbronner, E. (1954). *Helv. Chim. Acta* **37**, 913.
69. Heilbronner, E. (1964). *Tetrahedron Lett.* **29**, 1923.
70. Heilbronner, E. and Bock, H. (1968). "Das HMO-Modell und seine Anwendung", Verlag Chemie, Weinheim.
71. Heilbronner, E. and Straub, E. (1966). "Hückel Molecular Orbitals", Springer, New York.
72. Hess, B. A. and Schaad, L. J. (1971a). *J. Amer. Chem. Soc.* **93**, 305.
73. Hess, B. A. and Schaad, L. J. (1971b). *J. Amer. Chem. Soc.* **93**, 2413.
74. Hoffman, A. J. (1963). *Amer. Math. Monthly* **70**, 30.
75. Hoffmann, R., Imamura, A. and Zeiss, G. D. (1967). *J. Amer. Chem. Soc.* **89**, 5215.
76. Hosoya, H. (1971). *Bull. Chem. Soc. Japan* **44**, 2332.
77. Hosoya, H. (1972). *Theor. Chim. Acta* **25**, 215.
78. Hückel, E. (1931). *Z. Phys.* **70**, 204.
79. Hückel, E. (1932). *Z. Phys.* **72**, 628.
80. Kettle, S. F. A. and Reynolds, D. J. (1971). *Theor. Chim. Acta* **22**, 239.
81. Kettle, S. F. A. and Tomlinson, V. (1969a). *J. Chem. Soc.* 2002.
82. Kettle, S. F. A. and Tomlinson, V. (1969b). *J. Chem. Soc.* 2007.
83. Kettle, S. F. A. and Tomlinson, V. (1969c). *Theor. Chim. Acta* **14**, 175.
84. Kirchhoff, G. (1847). *Ann. Phys. Chem.*, **72**, 497.
85. Klasinc, L. and Trinajstić, N. (1971). *Tetrahedron* **27**, 4045.
86. König, D. (1936). "Theorie der endlichen und unendlichen Graphen", Akademischer Verlag, Leipzig.
87. Koutecký, J. (1966). *J. Chem. Phys.* **44**, 3702.
88. Lennard-Jones, J. E. and Coulson, C. A. (1939). *Trans. Far. Soc.* **35**, 811.
89. Longuet-Higgins, H. C. (1950). *J. Chem. Phys.* **18**, 265.
90. Longuet-Higgins, H. C. and Salem, L. (1959). *Proc. Roy. Soc.* **A251**, 172.
91. Mallion, R. B. (1969) "Nuclear Magnetic Resonance of Condensed Benzenoid Hydrocarbons", Thesis, University of Wales.
92. Margenau, H. and Murphy, G. M. (1956). "The Mathematics of Physics and Chemistry", 2nd Edit., Van Nostrand, Princeton.

93. Marimont, R. B. (1969). *Bull. Math. Biophys.* **31**, 255.
94. Mayot, M., Berthier, G. and Pullman, B. (1951). *J. Phys. Radium* **12**, 652.
95. McWeeny, R. (1952). *Proc. Phys. Soc.* **A65**, 839.
96. McWeeny, R. (1953). *Proc. Phys. Soc.* **A66**, 714.
97. Metcalf, B. W. and Sondheimer, F. (1971). *J. Amer. Chem. Soc.* **93**, 5271.
98. Meyer, E. (1970). *Angew. Chem.* **82**, 605.
99. Milun, N., Sobotka, Z. and Trinajstić, N. (1972). J. Org. Chem. **37**, 139.
100. Mowshowitz, A. (1972). *J. Comb. Theory.* **B12**, 177.
101. Perron, O. (1907). *Math. Ann.* **64**, 248.
102. Phelan, N. F. and Orchin, M. (1966). *J. Chem. Educ.* **43**, 571.
103. Platt, J. R. (1952). *J. Phys. Chem.* **56**, 238.
104. Ponstein, J. (1966). "Matrices in Graph and Network Theory", Thesis, University of Utrecht.
105. Popov, N. A. (1969). *Zhur. Strukt. Khim.* **10**, 533.
106. Rebane, T. K. (1965). *In* "Methods of Quantum Chemistry", (Ed. M. G. Veselov,) Academic Press, New York, London and San Francisco.
107. Rescigno, A. and Segre, G. (1965). *Bull. Math. Biophys.* **27**, 315.
108. Rouvray, D. H. (1972a). *Comptes Rend. Acad. Sci. Paris* **C274**, 1561.
109. Rouvray, D. H. (1972b). *Comptes Rend. Acad. Sci. Paris* **C275**, 657.
110. Rouvray, D. H. (1972c). *Comptes Rend. Acad. Sci. Paris* **C275**, 363.
111. Rouvray, D. H. (1973). *Amer. Sci.* **61**, 729.
112. Ruedenberg, K. (1954). *J. Chem. Phys.* **22**, 1878.
113. Ruedenberg, K. (1958). *J. Chem. Phys.* **29**, 1232.
114. Ruedenberg, K. (1961). *J. Chem. Phys.* **34**, 1884.
115. Rumer, G. (1932). *Nachr. Ges. Wiss. Göttingen, Math. Phys. Kl,* 337.
116. Rutherford, D. E. (1946). *Proc. Roy. Soc. Edinburgh* **A62**, 229.
117. Sachs, H. (1962). *Publ. Math. Debrecen* **9**, 270.
118. Sachs, H. (1963). *Publ. Math. Debrecen* **11**, 199.
119. Samuel, I. (1949). *Comptes Rend. Acad. Sci. Paris* **229**, 1236.
120. Samuel, I. (1958). "Méthode des polygones, procédé d'étude graphique des déterminants. Applications aux problèmes de chimie théorique", Thesis, University of Paris.
121. Schaad, L. J. and Hess, B. A. (1972). *J. Amer. Chem. Soc.* **94**, 3068.
122. Schmidtke, H. H. (1966). *J. Chem. Phys.* **45**, 3920.
123. Schmidtke, H. H. (1967). *Coord. Chem. Rev.* **2**, 3.
124. Schmidtke, H. H. (1968a). *Int. J. Quant. Chem. Symp.* **2**, 101.
125. Schmidtke, H. H. (1968b). *Theor. Chim. Acta* **9**, 199.
126. Schuster, P. (1965). *Theor. Chim. Acta* **9**, 278.
127. Shustorovich, E. M. (1963). *Zhur. Strukt. Khim.* **4**, 773.
128. Spialter, L. (1964). *J. Chem. Docum.* **4**, 269.
129. Streitwieser, A. (1961). "Molecular Orbital Theory for Organic Chemists", Wiley, New York.
130. Streitwieser, A. and Brauman, I. J. I. (1965). "Supplemental Tables of Molecular Orbital Calculations", Pergamon, Oxford.
131. Swain, C. G. and Thorson, W. R. (1959). *J. Org. Chem.* **24**, 1989.
132. Sylvester, J. J. (1878). *Amer. J. Math.* **1**, 64.
133. Trinajstić, N. (1971). *Record Chem. Prog.* **32**, 85.
134. Turner, J. (1968). *S.I.A.M. J. Appl. Math.* **16**, 520.
135. Urch, D. S. (1958). *J. Chem. Soc.* **4**, 4767.
136. Weiner, H. (1947). *J. Amer. Chem. Soc.* **69**, 17.
137. Wild, U., Keller, J. and Günthard, H. H. (1969). *Theor. Chim. Acta* **14**, 383.

138. Yen, T. F. (1971). *Theor. Chim. Acta* **20**, 399.
139. Zimmerman, H. E. (1966). *J. Amer. Chem. Soc.* **88**, 1564.
140. Zimmerman, H. E. (1971). *Accts. Chem. Res.* **4**, 272.
141. Živković, T. (1972). *Croat. Chem. Acta* **44**, 351.
142. Živković, T., Trinajstić N. and Randić, M. (1975). *Molec. Phys.* **30**, 517.
143. Zivković, T. and Trinajstić, N. (1969). *Can. J. Chem.* **47**, 697.
144. Daudel, R., Lefebvre, R. and Moser, C. (1959) "Quantum Chemistry", Interscience, New York, p. 539.

APPENDIX I

Proof of the equivalence of the topological and Hückel matrices

In simple molecular orbital theory the energy levels of the π-electrons in chemical species are given by the eigenvalues of the Hückel secular determinant thus:

$$\mathbf{H} = \mid \mathbf{M} - \epsilon_k'\mathbf{S} \mid = 0 \tag{A1}$$

where $\mathbf{M}$ is the Hamiltonian energy matrix, ϵ_k' is an eigenvalue, and $\mathbf{S}$ the overlap matrix. All entries in $\mathbf{S}$ are zero except those for nearest neighbouring atoms, for which the entry is β, the resonance integral. Now the matrices $\mathbf{M}$ and $\mathbf{S}$ may be resolved as follows:

$$\mathbf{M} = \alpha\mathbf{1} + \beta\mathbf{A} \tag{A2}$$

$$\cdot\mathbf{S} = \mathbf{1} + \mathbf{SA} \tag{A3}$$

where α is the Coulomb integral, and $\mathbf{A}$ the topological matrix for the species. This resolution will be strictly valid only if all the atoms in the species are topologically equivalent, i.e. if all the α terms and all the β terms are exactly equal to one another.

Substitution of the resolved expressions into equation (A1) yields the result

$$\mid \mathbf{H} \mid = \mid \alpha\mathbf{1} + \beta\mathbf{A} - \epsilon_k'(\mathbf{1} + \mathbf{S} \cdot \mathbf{A}) \mid \tag{A4}$$

$$= \mid \mathbf{A}(\beta - \epsilon_k'\mathbf{S}) + \mathbf{1}(\alpha - \epsilon_k') \mid \tag{A5}$$

$$= \left| \mathbf{A} - \left(\frac{\epsilon_k' - \alpha}{\beta - \epsilon_k'\mathbf{S}} \right)\mathbf{1} \right| = 0 \tag{A6}$$

The terms in the bracket of (A6) are known as the Hückel numbers and are usually given the symbol ρ; they will be determinate only when the expression $\beta - \epsilon_k'\mathbf{S}$ is not zero. In this case the ρ form the eigenvalues of $\mathbf{A}$, which is equivalent to saying that the spectrum of eigenvalues of $\mathbf{M}$ with respect to the metric $\mathbf{S}$ is expressed by the Hückel numbers. The eigenvalues of the Hückel matrix are closely related. They are however expressed in units of β relative to a zero energy level of α. Thus, if ρ is an eigenvalue of $\mathbf{A}$, the corresponding eigenvalue of $\mathbf{H}$ will be $\alpha + \beta\rho$.

The eigenvectors for corresponding eigenvalues of the matrices $\mathbf{A}$ and $\mathbf{H}$ will be identical. From equation (A2) it is evident that $\mathbf{H}$ and $\mathbf{A}$ will commute:

$$\mathbf{H} \cdot \mathbf{A} - \mathbf{A} \cdot \mathbf{H} = 0 \tag{A7}$$

If X is an eigenvector of A for the eigenvalue ρ we have

$$AX = \rho X. \tag{A8}$$

Similarly, for matrix H we may write

$$HX = (\alpha 1 + \beta A)X \tag{A9}$$

$$= \alpha X + \beta(AX) \tag{A10}$$

$$= \alpha X + \beta(\rho X) \tag{A11}$$

$$= (\alpha + \beta\rho)X. \tag{A12}$$

This establishes that X is the same for both A and H, since the eigenvalue in equation (A12) corresponds to the eigenvalue ρ for A.

APPENDIX II

A graph-theoretical proof of the Pairing Theorem

When discussing alternant molecules, it was demonstrated that the 'starring' procedure of Coulson and Rushbrooke[24] is equivalent to a graph-theoretical colouring process. This process is illustrated for the case of the naphthalene molecule in Fig. 25. The starred atoms (colour 1) are numbered successively, followed by the unstarred atoms (colour 2) also numbered successively. The colouring process results in the partitioning of the topological matrix for such a system as follows:

$$\left(\begin{array}{c|c} 0 & R \\ \hline R^t & 0 \end{array}\right)$$

Bringing this matrix to this form for any bipartite graph will not alter the eigenvalues, as these are invariant under interchange of rows and columns.

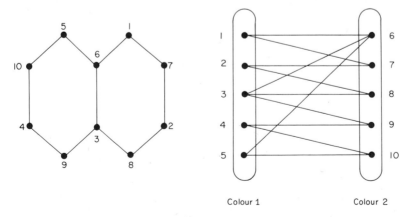

Colour 1   Colour 2

FIG. 25   Partitioning of the vertices of the graph representing a naphthalene molecule into two mutually exclusive sets.

To prove the theorem, we let an eigenvalue ϵ_k of this matrix have an associated eigenvector X thus:

$$
X = \begin{pmatrix} x_1 \\ \cdot \\ \cdot \\ \cdot \\ x_m \\ y_n \\ \cdot \\ \cdot \\ \cdot \\ y_s \end{pmatrix}
\tag{A13}
$$

where vertices of colour 1 have been labelled from 1 to m, and those of colour 2 have been labelled from n to s. Hence we may write

$$
\begin{pmatrix} 0 & R \\ R^t & 0 \end{pmatrix} \begin{pmatrix} x_1 \\ \cdot \\ \cdot \\ \cdot \\ x_m \\ y_n \\ \cdot \\ \cdot \\ \cdot \\ y_s \end{pmatrix} = R^t \begin{pmatrix} R \begin{pmatrix} y_n \\ \cdot \\ \cdot \\ \cdot \\ y_s \end{pmatrix} \\ x_1 \\ \cdot \\ \cdot \\ \cdot \\ x_m \end{pmatrix} = \epsilon_k \begin{pmatrix} x_1 \\ \cdot \\ \cdot \\ \cdot \\ x_m \\ y_n \\ \cdot \\ \cdot \\ \cdot \\ y_s \end{pmatrix}
\tag{A14}
$$

We now define a new vector Y as follows:

$$
Y = \begin{pmatrix} x_1 \\ \cdot \\ \cdot \\ \cdot \\ x_m \\ -y_n \\ \cdot \\ \cdot \\ \cdot \\ -y_s \end{pmatrix}
\tag{A15}
$$

For Y we may similarly write

$$
\begin{pmatrix} 0 & R \\ R^t & 0 \end{pmatrix}
\begin{pmatrix} x_1 \\ \cdot \\ \cdot \\ \cdot \\ x_m \\ -y_n \\ \cdot \\ \cdot \\ \cdot \\ -y_s \end{pmatrix}
= R^t
\begin{pmatrix} -y_n \\ \cdot \\ \cdot \\ \cdot \\ -y_s \\ x_1 \\ \cdot \\ \cdot \\ \cdot \\ x_m \end{pmatrix}
= R^t
\begin{pmatrix} y_n \\ \cdot \\ \cdot \\ \cdot \\ y_s \\ x_1 \\ \cdot \\ \cdot \\ \cdot \\ x_m \end{pmatrix}
\quad . \tag{A16}
$$

This demonstrates that $-\epsilon_k$ is also an eigenvalue for this topological matrix since

$$
\epsilon_k
\begin{pmatrix} -x_1 \\ \cdot \\ \cdot \\ \cdot \\ -x_m \\ y_n \\ \cdot \\ \cdot \\ y_s \end{pmatrix}
=
-\epsilon_k
\begin{pmatrix} x_1 \\ \cdot \\ \cdot \\ \cdot \\ x_m \\ -y_n \\ \cdot \\ \cdot \\ y_s \end{pmatrix}
\quad . \tag{A17}
$$

CHAPTER 8

Some Aspects of Graph Theory for Intermolecular Interactions in Chemical Physics

Jean Brocas

Faculty of Sciences,
Université Libre de Bruxelles, Bruxelles, Belgium

1   INTRODUCTION

The aim of statistical mechanics is to provide a bridge between the *microscopic* properties of systems containing a large number of particles and their *macroscopic* properties. For example, equilibrium statistical mechanics permits us, in principle at least, to obtain any thermodynamic quantity (pressure, energy, dielectric constant, etc) from the *assumed knowledge* of particle properties (intermolecular forces, dipole moments, etc.). The same is true for non-equilibrium statistical mechanics where transport coefficients and time dependent autocorrelation functions are of interest. One considers always the *thermodynamic limit*: N (number of particles) and V (volume) infinite; $N/V = \rho$ finite.

The tool playing the central role in equilibrium statistical mechanics is the *partition function*. From this quantity, any thermodynamic quantity may be obtained through well defined mathematical operations. The difficult problem is the derivation of the partition function from the microscopic properties. This may be done in a rigorous way in a very small number of cases, essentially when the system is made of independent, non interacting individuals as in a dilute gas or an harmonic solid.

If the particles may not be treated as independent, one has to develop approximate methods for the computation of the partition function. In doing this, one starts from the known solution of the unperturbed problem (independent particles) and one introduces the interactions as a perturbation. This scheme implies the choice of a *parameter* which tells us how strong the imperfections are.

Once this choice is made, one has to develop a consistent expansion of the partition function in the given parameter. Graph theoretic methods have been widely used in the solution of this type of problem.[17] The graphical representation serves as a tool to classify the contributions to the partition function according to their dependence on the expansion parameter and, as a result, the expansion coefficients may be formulated through topological rules, which are easy to visualize. The problem of finding the numerical values of these coefficients then remains to be solved: it rests on the *effective knowledge* of the microscopic properties (interaction potential) and on the elaboration of refined computational techniques.

In non-equilibrium, the general approach is essentially the same but the situation is even more difficult because one needs the time dependence of the *distribution function* to study the evolution towards equilibrium. The answer to this question is again relatively well-known for systems with independent particles[1] and may be used for the study of more complicated situations. Recently, Prigogine[19-21] and coworkers and Van Hove[28,29] have elaborated theories describing the approach to equilibrium of macroscopic systems which are valid for a very wide class of initial situations. In particular, the Prigogine theory rests upon a diagram technique which has some aspects in common with equilibrium theories but which widely differs in other respects.

In this contribution we will present the statistical mechanical approach to some problems in the field of chemical physics and we will show that the use of graphs is of crucial importance for their description. However, the graphs used in this field have a different meaning to those used in other areas such as topological representations of stereoisomerism.[11]

In Section 2, we shall recall the simplest possible non trivial case: the calculation of the partition function and of the equation of state of a classical monoatomic, one component gas. In order to avoid an abstract presentation of the topological concepts that we will need in the other sections, we introduce them step by step in the discussion of this well-known problem. We will see that the concentration of the gas is the parameter measuring the importance of the non-independence of the particles. The equation of state of the fluid will be obtained as a series expression using this parameter. The coefficients therein are expressed by simple graphical rules.

Section 3 will be devoted to the study of some applications of graph theory to the irreversible approach of a system to equilibrium. It will be shown that topological properties are still important but that they must be used with even more care than in equilibrium problems. It will be seen that some graphical concepts may be transferred *mutatis mutandis* from equilibrium to non-equilibrium, while others may not. For example, the theory of the non-equilibrium evolution given by Cohen, and which parallels very closely the equilibrium graph theory of pair correlation functions, is equivalent[4] to the dynamical theory of Prigogine and coworkers.[19] On the other hand, the semi-invariant formulation[6] is an attempt to extend the concept of connected graphs out of equilibrium but this may give rise to a misleading description of irreversible phenomena.

In Section 4, we will give an account of two examples where equilibrium graphs may be used to study the effect of imperfection on chemical equilibrium or on reaction rates. The first one requires the calculation of the partition function of a multicomponent gas,[16] taking into account that the imperfections modify the equilibrium concentrations of the species. This must be true if the distance for physical clustering is bigger than the distance for chemical reaction. If this condition is satisfied, and if the reaction does not modify the cluster concentration, one may also study the effect of imperfections on the rates. We will give a brief description of recent work in this field[9] and, finally, draw graphs which should be important for the study of these problems.

2   THE IMPERFECT GAS IN EQUILIBRIUM

2.1   Mayer graphs

We are interested in the macroscopic properties of a system of N classical and monoatomic particles in the volume V. The Hamiltonian of this system is given by

$$H_N = K_N + V_N \tag{2.1}$$

where the kinetic energy is

$$K_N = \frac{1}{2m} \sum_{i=1}^{N} p_i^2 \tag{2.2}$$

and the potential energy responsible for intermolecular interactions is

$$V_N = \sum_{i>j=1}^{N} v_{ij} \tag{2.3}$$

In these equations, m is the mass of each particle, p_i and v_{ij} represent, respectively, the impulsion of particle i and the interaction energy between particles i and j, which is a function of the distance $|\mathbf{r}_i - \mathbf{r}_j|$ between them. In statistical mechanics, one shows[13,14] that the thermodynamic properties of the system may be derived from the *partition function*:

$$Q_N = \frac{1}{h^{3N} N!} \int d\mathbf{p}_1 \ldots d\mathbf{p}_N \int d\mathbf{r}_1 \ldots d\mathbf{r}_N e^{-H_N/kT} \tag{2.4}$$

which is a function of V and T, the temperature of the system. In this expression, the position integrations are performed over the volume V of the system; k and h are respectively Boltzmann's and Planck's constants.

The integral over the impulsions is easy to calculate. One gets

$$Q_N = \frac{Z_N}{N! \Lambda^{3N}} \tag{2.5}$$

where one has put

$$\Lambda = \frac{h}{(2\pi mkT)^{1/2}} \tag{2.6}$$

and

$$Z_N = \int d\mathbf{r}_1 \ldots d\mathbf{r}_N e^{-V_N/kT} \tag{2.7}$$

for the configuration integral. This quantity is impossible to compute exactly, but one may try to classify the various contributions to Z_N according to certain criteria. Therefore, following Mayer and Mayer,[17] let us define

$$f_{ij} = e^{-v_{ij}/kT} - 1 \tag{2.8}$$

which allows us to write:

$$e^{-(V_N/kT)} = \prod_{i>j=1}^{N} e^{-(v_{ij}/kT)} = \prod_{i>j=1}^{N} (1 + f_{ij}) \tag{2.9}$$

This expression is a product of $N(N-1)/2$ factors, because the potential interaction V_N contains $N(N-1)^2/2$ terms, one for each pair ij. If we write the first terms of the expresison, we obtain:

$$e^{-(V_N/kT)} = 1 + (f_{12} + f_{13} + f_{23} + \dots)$$
$$+ f_{12}f_{13} + f_{12}f_{14} + \dots$$
$$+ f_{12}f_{34} + \dots \tag{2.10}$$

the whole expression containing $2(N(N-1)/2)$ terms. A conenient way to visualize this expression is to associate a graph to each term in the following way: each particle $1,2 \dots N$ is represented by a labelled point and each f_{ij} factor by a line joining particles i and j. For example, in Fig. 1 we have represented the term $f_{12}f_{45}f_{56}f_{67}f_{57}f_{46}$ of the expansion (2.1) for $N = 8$. Each graph containing a line between the particles i and j vanishes if these particles are far apart. This property is due to the fact that v_{ij} and f_{ij} are zero for big r_{ij} distances. Therefore, if the gas is perfect (no intermolecular interactions), the only term which survives in (2.10) is the first one and the only non-vanishing graph has no lines.

We may now define an *l-cluster* as being a connected graph with l point. For example, the graph of Fig. 1 contains four disjoint connected parts to which two 1-cluster, one 2-cluster and one 4-cluster correspond. We call ml the number of *l*-clusters of a given graph. In Fig. 1, we have $m_1 = 2, m_2 = 1, m_4 = 1$, the other m_l being zero. Evidently, one has

$$\sum_{l=1}^{N} lm_l = N \tag{2.11}$$

Of course, there exist in general various *l*-clusters differing from each other by the manner in which the *l*-particles are connected. Figure 2 gives all the possible 3-clusters. Let us define a *cluster integral* as

$$b_l(V,T) = \frac{1}{l!V} \int dr_1 \dots dr_l \, \mathcal{U}_{1,2 \dots l} \tag{2.12}$$

where the cluster sum $\mathcal{U}_{1,2\dots l}$ is the sum of the various *l*-clusters that can be

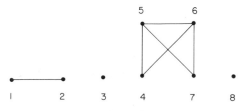

FIG. 1   A typical graph for $N = 8$.

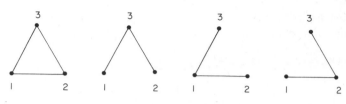

FIG. 2   All possible 3-clusters.

constructed with the particles $1, 2 \ldots l$. For example, the sum of the four graphs of Fig. 2 is $\mathcal{U}_{123}$.

We are now ready to analyse the configuration integral Z_N of (2.7) in terms of the cluster integrals b_l. We note that the configuration may be split according to the partition of the N particles in clusters. We have

$$Z_N = \sum_{\{m_l\}} Z(N, \{m_l\}) \tag{2.13}$$

where $Z(N,\{m_l\})$ is the contribution to Z_N of all the terms of (2.10) such that the particles are distributed in m_1 clusters of 1 particle, m_2 clusters of 2 particles, $\ldots$, m_l clusters of l particles and so on. The set of integers m_l determines $Z(N,\{m_l\})$ and the summation $\Sigma\{m_l\}$ is made over all the sets satisfying (2.11) (over all partitions).

The contribution $Z(N,\{m_l\})$ coming from a given partition $\{m_l\}$ can be obtained in the following way. Each particular term of $Z(N,\{m_l\})$ of m_1 integrals over one position vector for the 1-cluster, m_2 integrals over two position vectors for the 2-clusters, $\ldots m_l$ integrals over l position vectors for the l-clusters, and so on. The various terms contributing to $Z(N,\{m_l\})$ for fixed $\{m_l\}$ will differ from each other for two reasons. First, because a given labelled particle may belong to various clusters. Second, because each of the m_l integrals for the l-clusters may correspond to various ways of constructing an l-cluster (see Fig. 2 for $l = 3$). $Z_N(N,\{m_l\})$ is given, for fixed $\{m_l\}$, by a sum of terms corresponding to all the possible distributions of the particle labels among the clusters and to all the possible l-clusters.

The sum over all the possible ways of constructing an l-cluster merely replaces each of the m_l integrals over l-position vectors by $l! V b_l$. This is a consequence of the definition of b_l. Using (2.12), we get:

$$Z_N(N, \{m_l\}) = \sum_{\text{label}} \prod_{l=1}^{N} (l! V b_l)^{m_l} \tag{2.14}$$

where the symbol Σ_{label} means that we have still to sum over all the possible distributions of the particle labels among the clusters. Since each of these distributions gives the same value to $(l! V b_l)^{m_l}$ it is clear that we have only to count the number of terms in Σ_{label}. Let us suppose that we first permute the N particle labels. There are $N!$ such permutations. But the permutations of the l labels in an l-cluster integral do not give a new graph and should lead to an overcounting of the number of graphs (for example, commute the labels in Fig. 2). For this reason, one has to introduce a factor $1/l!$ for each l-cluster integral. Moreover, the interchange of the two sets of l-labels from one l-cluster integral to another does not give a new graph either. This introduces a factor $1/m_l!$ for the m_l l-cluster integrals. The equation (2.14) now reads:

$$Z(N, \{m_l\}) = N! \prod_{l=1}^{N} \frac{(Vb_l)^{m_l}}{m_l!} \qquad (2.15)$$

If we combine this equation with (2.13), we obtain the well known formula for the configuration integral

$$Z_N = N! \sum_{m_l} \prod_{l=1}^{N} \frac{(Vb_l)^{m_l}}{m_l!} \qquad (2.16)$$

which is, for the moment, only a convenient way to re-organize the series (2.10) in terms of the cluster integrals b_l. However the complexity of these quantities increases rapidly with l: b_2 contains one term which is f_{12}, b_3 has four terms (see Fig. 2), b_4 has 38 terms and so on. Now we shall see that equation (2.16) may still be simplified by introducing the concept of irreducible clusters. Therefore we need first to define the topological notion of *articulation point*; if, by cutting all the lines attached to a given point of a connected graph, one generates a disconnected one, then this point is called an articulation point. In Fig. 2, the first graph to the left contains no articulation points, but the point connected to two lines of the three other graphs is an articulation point. If a connected graph has no articulation points, it will be called a *star* or an *irreducible graph*. If it contains one or more articulation points, it will be called a *reducible graph*.

The definition of an irreducible l-cluster integral β_l follows closely the definition (2.12) of a cluster integral:

$$\beta_l = \frac{1}{l!V} \int d\mathbf{r}_1 \dots d\mathbf{r}_{l+1} \sigma_{1,2} \dots_{l+1} \qquad (2.17)$$

where $\sigma_{1,2\dots l+1}$ is now the sum of the various $l+1$ irreducible clusters that can be built with the particles $1,2,\dots l+1$. Figure 3 gives σ_{12} and σ_{123}.

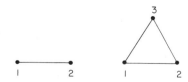

FIG. 3   Irreducible graphs for σ_{12} and σ_{123}.

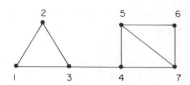

FIG. 4   Factorization of an irreducible cluster

We indicate briefly in an example that the cluster integral (2.12) may be expressed as a product of irreducible cluster integrals. We consider the particular term associated with the graph of Fig. 4 which contains two articulation points 3 and 4. Let us consider the contribution C to b_7 (V,T) [see (2.12)] of this graph:

$$C = \frac{1}{7!V} \int dr_1 \ldots dr_7 (f_{12}f_{13}f_{23})(f_{34})(f_{45}f_{56}f_{57}f_{67}f_{47})$$

where the various parentheses correspond to cutting the graph through its articulation points. We introduce new integration variables which are the coordinates of 3 and the relative coordinates r_{i3} of the other particles with respect to 3:

$$C = \frac{1}{7!V} \int dr_{13} \int dr_{23} \int dr_{43} \int dr_{53} \int dr_{63} \int dr_{73} \int dr_3 \, (\text{idem}) \qquad (2.18)$$

The integral over dr_3 gives V and the quantity

$$C_1 = \int dr_{13} \int dr_{23} f_{12} f_{13} f_{23}$$

may be factored out because the variables r_{13} and r_{23} do not appear in the other parentheses. It is easy to show from (2.17) that

$$C_1 = 2!\beta_2$$

which puts (2.18) in the form

$$C = \frac{2!\beta_2}{7!} \int dr_{34} \int dr_{45} \int dr_{46} \int dr_{47} (f_{34})(f_{45}f_{56}f_{57}f_{67}f_{47})$$

$$= \frac{2!\beta_2\beta_1}{7!} \int dr_{45} \int dr_{46} \int dr_{47} (f_{45}f_{56}f_{57}f_{67}f_{47})$$

because $\int dr_{34} f_{34}$ factors out again to give β_1. If we had considered b_f instead of C, the product $f_{45}f_{56}f_{57}f_{67}f_{47}$ which still appears in C would have been replaced by the irreducible cluster sum σ_{4567}, then the contribution C should have taken the value

$$\frac{1}{7!} (2!\beta_2)(\beta_1)(3!\beta_3)$$

It is then easy to convince oneself that the product

$$\frac{1}{l!} \prod_{j=1}^{l-1} (j!\beta_j)^{n_j}$$

(2.19)

is the contribution to b_l of all the l-clusters that can be reduced into n_1 irreducible 2-cluster integrals, n_2 irreducible 3-cluster integrals ... n_j irreducible $(j+1)$-cluster integrals. We should now compute the number of l-clusters satisfying the condition. This difficult problem has been solved[17] and this number is:

$$\frac{l!}{l^2} \prod_{j=1}^{l-1} \left(\frac{l}{j!}\right)^{n_j} \frac{1}{n_j!}$$

so that the contribution to b_l (2.19) reads:

$$\frac{1}{l^2} \prod_{j=1}^{l-1} \frac{(l\beta_j)^{n_j}}{n_j!}$$

which is still to be summed over the various sets of n_j to give:

$$b_l = \frac{1}{l^2} \sum_{\{n_j\}} \prod_{j=1}^{l-1} \frac{(l\beta_j)^{n_j}}{m_j!}$$

(2.20)

In this expression and in (2.19) the n_j must satisfy the condition

$$l = \sum_{j=1}^{l-1} jn_j + 1$$

2.2   Equation of state

As an illustration, we shall now recall the well known expansion of the pressure p of an imperfect gas as a function of its density ρ. As a definition of the so called virial coefficients B_r we write:

$$\frac{p}{\rho kT} = 1 + B_2\rho + B_3\rho^2 + \dots$$

(2.21)

where the first term corresponds to the perfect gas approximation.

In order to evaluate the coefficients B_r in terms of microscopic quantities we need a microscopic expression for the pressure. We first define the *grand partition function*:

$$\Theta(T,V,N) = \sum_{N \geqslant 0} Q_N \lambda^N$$

(2.22)

where $\lambda = e^{\mu/kT}$ and μ is the chemical potential. Using (2.5), one has

$$\Theta(T,V,N) = \sum_{N \geqslant 0} \frac{Z_N}{N!} \left(\frac{\lambda}{\Lambda^3}\right)^N$$

(2.23)

where the quantity λ/Λ^3 is the activity z of the system.

In terms of the statistical grand partition function, one could show that the thermodynamic variables p and ρ are given by[13]

$$p = \frac{kT}{V} \ln \Theta \tag{2.24}$$

$$\rho = \frac{kT}{V} \left(\frac{\partial \ln \Theta}{\partial \mu} \right)_{T,V} \tag{2.25}$$

If one puts the result (2.16) into the expression (2.23), one obtains

$$\Theta = \sum_{N \geqslant 0} \sum_{\{m_l\}} \prod_{l=1}^{N} \frac{(Vb_l)^{m_l}}{m_l!} z^N$$

and, since the sum over $\{m_l\}$ verifies (2.11) this may be written as:

$$\Theta = \prod_{l=1}^{\infty} \sum_{m_l \geqslant 0} \frac{(Vb_l z^l)^{m_l}}{m_l!}$$

the sum over m_l being, this time, unrestricted. For given l, the sum over m_l is identical with the Taylor expansion of $e^{Vb_l z^l}$ and, therefore

$$\ln \Theta = \ln \prod_{l=1}^{\infty} e^{Vb_l z^l} = V \sum_{l=1}^{\infty} b_l z^l$$

From this expression and the formulae (2.24) and (2.25), one gets easily p/kT and ρ as powers series in the activity:

$$\frac{p}{kT} = \sum_{l=1}^{\infty} b_l z^l \tag{2.26}$$

$$\rho = \sum_{l=1}^{\infty} l b_l z^l \tag{2.27}$$

since

$$\frac{\partial}{\partial \mu} = \frac{z}{kT} \cdot \frac{\partial}{\partial z}$$

In order to get p/kT as a power series in the density ρ, we first express the activity as a series in ρ:

$$z = \rho + \sum_{n=2}^{\infty} a_n \rho^n \tag{2.28}$$

where the coefficients a_n are obtained by replacing z given by (2.28) in (2.27) and by expressing that the coefficient of $\rho^n (n \geqslant 2)$ in the right hand side is equal to zero. This yields a set of relations between the a_n and the b_l

$$a_2 = -2b_2$$
$$a_3 = -3b_3 + 8b_2^2$$
$$a_4 = -4b_4 + 30b_2 b_3 - 40b_2^3$$

and the formula for z obtained in this way has to be carried to (2.26). One gets of course p/kT as a power series in ρ where the coefficients are expressible in terms of the b_l:

$$\frac{p}{kT} = \rho[1 - b_2 \rho + (4b_2^2 - 2b_3)\rho^2 + \ldots] \qquad (2.29)$$

But according to (2.20), one must have

$$b_2 = \frac{1}{2}\beta_1$$

$$b_3 = \frac{1}{9}\left[3\beta_2 + \frac{(3\beta_1)^2}{2!}\right] = \frac{\beta_2}{3} + \frac{\beta_1^2}{2}$$

and, when these expressions are substituted for the b_l it turns out that the coefficient of ρ^{l+1} in (2.29) is $-(l/l+1)\beta_l$:

$$\frac{p}{\rho kT} = 1 - \sum_{l=1} \frac{l}{l+1}\beta_l \rho^l \qquad (2.30)$$

This very nice result has been proved to be true for arbitrary l.[15] It indicates that the topological nature of the terms of the configuration integral (2.7) is of crucial importance for the determination of the physical properties of imperfect gases. The configuration integral itself may be expressed in terms of the cluster integrals b_l as in (2.16). These quantities appear also as coefficients in the activity expansions of the pressure and the number density whereas the irreducible cluster integrals are needed for the density expansion of the pressure. Of course, comparison of (2.21) and (2.30) yields the statistical expression for the virial coefficients B_n:

$$B_{n+1} = -\frac{n}{n+1}\beta_n \qquad (2.31)$$

2.3   Semi-invariants

For further purposes we will also need *semi-invariants of Thiele or cumulants*. These quantities may be introduced by considering the quantity:[5]

$$\ln \frac{1}{V^N}\int d\mathbf{r}_1 \ldots d\mathbf{r}_N\, e^{-(V_N/kT)} = \sum_{n=1}^{\infty} \frac{M_n}{n!}\left(-\frac{1}{kT}\right)^n \qquad (2.32)$$

where the semi-invariant of order n, M_n may be obtained by expanding the exponential and the logarithm in a Taylor series of the parameter $1/kT$. In this way, the first semi-invariants are shown to be:

$$M_1 = \langle V_N \rangle$$
$$M_2 = \langle V_N^2 \rangle - \langle V_N \rangle^2$$
$$M_3 = \langle V_N^3 \rangle - 3 \langle V_N^2 \rangle \langle V_N \rangle + 2 \langle V_N \rangle^3 \tag{2.33}$$

where $\langle V_N^i \rangle$ means

$$\frac{1}{V^N} \int dr_1 \dots dr_N V_N^i$$

Because of the thermodynamic relation[13] between the free energy F and the partition function Q_N

$$F = -kT \ln Q_N \tag{2.34}$$

it is easy to realise that the left hand side may be written:

$$\ln \frac{Z_N}{V^N} = \ln \frac{Q_N \Lambda^{3N} N!}{V^N} = -\frac{1}{kT} (F - F_0) \tag{2.35}$$

where

$$F_0 = -kT \ln \frac{V^N}{N! \Lambda^{3N}}$$

is the free energy of the perfect gas for which

$$Q_N = \frac{V^N}{N! \Lambda^{3N}}.$$

It is possible to show[5] that the semi-invariant expansion (2.32) may be reorganized in powers of the density to yield:

$$-\frac{1}{NkT} (F - F_0) = \sum_{l=1}^{\infty} \frac{\beta_l \rho^l}{l+1} \tag{2.36}$$

This result is not obtainable immediately but it is easy to convince oneself that it reduces to the density expansion of the pressure (2.30) if one uses the well-known thermodynamic relation

$$p - p_0 = -\left(\frac{\partial (F - F_0)}{\partial V} \right)_T \tag{2.37}$$

Comparison of the expressions (2.32), (2.35) and (2.37) indicate that there exists a close relation between the semi-invariants and the irreducible cluster integrals β_l. In fact, if one replaces V_N by $\sum v_{ij}$ in (2.33), one obtains expressions which may be represented by multigraphs with vertices (one for each particle $i,j, \ldots$) and lines between them (a line between i and j for each v_{ij} factor). The only difference between these graphs and the Mayer graphs is due to the fact that the points i and j may here be connected by more than one line because M_n, containing $\langle V_N^n \rangle$, has a term proportional to $\int d\mathbf{r}_i \ldots d\mathbf{r}_N (v_{ij})^n$. If the expressions (2.33) for M_n are analysed with this graph technique, it appears indeed that only the irreducible graphs contribute to the semi-invariants.

3   IMPERFECT GAS OUT OF EQUILIBRIUM

3.1   The Prigogine theory

We will now discuss some graph theoretical features in non-equilibrium classical statistical mechanics.

The non-equilibrium situation differs from the equilibrium one by the fact that the distribution function $\rho_N(\mathbf{r}_N,\mathbf{p}_N,t)$ — which gives the probability of finding the N particles with positions $\mathbf{r}_N$ and impulsions $\mathbf{p}_N$ — now depends on the time t. Its evolution is governed by the Liouville equation[26]

$$i \frac{\partial \rho_N}{\partial t} = L_N \rho_N \tag{3.1}$$

where $L_N = L_0 + \lambda \delta L_N$, the Liouville operator, has been separated in an unperturbed part

$$L_0 = -i \sum_{k=1}^{N} \frac{\partial H_N}{\partial \mathbf{p}_k} \frac{\partial}{\partial \mathbf{r}_k} = -i \sum_{k=1}^{N} \mathbf{v}_k \cdot \frac{\partial}{\partial \mathbf{r}_k} \tag{3.2}$$

and a perturbation

$$\lambda \delta L_N = i\lambda \sum_{k=1}^{N} \frac{\partial H_N}{\partial \mathbf{r}_k} \cdot \frac{\partial}{\partial \mathbf{p}_k} = i\lambda \sum_{k=1}^{N} \frac{\partial V_N}{\partial \mathbf{r}_k} \cdot \frac{\partial}{\partial \mathbf{p}_k}$$

For long times, the distribution function reaches its equilibrium value:

$$\rho_N^{eq} = \frac{e^{-H_N/kT}}{Q_N} \tag{3.3}$$

where Q_N is given by (2.4). The problem of the compatibility of this irreversible evolution with the reversible equation (3.1) is a very crucial one. Non-equilibrium statistical mechanics will enable us to elucidate this paradox; the reader will find further details about this point in the theory in the articles by Prigogine[19,22] and Résibois.[24] We will here recall only those results of this theory that are needed for our special discussion; for a more precise idea of the present status of this work, the reader is referred to the note added in proof.

The formal solution of equation (3.1) is

$$\rho_N(t) = e^{-iL_Nt}\rho_N(0) \tag{3.4}$$

and may also be formulated in terms of the resolvant operator $(L_N - z)^{-1}$

$$\rho_N(t) = -\frac{1}{2\pi i} \oint_c dz\, e^{-izt} \frac{1}{L_N - z} \rho_N(0) \tag{3.5}$$

where z is a complex variable and where c is a parallel to the real axis and a large semi-circle in the lower half plane.

The resolvant may be expanded as a series in the coupling parameter λ

$$\frac{1}{L_N - z} = \frac{1}{L_0 - z} \sum_{n=0}^{\infty} \left(-\lambda\delta L_N \frac{1}{L_0 - z}\right)^n$$

and, under this form, it appears as a sequence of free propagations of the particles (described by $1/L_O - Z$) interrupted by interactions (described by $\lambda\delta L_N$).

Let us define the matrix elements of an operator A by the equation

$$\langle k \mid A \mid k' \rangle = \frac{1}{V^N} \int dr_1 \ldots dr_N \exp\left(i \sum_{l=1}^{N} k_s \cdot r_s\right) A \exp\left(-i \sum_{m=1}^{N} k_m r_m\right) \tag{3.6}$$

where k and k' represent the sets k_s and k'_s. In this notation the central operator of the theory is the collision operator

$$\psi(z) = \langle 0 \mid -\lambda\delta L_N \sum_{n=1}^{\infty} \left(-\lambda\delta L_N \frac{1}{L_0 - z}\right)^n \mid 0 \rangle_{irr}$$

where the subscript 'irr' means that the only contributions that are retained must satisfy the *dynamical irreducibility condition*. This condition expresses that, in a collision, the particles are infinitely far apart at the *initial* time, then interact and separate infinitely far apart at the *final* time. It excludes explicitly situations where the particles should be infinitely separated at *intermediate* times.

In Fig. 5 we give some contributions to the resolvant in order to illustrate the dynamical irreducibility condition. These graphs should not be confused with equilibrium graphs: the labelled lines represent the free movement of the particles under the influence of the unperturbed Hamiltonian K_N, the dots are interactions between the particles due to the perturbation V_N. Moreover the *time* appears explicitly in these graphs: each event anterior to another is drawn at the right of this one. It is seen that the graphs (a) and (b) do not satisfy the dynamical irreducibility condition because the intermediate state indicated by a

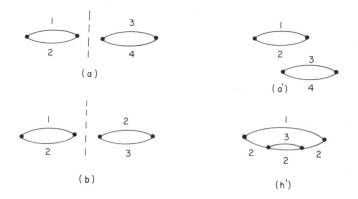

FIG. 5   Some dynamical graphs of the Prigogine theory.

dotted vertical line corresponds to infinitely separated particles. Such a state does not exist in (a') and (b') which are indeed dynamically irreducible.

It may be shown[19,21] that, for long times, the evolution equation for the velocity distribution function $\rho_o(t)$ may be written

$$\frac{\partial \rho_0(t)}{\partial t} = \Omega \psi \rho_0(t) \tag{3.7}$$

the operator $\Omega \psi$ being a well-defined function of the operator $\psi(z)$. This equation means that the evolution of the velocities is governed by the collisions through the collision operator $\psi(o)$. The fact that the collisions are *non-instantaneous* is contained in the operator Ω. It has also been shown that the equilibrium velocity distribution function is a stationary solution of (3.7).

3.2   Semi-invariant formulation compared to the Prigogine theory

The velocity distribution function at time t is related to its value at $t = 0$ through the equation

$$\rho_0(t) = \langle 0|S_{-t}^N|0\rangle \rho_0(0) \tag{3.8}$$

where we have already neglected the destruction of initial correlation, an approximation which is valid for sufficiently long times[19,21] We have noted

$$\langle 0|S_{-t}^N|0\rangle = \langle 0|e^{-iL N t}|0\rangle \tag{3.9}$$

If we take the logarithm of this expression and use (3.6), we obtain an expression which looks like (2.32), apart from the fact that $1/kT$ has to be replaced by (it) and that the L_N must be substituted for V_N. For this reason it seems natural to define non-equilibrium semi-invariants $\tilde{M}_n$ in a way which parallels (2.32)

$$\ln \langle 0 \mid S^N_{-t} \mid 0 \rangle \equiv \sum_{n=0}^{\infty} \frac{\tilde{M}_n}{n!} (-it)^n \tag{3.10}$$

We have now to analyse the properties of the $\tilde{M}_n$ and, therefore, we will use tentatively the properties of the equilibrium semi-invariants. In fact, we will see that such a procedure needs great care because some of the equilibrium properties may not be translated out of equilibrium.[3]

To understand this let us first write the left hand side of (3.9) in an equivalent form

$$\langle 0 \mid S^N_{-t} \mid 0 \rangle = \langle 0 \mid T \exp \left\{ -i\lambda \int_0^t \delta L_N(t') dt' \right\} \mid 0 \rangle \tag{3.11}$$

where

$$\delta L_N(t) = \exp[-iL_0 t] \, \delta L_N \, \exp[iL_0 t]$$

and where Dyson's chronological operator (see for example Schweber, Hoffman and Bethe[25]) T rearranges the time labels of a sequence of operators in their chronological order.

$$T(\delta L_N(t_1)\delta L_N(t_2) \ldots \delta L_N(t_k) = \delta L_N(t_s)\delta L_N(t_r) \ldots \delta L_N(t_n)$$

with

$$t_s > t_r > \ldots > t_n$$

Using (3.10) and (3.11) and taking into account that $\langle 0|\delta L_N(t)|0 \rangle = 0$ one gets the first few non-equilibrium semi-invariants:

$$\tilde{M}_1(t) = 0$$

$$\tilde{M}_2(t) = \lambda^2 \int_0^t dt_i \int_0^t dt_j \, \langle 0 \mid T\delta L_N(t_i)\delta L_N(t_j) \mid 0 \rangle$$

$$\tilde{M}_3(t) = \lambda^3 \int_0^t dt_i \int_0^t dt_j \int_0^t dt_k \, \langle 0 \mid T\delta L_N(t_i)\delta L_N(t_j)\delta L_N(t_k) \mid 0 \rangle$$

$$\tilde{M}_4(t) = \lambda^4 \int_0^t dt_i \int_0^t dt_j \int_0^t dt_k \int_0^t dt_l$$

$$\{ \langle 0 \mid T\delta L_N(t_i)\delta L_N(t_j)\delta L_N(t_k)\delta L_N(t_l) \mid 0 \rangle$$

$$- 3 \langle 0 \mid T\delta L_N(t_i)\delta L_N(t_j) \mid 0 \rangle \langle 0 \mid T\delta L_N(t_k)\delta L_N(t_l)|0 \rangle$$

$$\tag{3.12}$$

and we are now going to examine $\tilde{M}_4(t)$. We first eliminate the T operators; this

FIG. 6   Two contributions to (3.12).

merely orders the t integrations and introduces numerical factors. We obtain:

$$\tilde{M}_4(t) = 4!\lambda^4 \int_0^t dt_i \int_0^{t_i} dt_j \int_0^{t_j} dt_k \int_0^{t_k} dt_l$$

$$\langle 0 | \delta L_N(t_i) \delta L_N(t_j) \delta L_N(t_k) \delta L_{N'}(t_l) | 0 \rangle$$

$$- 3.2!2!\lambda^4 \int_0^t dt_i \int_0^{t_i} dt_j \langle 0 | \delta L_N(t_i) \delta L_N(t_j) | 0 \rangle$$

$$\int_0^t dt_k \int_0^{t_k} dt_l \langle 0 | \delta L_N(t_k) \delta L_N(t_l) | 0 \rangle$$

We now consider two specific contributions to the preceding expressions $\tilde{M}'_4(t)$ and $\tilde{M}''_4(t)$ which are associated respectively to the graphs (a) and (b) of Fig. 6. In these drawings, we adopt a convention which is similar to the one used at the end of Section 2 except that a line between two vertices s and t here represents

$$\delta L^{(st)} = \frac{\partial v_{st}}{\partial r_s} \cdot \frac{\partial}{\partial p_s}$$

and not v_{st} itself. The graph (a) is thus a disconnected graph while (b) is a connected but a reducible one. It is expected that neither (a) nor (b) contribute to $\tilde{M}_4(t)$ since neither one of them is connected *and* irreducible.

The explicit expressions for these contributions are:

$$\tilde{M}'_4(t) = 4!\lambda^4 \int_0^t dt_i \int_0^{t_i} dt_j \int_0^{t_j} dt_k \int_0^{t_k} dt_l (A + B)$$

$$- 6\lambda^4 2!2! \int_0^t dt_i \int_0^{t_i} dt_j \langle 0 | \delta L^{(12)}(t_i) \delta L^{(12)}(t_j) | 0 \rangle$$

$$\int_0^t dt_k \int_0^{t_k} dt_l \langle 0 | \delta L^{(34)}(t_k) \delta L^{(34)}(t_l) | 0 \rangle \qquad (3.13)$$

where:

$$A = \delta L^{(12)}(t_i)\delta L^{(12)}(t_j)\delta L^{(34)}(t_k)\delta L^{(34)}(t_l)$$
$$+ \delta L^{(12)}(t_i)\delta L^{(34)}(t_j)\delta L^{(12)}(t_k)\delta L^{(34)}(t_l) \qquad (3.14)$$
$$+ \delta L^{(12)}(t_i)\delta L^{(34)}(t_j)\delta L^{(34)}(t_k)\delta L^{(12)}(t_l)$$

and B is the same expression except that one has to exchange the role of (12) and (34) everywhere.

The expression for $M_4''(t)$ is obtained from (3.13) and (3.14) simply by replacing index 4 by 2.

We will now verify that $\tilde{M}_4''(t)$ does not contribute to $\tilde{M}_4(t)$ as is expected from equilibrium arguments. In contradiction with these arguments, we will see that $\tilde{M}_4''(t)$ does not vanish and contributes to $\tilde{M}_4(t)$.

The vanishing of $\tilde{M}_4''(t)$ will not be demonstrated mathematically (see Brocas and George, 1967) but will merely be suggested. It comes from the composition of the two terms of (3.13). In the second term of this expression the two interactions $\delta L^{(12)}$ are ordered with respect to each other and the same is true for the two interactions $\delta L^{(34)}$. But no ordering is prescribed for $\delta L^{(12)}$ with respect to $\delta L^{(34)}$ In the first term of (3.13) all the orderings of the four events (two $\delta L^{(12)}$ and two $\delta L^{(34)}$) are represented [see (3.14)]. Since $\delta L^{(12)}$ and $\delta L^{(34)}$ commute, one may put the $\delta L^{(12)}$ before the $\delta L^{(34)}$ in each term of A and B and see that the two terms of (3.13) compensate exactly.

The same type of operation is impossible for $M_4''(t)$ because $\delta L^{(12)}$ and $\delta L^{(23)}$ do not commute: they have particle 2 in common. For this reason, $\tilde{M}_4''(t)$ does not vanish and the equilibrium prediction is not true out of equilibrium.

This example illustrates the fact that topological properties which are extremely useful in equilibrium problems may be very difficult to translate out of equilibrium. The main difficulty of non-equilibrium expressions is that they involve operators in specific chronological order. This aspect is not taken into account by the equilibrium irreducibility condition but it is included in the dynamical irreducibility condition which tells us not only about the interactions between particles but also about the chronological order of these events. This point is clearly illustrated by Fig. 5 where graphs (a), (b) and (a'), (b') correspond to different orderings of the events contained respectively in the graphs (a), (b) of Fig. 6.

A more detailed discussion of this problem is a very difficult task. The reader interested in these aspects will find further details elsewhere.[2,19,23]

3.3   Cohen's formulation compared to the Prigogine theory

We now discuss another example where the translation of the equilibrium topological structure has been possible. This example is due to Cohen[7] who studied the approach to equilibrium of the *one-particle velocity distribution function.* $\varphi^{(1)}$ of an homogeneous fluid. A formal equation for the evolution of this quantity may be obtained by integrating the Liouville equation (3.1) over $N - 1$ velocities and positions. One gets:

$$\frac{\partial \varphi^{(1)}(\mathbf{v}_1, t)}{\partial t} = -i \int d\mathbf{r}_2 \int d\mathbf{p}_2 \, \delta L^{(12)} f^{(2)}(\mathbf{r}_1, \mathbf{p}_1; \mathbf{r}_2, \mathbf{p}_2, t) \tag{3.15}$$

where $f^{(2)}$ represents the *two-particle distribution function.* This equation is still useless since $\varphi^{(1)}$ is related to $f^{(2)}$; $f^{(2)}$ itself may be expressed as a function of $f^{(3)}$, and so on. One gets, in this way, an hierarchy of equations where each distribution function is expressed as a function of the distribution function containing one particle more.

We give a brief account of the procedure used by Cohen in order to express $f^{(2)}$ as a function of the one particle velocity distribution function only, in order to understand the role of graph theory in its formulation.

The starting point is the so-called *n*-particle $(n \leqslant N)$ streaming operator

$$S_{-t}^{(12 \cdots n)} = S_{-t}^{(n)} = \exp(-iL_n t) \tag{3.16}$$

where L_n is defined for n particles in a way analogous to (3.2) and (3.3). It is easy to verify that $S_{-t}^{(1)} = 1$ for an homogeneous system: with one particle $\delta L_n = 0$ and L_n contains only one position derivative. Each $S_{-\tau}^n$ is decomposed in dynamical Ursell functions $\mathscr{U}_{-\tau}^{(12 \cdots s)}$

$$S_{-\tau}^{(1)} = 1$$
$$S_{-\tau}^{(12)} = \mathscr{U}_{-\tau}^{(12)} + 1 \tag{3.17}$$
$$S_{-\tau}^{(123)} = \mathscr{U}_{-\tau}^{(123)} + \mathscr{U}_{-\tau}^{(12)} + \mathscr{U}_{-\tau}^{(13)} + \mathscr{U}_{-\tau}^{(23)} + 1$$

and so on. The relation that is constructed here defines the $\mathscr{U}_{-\tau}^{(12 \cdots s)}$ functions. It parallels completely the Ursell expansion of equilibrium: the relation between the streaming operators and the dynamical Ursell functions is analogous to the relation between the $e^{-Hn/kT}$ and the equilibrium Ursell functions.[27] The only major difference is that (3.17) involves operators instead of functions.

The relation for arbitrary n is:

$$S_{-\tau}^{(n)} = \sum_{\{n_i\}} \sum_{\text{lab}} \prod_{i=1} \mathscr{U}_{-\tau}^{(n_i)} \tag{3.18}$$

where the symbols mean that we have to sum over all the partitions $\{n_i\}$ of the n particles, to distribute the labels of the particles among the parts in all possible ways and to perform the product of the dynamical Ursell functions corresponding to each part.

The next step involves the new concept of Husimi trees[12] that we define now. Let us first consider a Cayley tree which is a graph with no cycles. One may replace each of its lines by a triangle, or a given polygon and generate in this way cacti or pure Husimi trees respectively. An important restriction is that no line lies on more than one polygon. Mixed Husimi trees are obtained if different polygons are substituted for the lines of the original Cayley tree.

The Husimi operators for particles $(12 \ldots n)$, i.e. $\mathscr{V}_{-\tau}^{(12 \ldots n)}$, are defined by the following set of relations:

$$\mathscr{U}_{-\tau}^{(12)} = \mathscr{V}_{-\tau}^{(12)}$$

$$\mathscr{U}_{-\tau}^{(123)} = \mathscr{V}_{-\tau}^{(123)} + \mathscr{V}_{-\tau}^{(12)}\mathscr{V}_{-\tau}^{(23)} + \mathscr{V}_{-\tau}^{(12)}\mathscr{V}_{-\tau}^{(13)} + \mathscr{V}_{-\tau}^{(13)}\mathscr{V}_{-\tau}^{(23)}$$

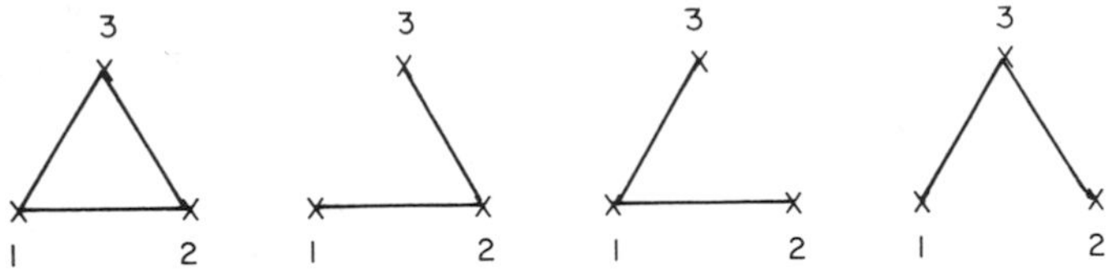

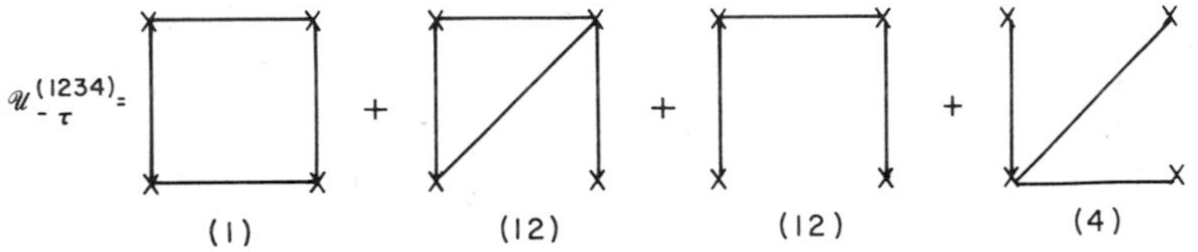

FIG. 7   The Husimi operators.

which are analogous to the equilibrium relations between the Ursell functions and the star functions.[27] In Fig. 7, we have used crosses instead of dots because these graphs have a significance which is slightly different from the usual one. For example, in equilibrium theory, the equilibrium star function σ_{1234} is given in Fig. 8. Moreover, the number between parentheses under each graph indicates the number of different labellings of the particles that can be realized for the graph under consideration.

We must note that, because of the *operator* character of the quantities, these do not commute with each other if they have at least one particle in common. For this reason, one must choose a standard order for these operators. One decides to put at the extreme left the factor which contains the smallest particle label and, going to the right, we put the factor containing the smallest particle label in ascending order. This problem does not appear in (3.17) since a given particle does not appear in more than one factor of a product of $\mathscr{U}_{-\tau}$ operators.

We are now ready to write down the time-evolution of the one particle velocity distribution function

$$\frac{\partial \varphi^{(1)}(v_1, t)}{\partial t} \quad \sum_{n=1}^{\infty} \left(\frac{\partial \varphi}{\partial t}\right)^{[n]} \tag{3.19}$$

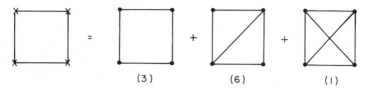

FIG. 8   The equilibrium star function σ_{1234}.

the contribution

$$\left(\frac{\partial \varphi}{\partial t}\right)^{[n]}$$

involving n particles being:

$$\left(\frac{\partial \varphi}{\partial t}\right)^{[n]} = \frac{-iC^{n-1}}{(n-2)!} \lim_{\tau \to \infty} \int dp_2 \int dr_2 \ldots \int dp_n \int dr_n \delta L^{(12)} B_{-\tau}^{(12 \ldots n)}$$

$$\prod_{i=1}^{n} \varphi^{(1)}(v_i,t) \tag{3.20}$$

The expressions for the first few $B_{-\tau}^{(12 \cdots n)}$ operators are given in Fig. 9.

$$B_{-\tau}^{(12)} = \mathscr{V}(12)_{-\tau}$$

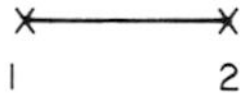

$$B_{-\tau}^{(123)} = \mathscr{V}(123)_{-\tau} + \mathscr{V}(13)_{-\tau} \mathscr{V}(23)_{-\tau}$$

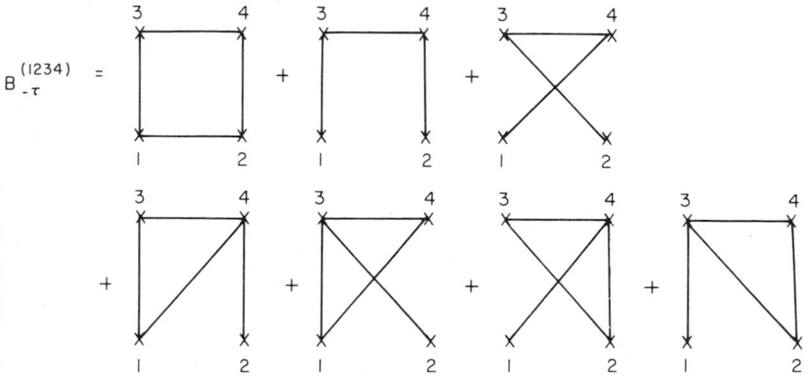

FIG. 9   Graphical expressions for the $B_{-\tau}^{(1234)}$ operators.

The important difference between the $\mathcal{U}_{-\tau}^{(12\cdots n)}$ operators of Fig. 6 and the $B_{-\tau}^{(12\cdots n)}$ operators of Fig. 9 lies in the fact that the $\mathcal{U}_{-\tau}^{(12\cdots n)}$ contain two types of contributions (1,2-reducible Husimi trees *and* 1,2-irreducible Husimi trees) while $B_{-\tau}^{(12\cdots n)}$ contain only 1,2-irreducible Husimi trees. These are *rooted* Husimi trees consisting of a chain of polygons, having no more than one vertex in common with each other. The two ends of the 1,2 chain are two polygons containing either particle 1 or particle 2. If the chain consists of one polygon, it contains both 1 and 2. In contrast with the 1,2-reducible Husimi trees, the 1,2-irreducible Husimi trees may not contain *appending parts* attached to the 1,2-chain.

By comparing the equations (3.15) and (3.19)–(3.20) one sees that the $B_{-\tau}^{(12 \cdots n)}$ expressions are the coefficients of the concentration expansion of the non-equilibrium two particle distribution function $f^{(2)}(\mathbf{r}_1,\mathbf{p}_1;\mathbf{r}_2,\mathbf{p}_2,\tau)$. It is striking that the final expression for this function may be obtained by a topological rule which is very closely related to the corresponding rule for equilibrium quantities. In principle,* at least, it is easy to write down the explicit expression of $B_{-\tau}^{(12\cdots n)}$ for arbitrary n and this provides a generalization of the Boltzmann equation to arbitrary order in the density.

The relation between this derivation and the kinetic equations that have been obtained by Prigogine and his collaborators[19] is a very intriguing point, because, as we have mentioned before, the dynamical irreducibility condition of the Prigogine theory is of a quite different nature to the equilibrium irreducibility condition of the Cohen formulae. However it has been proved[4] that the two formulations are completely equivalent. The details of the demonstration are rather involved and will not be reproduced here. We shall again simply try to give a general idea of the arguments that are used, because it might give a better understanding of the relation between topology and dynamical irreducibility.

We start from the formula:

$$\mathcal{U}_{-\tau}^{(12\cdots n)} = B_{-\tau}^{(12\cdots n)} + R_{-\tau}^{(12\cdots n)} \tag{3.21}$$

which defines $R_{-\tau}^{(12\cdots n)}$ as the sum of all 1,2-reducible Husimi trees, according to Figs. 7 and 9 and to the comments following Fig. 9. The first $R_{-\tau}^{(12\cdots n)}$ operators are given in Fig. 10, and a similar formula is used for $R_{(124)}^{(1234)}$.

The formulae given in Fig. 10 clearly illustrate the fact that each 1,2-reducible Husimi tree (chain) on n points may be decomposed into two parts: a 1,2-irreducible Husimi tree on n' points ($n' < n$) and appending parts attached to the main 1,2-chain: for clarity we have drawn the 1,2 chain in heavy lines. Moreover we have written $R_{(12\cdots n)}^{(12\cdots n)}$ for the sum of all the 1,2 irreducible Husimi trees with n points whose 1,2 chain consists of the n' points $1,2,\ldots n'$.

---

*We do not discuss here the difficulties involved with the use of this type of expression. It is well known that divergencies appear in the virial expansion of transport coefficients which are related to special dynamical effects (details and references may be found for example in Ernst, Haines and Dorfman[10]).

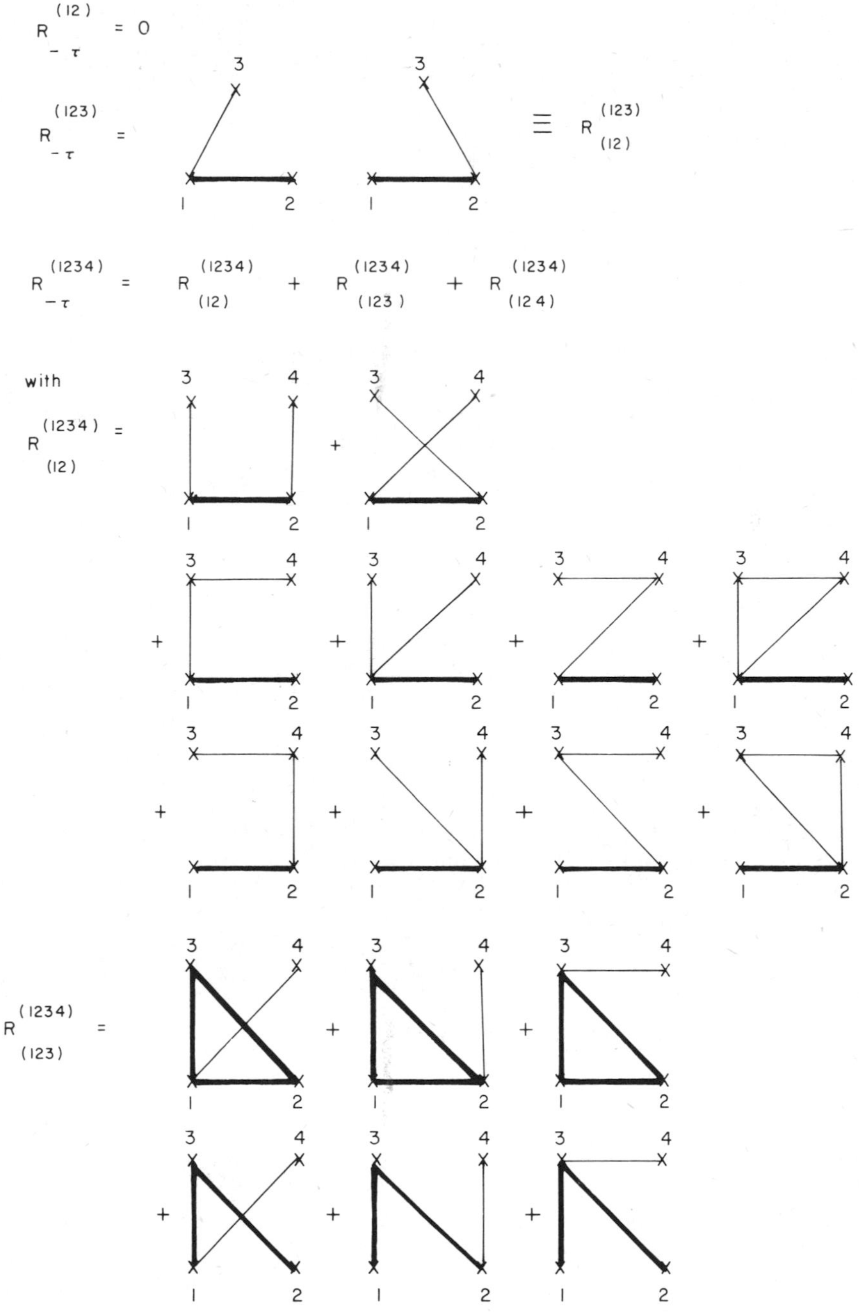

$$R_{-\tau}^{(12)} = 0$$

$$R_{-\tau}^{(123)} = \quad \equiv \quad R_{(12)}^{(123)}$$

$$R_{-\tau}^{(1234)} = R_{(12)}^{(1234)} + R_{(123)}^{(1234)} + R_{(124)}^{(1234)}$$

with

$$R_{(12)}^{(1234)} =$$

$$R_{(123)}^{(1234)} =$$

FIG. 10   1,2-reducible Husimi trees.

The formulae of Fig. 10 may be generalized as

$$R_{-\tau}^{(12\,\cdots\,n)} = \sum_{n'=2}^{n-1} \sum_{\text{lab}} R_{\{1\,2\,\cdots\,n'\}}^{\{1\,2\,\cdots\,n\}}$$

(3.22)

and the sum over the labelling takes into account that there are $(n-2)!/(n'-2)!(n-n')!$ ways to select the particles 1,2 and $n'-2$ others among the $n-2$ particles differing from 1 and 2.

We now give two very important rules:

Rule 1: We decompose $R_{\{1\,2\,\cdots\,n'\}}^{(1\,2\,\cdots\,n)}$ according to the topology of the 1,2-chain on which it is built. For example, $R_{\{1\,2\,3\}}^{\{1\,2\,3\,4\}}$ contains terms built on $\mathscr{V}_{-\tau}^{(1\,2\,3)}$ or on $\mathscr{V}_{-\tau}^{(1\,3)}\mathscr{V}_{-\tau}^{(2\,3)}$. We remark that for any topology the number of terms is the same. This comes from the fact that any 1,2-chain built on the particles $1,2,\ldots n'$ may be 'decorated' by the appending parts in the same way. It implies that, if we want to obtain $R_{\{1\,2\,\cdots\,n'\}}^{\{1\,2\,\cdots\,n\}}$ we may decorate the *sum* $B_{-\tau}^{(1\,2\,\cdots\,n')}$ of all 1,2-irreducible Husimi trees with n' points instead of decorating separately each 1,2-irreducible Humsimi tree.

Rule 2: It concerns specifically the decoration by the appending parts. The points which connect the 1,2-chain to the appending parts are called *articulation points*. Consider one of the appending parts: it is attached to the 1,2-chain through the point l and contains, apart from l, m_l other particles ($m_l \leqslant n-n'$). For example, in $R_{\{1\,2\}}^{\{1\,2\,3\,4\}}$ the appending part attached to the 1,2-chain $\mathscr{V}_{-\tau}^{(1\,2)}$ through the articulation point 1 and containing the particles 1, 3 and 4 may be $\mathscr{V}_{-\tau}^{(1\,3\,4)}$ or $\mathscr{V}_{-\tau}^{(1\,3)}\mathscr{V}_{-\tau}^{(3\,4)}$ or $\mathscr{V}_{-\tau}^{(1\,4)}\,\mathscr{V}_{-\tau}^{(3\,4)}$ or $\mathscr{V}_{-\tau}^{(1\,3)(1\,4)}$ whose sum is simply $\mathscr{U}_{-\tau}^{(1\,3\,4)}$. In general, to add the appending part containing m_l particles plus the articulation point 1 is thus equivalent to multiplication by the operator $\mathscr{U}_{-\tau}^{(m_l,\,l)}$

We may write down explicitly the consequences of these two rules and obtain in this way the form of the 1,2-reducible Husimi trees:

$$R_{\{1\,2\}}^{\{1\,2\,3\}} = B_{-\tau}^{(1\,2)}[\mathscr{U}_{-\tau}^{(1\,3)} + \mathscr{U}_{-\tau}^{(2'\,3)}]$$

$$R_{\{1\,2\}}^{\{1\,2\,3\,4\}} = B_{-\tau}^{(1\,2)}[\mathscr{U}_{-\tau}^{(1\,3)}\mathscr{U}_{-\tau}^{(2\,4)} + \mathscr{U}_{-\tau}^{(1\,4)}\mathscr{U}_{-\tau}^{(2\,3)} + \mathscr{U}_{-\tau}^{(1\,3\,4)} + \mathscr{U}_{-\tau}^{(2\,3\,4)}]$$

$$R_{\{1\,2\,3\}}^{\{1\,2\,3\,4\}} = B_{-\tau}^{(1\,2\,3)}[\mathscr{U}_{-\tau}^{(1\,4)} + \mathscr{U}_{-\tau}^{(2\,4)} + \mathscr{U}_{-\tau}^{(3\,4)}]$$

(3.23)

and a similar expression for $R_{\{1\,2\,4\}}^{\{1\,2\,3\,4\}}$.

We are faced with the following situation: in equation (3.21) $U_{-\tau}^{(12\,\cdots\,n)}$ is expressed in terms of $B_{-\tau}^{(12\,\cdots\,n)}$ and of $R_{-\tau}^{(12\,\cdots\,n)}$. This last quantity, through (3.22) and (3.23), is given in terms of $B_{-\tau}^{(12\,\cdots\,n')}$ and of $U_{-\tau}^{(l,m\,l)}$. If one eliminates $R_{-\tau}^{(12\,\cdots\,n)}$ or $R_{\{1\,2\,\cdots\,n'\}}^{(12\,\cdots\,n)}$ from these three equations, one gets expressions of $B_{-\tau}^{(12\,\cdots\,n)}$ as function of $U_{-\tau}^{(l,m\,l)}$ quantities. Since these expressions have to be put in the integral (3.20), all the particles, except 1 and 2, play the same role and all the terms of $B_{-\tau}^{(12\,\cdots\,n)}$ which differ merely by the interchange of the dummy particles $3,4\ldots n$ give the same contribution. Taking this remark into account, the general expression for $R_{(12\ldots n')}^{(12\ldots n)}$ may be given:

$$R_{\{12\ldots n'\}}^{\{12\ldots n\}} = B_{-\tau}^{(12\ldots n')}\, m!\, \sum_{\{m_k\}}'\ \prod_{k=1}^{n'} \frac{\mathscr{U}_{-\tau}^{(k,\,m\,k)}}{m_k!} \tag{3.24}$$

where the combinatorial factor $m!/\pi m_k!$ counts the number of equivalent contributions that may be obtained by interchange of the dummy particles whereas the symbol $\Sigma'\{m_k\}$ indicates that one has to divide in all possible ways the $m = n - n'$ particles which do not appear in $B_{-\tau}^{(12\ldots n')}$ into n' groups (some of them being eventually empty).

In order to understand the relation between the equilibrium irreducibility and the dynamical irreducibility we must introduce the wave vector language of the Prigogine formalism into the expressions derived by Cohen. Using the notation (3.6), one gets from (3.20):

$$\left(\frac{\partial\varphi}{\partial t}\right)^{[n]} = \frac{(-i)N^{n-1}}{(n-2)!}\ \lim_{\tau\to\infty}\ \int dp_2 \ldots \int dp_n\ \langle 0|\,\delta L^{(12)}B_{-\tau}^{(12\ldots n)}\,|0\rangle$$

$$\prod_{i=1}^{n} \varphi^{(1)}(v_i,t) \tag{3.25}$$

the right hand side of which derives from (3.21):

$$\int dp_2 \ldots \int dp_n\ \langle 0\,|\,\delta L^{(12)}B_{-\tau}^{(12\ldots n)}\,|0\rangle$$

$$= \int dp_2 \ldots dp_n\ \Big[\langle 0\,|\,\delta L^{(12)}\mathscr{U}_{-\tau}^{(12\ldots n)}\,|0\rangle$$

$$- \sum_{n'=2}^{n-1} \frac{(n-2)!}{(n'-2)!(n-n')!}\,\langle 0|\,\delta L^{(12)}R_{\{12\ldots n'\}}^{\{12\ldots n\}}\,|0\rangle\Big] \tag{3.26}$$

In (3.26) the combinatorial factor comes from the fact that the dummy particles play again the same role (see the arguments following [3.22]). We have still to calculate the last term of this expression:

$$\int dp_2 \ldots \int dp_n\ \langle 0|\,\delta L^{(12)}R_{\{12\ldots n'\}}^{\{12\ldots n\}}\,|0\rangle$$

$$= \sum_{u,v,w,s} \int dp_2 \ldots \int dp_n\ \langle 0|\,\delta L^{(12)}B_{-\tau}^{(12\ldots n')}\,|u\rangle$$

$$\sum_{\{m_k\}} m!\ \frac{\langle u\,|\mathscr{U}_{-\tau}^{(1,m_1)}\,|v\rangle\langle v\,|\mathscr{U}_{-\tau}^{(2,m_2)}\,|w\rangle\ldots\langle s\,|\mathscr{U}_{-\tau}^{(l,m_l)}\,|0\rangle}{\prod_{k=1}^{n'} m_k!}$$

$$\tag{3.27}$$

The two equations (3.26) and (3.27) allow us to obtain any $\int dp_2 \ldots \int dp_n$ $\langle 0 | \delta L^{(12)} B_{-\tau}^{(12\cdots n)} | 0 \rangle$ in terms of $\int dp_2 \ldots \int dp_n \langle u | \mathcal{U}_{-\tau}^{(l,ml)} | v \rangle$ factors. We must notice that, in principle, every wave vector u, v, w, s, etc... appears in the product of $\mathcal{U}_{-\tau}^{(l,ml)}$ of (3.27). However, because of the fact that each $\mathcal{U}_{-\tau}^{(l,ml)}$ factor has only particle l in common with $B_{-\tau}^{(12\cdots n')}$ and of the fact that any pair of $\mathcal{U}_{-\tau}^{(l,ml)}$ factors has no particle in common, the only term in the sum over the wave vectors that remains is the term $u = v = w = \ldots = s = 0$. This property is a consequence of both the translational invariance of the system and the concept of articulation point. It allows us to write:

$$\int dp_2 \ldots \int dp_n \langle 0 | \delta L^{(12)} R\{^{1\,2}_{1\,2} \cdots ^n_{n'}\} | 0 \rangle$$

$$= \int dp_2 \ldots \ dp_n \langle 0 | \delta L^{(12)} B_{-\tau}^{(12\cdots n')} | 0 \rangle \, m!$$

$$\sum_{\{m_k\}} \prod_{k=1}^{n'} \frac{\langle 0 | \mathcal{U}_{-\tau}^{(k,m\,k)} | 0 \rangle}{m_k!} \qquad (3.28)$$

The physical meaning of this result is double. In equilibrium language we know that the left hand side of (3.26) is 1,2-irreducible. Therefore, from the first term of the right hand side of (3.26) which is not, we have to subtract the 1,2 reducible contribution. In non-equilibrium language, we know that the left hand side is a function of an operator which satisfies the dynamical irreducibility condition [see (3.7)]. The first term of (3.26) does not satisfy this requirement. Therefore, one has to subtract from it contributions which are not only 1,2 reducible in the equilibrium sense but also *dynamically reducible*. This only indicates what happens. We have still to demonstrate that the building up of $\Omega\psi$ in equation (3.28) is equivalent to the subtraction of the 1,2-reducible contribution. This point is dealt with below.

We start from the definition (3.9) and from equations (3.4) and (3.5) which imply:

$$\langle 0 | S_{-\tau}^N | 0 \rangle = - \frac{1}{2\pi i} \oint_c dz \, e^{-iz\tau} \langle 01 | \frac{1}{L_N - z} | 0 \rangle \qquad (3.29)$$

and, if one uses the perturbation expansion of $(L_N - z)^{-1}$ and the definition of $\psi(z)$ one gets:

$$\langle 0 | S_{-\tau}^N | 0 \rangle = \frac{1}{2\pi i} \oint_c dz \, e^{-iz\tau} \frac{1}{z + \psi(z)}$$

If one is interested in the long time behaviour, then one expects that the only zero of $z + \psi(z)$ that contributes significantly to the integral is:[21]

$$z_0 = -\psi(z_0) = -\Omega\psi$$

therefore, one obtains:

$$\lim_{\tau \to \infty} \langle 0 | S_{-\tau}^N | 0 \rangle = \lim_{\tau \to \infty} \left\{ \frac{e^{-iz\tau}}{1 + \dfrac{\partial \psi(z)}{\partial z}} \right\}_{z=z_0}$$

and, after taking the logarithmic derivative of both sides with respect to τ we have for $\tau \to \infty$

$$\langle 0 | \tau_\tau S_{-\tau}^N | 0 \rangle \langle 0 | S_{-\tau}^N | 0 \rangle^{-1} = -iz_0 = i\Omega\psi$$

Using now the Liouville equation (3.1) and multiplying both sides to the right by $\langle 0 | S_{-\tau}^N | 0 \rangle$ we get finally;

$$\lim_{\tau \to \infty} \lambda \langle 0 | \delta L_N S_{-\tau}^N | 0 \rangle = -\Omega\psi \lim_{\tau \to \infty} \langle 0 | S_{-\tau}^N | 0 \rangle \qquad (3.30)$$

It is immediately seen that the structure of this equation is very similar to the one of (3.27): we have to prove the equivalence of a part of $\Omega\psi$ and $\langle 0 | \delta L^{(12)} B_{-\tau}^{(12 \cdots n)} | 0 \rangle$ which play indeed an equivalent role in the two equations. Let us now make this statement more precise.

We call $(\Omega\psi)^{(12 \cdots n)}$ the part of $\Omega\psi$ involving the particles $1, 2 \ldots, n$. It may be shown[2] that $(\Omega\psi)^{(12 \cdots n)}$ is proportional to V^{1-n} and, therefore, any integral $\int dp_2 \ldots \int dp_n (\Omega\psi)^{(12 \cdots n)}$ provides the contributions of $\Omega\psi$ which are proportional to c^{n-1}; in this sense $(\Omega\psi)^{(12 \cdots n)}$ are the *virial coefficients* of $\Omega\psi$.

According to the equations (3.8) and (3.19), we may write:

$$\left(\frac{\partial \varphi}{\partial t} \right)^{[n]} = iN^{n-1} \int dp_2 \ldots \int dp_n (\Omega\psi)^{(12 \cdots n)} \prod_{i=1}^{n} \varphi_{(v_i, t)}^{(1)} \qquad (3.31)$$

because of the factorization of the N particle velocity distribution function[19] and of the equivalent role of the dummy particles.

We multiply both sides of (3.30) by

$$\prod_{i=1}^{N} \varphi^{(1)}(v_i, t)$$

and we integrate over all momenta except p_1. We then extract the contributions proportional to V^{1-n} of each member. We get:

$$-i\lambda \int dp_2 \ldots dp_N \langle 0 | \delta L_N S_{-\tau}^N | 0 \rangle^{[n]} \prod_{i=1}^{N} \varphi^{(1)}(v_i, t)$$

$$= i \sum_{n'=2}^{n} N^{n'-1} \int dp_2 \ldots \int dp_N (\Omega\psi)^{(12 \cdots n')}$$

$$\langle 0 | S_{-\tau}^N | 0 \rangle^{[n-n'+1]} \prod_{i=1}^{N} \varphi^{(1)}(v_i, t) \qquad (3.32)$$

where $A^{[n]}$ denotes that we extract the contributions of order V^{1-n} from the quantity A.

We now analyse separately both sides of (3.32) and first the left hand one:

$$\Gamma_n = -i\lambda \int dp_2 \ldots \int dp_N \langle 0 | \delta L_N S_{-\tau}^N | 0 \rangle^{[n]} \prod_{i=1}^{N} \varphi^{(1)}(v_i,t)$$

where we insert the non-equilibrium Ursell expansion (3.18). But δL_N has to contain the only non dummy particle, i.e. particle 1 and a dummy particle, say 2. It is easy to show[4] that the only non vanishing term of (3.18) has to contain both particles 1 and 2 in the same $\mathcal{U}_{-\tau}$ operator and, among those,

$$\Gamma_n = -i\lambda(N-1) \frac{(N-2)!}{(N-n)!(n-2)!} \int dp_2 \ldots \int dp_n$$

$$\langle 0 | \delta L^{12} \mathcal{U} \binom{12}{-\tau} \cdots n) | 0 \rangle \prod_{i=1}^{n} \varphi^{(1)}(v_i,t)$$

contributes effectively to the order V^{1-n}. The origin of the combinatorial factor is the same as in (3.24).

We now analyse the right hand side of (3.32). To do that we use the cluster expansion (3.18) again in order to explicit $\langle 0|S_\tau^n|0\rangle$. It is readily shown[4] that each factor of the $\mathcal{U}_{-\tau}$ product has to contain at least one of the particles $1, 2, \ldots n'$ of the $(\Omega\psi)^{(12\ldots n')}$ factor, otherwise it vanishes. On the other hand, if one $\mathcal{U}_{-\tau}$ factor contains more than one particle, the contribution where it appears is negligible in the thermodynamic limit $N \to \infty$, $V \to \infty$, N/V finite. We obtain for the right hand side of (3.32):

$$\bar{\Gamma}_n = \sum_{n'=2}^{n} N^{n'-1} \int dp_2 \ldots \int dp_n (i\Omega\psi)^{(12\ldots n')}$$

$$\times \sum_{\{m_k\}}' \frac{(N-n)!}{(N-n'-m)! \prod_{k=1}^{n'} m_k!} \prod_{k=1}^{n'} \langle 0 | \mathcal{U} \binom{k, m_k}{-\tau} | 0 \rangle$$

$$\prod_{i=1}^{n} \varphi^{(1)}(v_i,t)$$

It is seen that the virial expansion of the product $(\Omega\psi)\langle 0|S_{-\tau}^N 0\rangle$ is equivalent to the procedure of decorating the 1,2-irreducible chain $B^{(12\ldots n')}$ with the appending parts $\mathcal{U}_{-\tau}^{(k, m_k)}$ [see equations (3.26) and (3.27)]. This rather unexpected property explains why it is possible, in this particular case, to use the equilibrium language to describe non-equilibrium situations. The equality of

Γ_n and $\bar{\Gamma}_n$ implies:

$$\frac{1}{(n-2)!} \int dp_2 \ldots \int dp_n \langle 0 \mid \delta L^{(12)} \, \mathscr{U} \, {\textstyle \binom{12}{-\tau}} \ldots n) \mid 0 \rangle \prod_{i=1}^{n} \varphi^{(1)}(v_i,t)$$

$$= \int dp_2 \ldots \int dp_n \left[(\Omega\psi)^{(12 \ldots n)} + \right.$$

$$\left. \sum_{n'=2}^{n-1} (\Omega\psi)^{(12 \ldots n')} \sum_{\{m_k\}}' \sum_{k=1}^{n} \langle 0 \mid \mathscr{U} \, {\textstyle_{-\tau}^{(k,m}}_k) \mid 0 \rangle \right] \prod_{i=1}^{n} \varphi^{(1)}(v_i,t)$$

which parallels completely the equations (3.26) and (3.27) and shows that

$$\lim_{\tau \to \infty} \frac{\lambda}{(n-2)!} \int dp_2 \ldots \int dp_n \langle 0 \mid \delta L^{(12)} B_{-\tau}^{(12} \ldots n) \mid 0 \rangle \prod_{i=1}^{n} \varphi^{(1)}(v_i,t)$$

$$= - \int dp_2 \ldots \int dp_n (\Omega\psi)^{(12 \ldots n)} \prod_{i=1}^{n} \varphi^{(1)}(v_i,t)$$

If this last equality is compared to the relations (3.25) and (3.31) which express the virial expansion of the time evolution of $\varphi(1)(v_1,t)$ respectively in Cohen's formalism and in the Prigogine master equation theory, then it becomes obvious that these two formulations are equivalent.

We may close this section by concluding that the equilibrium formulations are usually not adapted to the study of non-equilibrium evolution. If one wants to extend equilibrium concepts out of equilibrium one has first to test the validity of the translation. This might be a very useful approach, since it allows the use of concise graph theoretical visualizations.

4   CHEMICAL APPLICATIONS

The use of graphs may also provide interesting information in chemical situations. In particular, they may serve to appreciate the effect of the non-ideality of the gas phase on the chemical equilibrium or on the reaction rates of imperfect gases. We will first examine the chemical equilibrium and we will turn to the reaction rates later.

4.1   Chemical equilibrium

The condition for chemical equilibrium is known to be

$$\sum_i \nu_i \mu_i = 0 \tag{4.1}$$

where ν_i is the stoichiometric coefficient of species i and μ_i its chemical potential. The bridge between this quantity and the microscopic properties of the system is provided by the formula[13]

$$-\frac{\mu_i}{kT} = \left(\frac{\partial}{\partial N_i} \ln \bar{Q}_{\{N_i\}}\right)_{V,T,N_1 \ldots N_{i-1}, N_{i+1} \ldots}$$ (4.2)

where the partition function $\bar{Q}_{\{N_i\}}$ refers to a system containing N_1 particles of species 1, N_2 of species 2, etc. . . . . and the bar indicates that each species has also internal degrees of freedom.

If we consider a system with only one kind of monoatomic particles, this formula reduces to:

$$-\frac{\mu}{kT} = \left(\frac{\partial \ln Q_N}{\partial N}\right)_{V,T}$$ (4.3)

If the system is a perfect gas:

$$Q_N = \frac{V^N}{N!\left(\dfrac{h^2}{2\pi m kT}\right)^{\frac{3N}{2}}}$$ (4.4)

(see (2.5), (2.6) and (2.7) and the chemical potential is simply

$$\frac{\mu}{kT} = \ln \frac{h^3}{(2\pi m kT)^{3/2}} + \ln \rho$$ (4.5)

If we are interested in the effect of the imperfections, the density ρ must be replaced by the activity z. The statistical expression for this quantity has been given in (2.28) where the first term refers to the perfect gas situation and the others are due to imperfections. In order to obtain μ as a function of microscopic quantities (intermolecular forces) one could replace the a_n of (2.28) by their expressions in terms of β_l, using the same scheme as for (2.30). A simpler way is provided by the formula $F = pV - \mu N$ where p and F are known through (2.30) and (2.36).

Of course, in chemical situations, one has to allow for the existence of internal degrees of freedom and of many molecular species. In this case, for a perfect gas, formula (4.5) must be modified in the following way:

$$\frac{\mu_i}{kT} = \ln \frac{h^3}{(2\pi m_i kT)Z_i^{\text{int}}} + \ln \rho_i$$ (4.6)

where subscript i refers to the ith molecular species and Z_i^{int} is the one particle partition function for internal degrees of freedom. We do not make its expression explicit here; a detailed discussion of this problem may be found in textbooks on statistical thermodynamics (see for example Münster[18]).

If one wants to take the imperfections into account, one replaces ρ_i by z_i; the activity of i; but the relation between the activity and the microscopic properties becomes more complicated than in the one component case. The solution of the problem when two species i and j are present has been given by Mayer[16]. We reproduce his result here, having simply added the effect of the internal degrees of freedom.

$$\frac{\mu_i}{kT} = \ln \frac{h^3 \rho_i}{(2\pi m_i kT)^{3/2} Z_i^{int}} - \frac{1}{x_i} \sum_{k=0}^{\infty} \sum_{\kappa=0}^{\infty} k\beta_{k,\kappa} x_i^k x_j^\kappa \frac{1}{v^{k+\kappa-1}}$$

$$\frac{\mu_j}{kT} = \frac{\ln h^3 \rho_i}{(2\pi m_j kT)^{3/2} Z_j^{int}} - \frac{1}{x_j} \sum_{k=0}^{\infty} \sum_{\kappa=0}^{\infty} \kappa\beta_{k,\kappa} x_i^k x_j^\kappa \frac{1}{v^{k+\kappa-1}} \qquad (4.7)$$

where x_i is the mole fraction of species i, and v is the volume per molecule in the system. In this expression $\beta_{k,\kappa}$ refers to the irreducible cluster integral with k particles of species i and κ of species j. In particular $\beta_{k,0}$ and $\beta_{0,k}$ are identical in form to β_{k-1}/k defined in equation (2.17).

The problem of writing down the contributions corresponding to $\beta_{k,\kappa}$ becomes more difficult to treat, because two kinds of particles must be disposed on the graph. For example, the three graphs for $\beta_{2,2}$ correspond to the *same* contributions if particles 1, 2, 3 and 4 are of the same kind but to two different contributions if 1 and 2 are of kind i and 3 and 4 of kind j. This is shown in Fig. 11 where different lines have been used according to whether they are of the type ii, ij or jj.

FIG. 11   Irreducible graphs with two different particles.

The two equations (4.7) express the chemical potential in terms of imperfections due to molecular interactions. Through graphical representations it is possible in principle to determine the magnitude of these effects on the chemical potential. At high pressure, as for example in the ammonia synthesis, the chemical potential may be modified in a very important way[8] and, as a result, the equilibrium concentrations are different from the ones that should be computed from the first term of (4.7) alone.

4.2   Reaction rates

The same type of treatment has been applied to the study of bimolar reaction rates.[9] The authors were interested in the effect of the non ideality of the gas on the initial rate and we give a brief account of their physical cluster approach to this problem. In this approximation the effect is only due to the reactants A and B and not to the products, the concentrations of which are assumed to be small. The reaction scheme is supposed to be the following:

$$A + B \xrightarrow{k_1} C + D$$
$$A_2 + B \xrightarrow{k_2} C + D + A$$
$$A + AB \xrightarrow{k_3} C + D + A$$
$$A + B_2 \xrightarrow{k_4} C + D + B$$
$$AB + B \xrightarrow{k_5} C + D + B$$
$$A + A + B \xrightarrow{k_6} C + D + A$$
$$A + B + B \xrightarrow{k_7} C + D + B \tag{4.8}$$

but the same ideas could be used for more complicated schemes. However, higher order corrections have to be introduced if steps of higher order than termolecular are permitted.

The assumption underlying this description is that we have physical interaction between A and B before they 'react' chemically. The physical interaction changes the effective number of molecules available for the reaction. This implies that the reaction distance is smaller than the distance for clustering. This point of view has also implicitly been adopted in the treatment of the effect of the imperfections on equilibrium properties.

We may write the rate equation for the evolution of ρ_A, the total density of A:

$$-\frac{d\rho_A}{dt} = k_1 n_A n_B + k_2 n_{A_2} n_B + k_3 n_A n_{AB} \tag{4.9}$$

where n_A, n_{AB} etc. represent the number density of monomer A, dimer AB etc. .... The relation between the total concentrations and the number densities of the cluster species are:

$$\rho_A = n_A + 2n_{A_2} + n_{AB} + \ldots \tag{4.10}$$

The grand canonical partition function (2.22) may be generalized to the case where there are various species.[13]

$$\Theta = 1 + Q_A \lambda_A + Q_B \lambda_B + Q_{A_2} \lambda_{A_2} + Q_{2A} \lambda_A^2$$
$$+ Q_{AB} \lambda_{AB} + Q_{A+B} \lambda_A \lambda_B + Q_{B_2} \lambda_{B_2} + Q_{2B} \lambda_B^2 + \ldots \tag{4.11}$$

if one keeps terms up to second order in the activities λ of monomers and up to first order in the activity of dimers. This is consistent since, because of the equilibrium condition (4.1) between monomers and dimers and the relation between activity and chemical potential $\lambda = e^{\mu/kT}$, one has:

$$\lambda_{A2} = \lambda_A^2 \qquad \lambda_{AB} = \lambda_A \lambda_B \qquad \lambda_{B2} = \lambda_B^2 \tag{4.12}$$

In (4.11), Q_A means the canonical partition function for one monomer A in the volume V, Q_{AB} is for one AB dimer, Q_{A_2} for one A_2 dimer, Q_{2A} or Q_{A+B} for two A monomers or for one monomer A and one B, etc. . . . .

The microscopic expression for the number density of each species follows from

$$n_A = \frac{\lambda_A}{V} \left(\frac{\partial \ln \Theta}{\partial \lambda_A} \right)_{T,V,\lambda_B,\lambda_{AB} \cdots} \tag{4.13}$$

and yields

$$V n_A = Q_A \lambda_A + (2Q_{2A} - Q_A^2)\lambda_A^2 + (Q_{A+B} - Q_A Q_B)\lambda_A \lambda_B + \ldots \tag{4.14}$$

and similar expressions for $n_B; n_{A2}; n_{AB}$ etc.

Using (4.10), it is easy to get

$$V\rho_A = Q_A \lambda_A + (2Q_{A_2} + 2Q_{2A} - Q_A^2)\lambda_A^2 + (Q_{AB} + Q_{A+B} - Q_A Q_B)\lambda_A \lambda_B \tag{4.15}$$

which, because of the use of (4.12) contains only λ_A and λ_B. Since one has a similar expression for ρ_B, it is possible to invert these equations and to express λ_A and λ_B as functions of ρ_A and ρ_B; the result is given below for the activity of A

$$\lambda_A = \frac{V}{Q_A} \rho_A + \frac{V^2}{Q_A^3} (Q_A^2 - 2Q_{A_2})\rho_A^2$$

$$+ \frac{V^2}{Q_A^2 Q_B} (Q_A Q_B - Q_{AB} - Q_{A+B})\rho_A \rho_B + \ldots$$

This may be introduced into the set of equations (4.14) and, as a result, one obtains n_A, n_B, n_{AB}, etc. . . . as functions of ρ_A and ρ_B:

$$n_A = \rho_A - 2\left(\frac{VQ_{A_2}}{Q_A^2} \right)\rho_A^2 - \left(\frac{VQ_{AB}}{Q_A Q_B} \right)\rho_A \rho_B + \ldots$$

$$n_B = \rho_B - 2\left(\frac{VQ_{B_2}}{Q_B^2} \right)\rho_B^2 - \left(\frac{VQ_{AB}}{Q_A Q_B} \right)\rho_A \rho_B + \ldots \tag{4.16}$$

and finally, (4.9) may be put into the form

$$-\frac{d\rho_A}{dt} = k_1 \rho_A \rho_B (1 + \xi_A \rho_A + \xi_B \rho_B + \ldots) \tag{4.17}$$

where:

$$\xi_A = \frac{V}{Q_A^2 Q_B}\left[\frac{k_2}{k_1}Q_B Q_{A_2} + \frac{k_3}{k_1}Q_A Q_{AB} + \frac{k_6}{k_1}\frac{Q_A^2 Q_B}{V} - 2Q_B Q_{A_2}\right.$$
$$\left. - Q_A Q_{AB}\right] \tag{4.18}$$

and a similar expression for ξ_B. To understand the meaning of this equation let us consider the case where only the first step of (4.8) occurs ($k_2 = k_3 \ldots = k_7 = 0$). The first term of (4.17) represents the rate of step k_1 if there is no cluster formation [$\rho_A = n_A$; $\rho_B = n_B$; see (1.10)]. The two last terms of the factor ξ_A account for the fact that because of the presence of dimers A_2, AB, B_2 the effective numbers of monomers n_A and n_B are smaller than ρ_A and ρ_B respectively.

This physical cluster procedure may be generalized to higher order in ρ_A and ρ_B. This extension may be useful even if only bi- or termolecular steps are taken into account. It cannot be avoided if higher order steps had been considered in (4.8). As a result, one should obtain new expressions for ξ_A where Q factors containing three, four, etc. . . . particles should appear.[13] In this case, of course, one should again use graphs to evaluate the various Q factors.

In Fig. 12 we give as an example the graphical expression for $Q_{2A,B}$ but analogous expressions may be easily obtained for higher order Q factors.

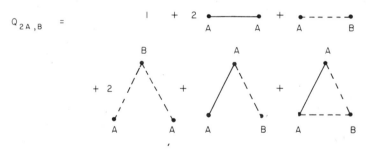

FIG. 12   Graph expression for $Q_{2A,B}$.

5   CONCLUSIONS

We have only presented here some selected applications of graph theory in statistical mechanics. Graphs have been used extensively in various other fields such as solid state physics, plasma physics, theory of electrolytic solution, etc. . . . where they serve to describe both equilibrium and non-equilibrium solutions.

The status of this tool varies widely according to the degree of elaboration of the field where it is used. For example, in the study of the equilibrium properties of a classical gas, reviewed in Section 2, very detailed topological properties are needed if one wants to find consistent approximations such as the hypernetted chain[30] one, or if one wants to have graphical expressions for complicated objects, such as 2,3 . . . -particle correlation functions.[27]

On the other hand, in problems such as the effect of imperfections on equilibrium constants or reaction rates, the situation is much less favourable. The description is so complicated that one is forced to be happy with qualitative or semi-quantitative indications. In this respect, graphs may be a useful tool because they may serve to *classify* the various effects even if one is unable to do more than estimate their order of magnitude. This is, we think, the spirit in which Section 4 should be understood.

A very important point is to know whether or not equilibrium graphs should be used out of equilibrium. They certainly give a part of the answer, since they describe what happens — interaction or no interaction between groups of particles — but they forget about the order of events which, as we have seen in Section 3, is of crucial importance and has, in principle, to be expressed by the dynamical irreducibility condition.[19] Actually, this condition is also of a topological nature, but here the time is involved in the graph. However, even with this restriction, it is possible to formulate a theory of the approach to equilibrium which uses the language of equilibrium graphs. The advantage of this possibility is that it gives a topological rule for the generalization of the Boltzmann equation. In other cases the use of equilibrium concepts may be more dangerous. For example, the translation of (2.32) out of equilibrium gives (3.10). Since M_n does not depend on kT, one should expect that $\tilde{M}_n$ should not depend on the time. This statement is not true.[3] It is a striking example of the care that is needed for this type of analogy.

We hope that we have indicated by these examples that graphs, which are so fascinating for mathematicians, are also of great interest to physical chemists, even for some practical aspects of their activity.

ACKNOWLEDGEMENTS

This work was elaborated in the department of Professor I. Prigogine who suggested several of the ideas presented here. I would like to thank him for his interest and his encouragement.

REFERENCES

1. Boltzmann, L. (1910). "Vorlesungen über Gastheorie", Liepzig, Barth.
2. Brocas, J. (1967). *In*: "Advances in Chemical Physics", vol. XI, Interscience, New York.
3. Brocas, J. and George, C. (1967). *Physica*, **34**, 515.

4. Brocas, J. and Résibois, P. (1966). *Physica,* 32, 1050.
5. Brout, R. and Carruthers, P. (1963). "Lectures on the Many Electron Problem", Interscience, New York.
6. Carruthers, P. and Dy, K. S. (1966). *Phys. Rev.,* 147, 214.
7. Cohen, E. G. D. (1962). *Physica,* 28, 1025.
8. Denbigh, K. (1961). "The Principles of Chemical Equilibrium", Cambridge Univ. Press.
9. Emptage, M. R. and Ross, J. (1969). *J. Chem. Phys.,* 51, 252.
10. Ernst, M. H., Haines, L. K. and Dorfman, J. R. (1969). *Rev. Mod. Phys.,* 41, 296.
11. Gielen, M. (1971). Chapter 9, this volume.
12. Harary, F. and Uhlenbeck, G. E. (1953). *Proc. Nat. Acad. Sci. U.S.A.,* 39, 315.
13. Hill, T. (1956). "Statistical Mechanics", McGraw Hill, New York.
14. Huang, K. (1965). "Statistical Mechanics", John Wiley, New York.
15. Kahn, B. (1938). Dissertation (Utrecht).
16. Mayer, J. E. (1939). *J. Chem. Phys.,* 43, 71.
17. Mayer, J. E. and Mayer, M. G. (1940). "Statistical Mechanics", John Wiley, New York.
18. Münster, A. (1969). "Statistical Thermodynamics", Springer Verlag, Academic Press.
19. Prigogine, I. (1962). "Non-equilibrium Statistical Mechanics", Interscience, New York.
20. Prigogine, I. and Balescu, R. (1959). *Physica,* 25, 281 and 302.
21. Prigogine, I. and Résibois, P. (1961). *Physica,* 27, 693.
22. Prigogine, I., George, C. and Henin, F. (1969). *Physica,* 45, 418, and references cited therein.
23. Résibois, P. (1963). *J. Math. Phys.,* 4, 166.
24. Résibois, P. (1965). *In* "Physics of Many Particles Systems", (Meeron, editor). Gordon and Breach, New York.
25. Schweber, S. S., de Hoffman, F. and Bethe, H. A. (1956). "Mesons and Fields", vol. I, 194. Row, Peterson and Co., New York.
26. Tolman, R. C. (1938). "The Principles of Statistical Mechanics", Oxford, London.
27. Uhlenbeck, G. E. and Ford, G. (1962). "Theory of Linear Graphs" in "studies in Statistical Mechanics", vol. I, North Holland, Amsterdam.
28. Van Hove, L. (1955). *Physica,* 21, 517.
29. Van Hove, L. (1957). *Physica,* 23, 441.
30. Van Leeuwen, J. M. J., Groeneveld, J. and de Boer, J. (1959). *Physica,* 25, 792.

NOTE ADDED IN PROOF

In this note we want to give a brief account of some recent work related to graph theory for intermolecular interactions in chemical physics. The following examples are far from exhaustive but illustrate merely some typical trends in this domain.

A. The non-equilibrium properties and, more specifically, the transport properties of the so-called van der Waals fluid have been extensively investigated recently. Such fluids are characterized by an intermolecular potential

$$v_{ij} = v^s(|r_i - r_j|) + \gamma^3 v^L(\gamma|r_i - r_j|)$$

where v^s is the short range of the potential and $\gamma^3 v^L$ the long range part. Here γ is the inverse range of v^L and is supposed to be small. Therefore, the properties of the system may be expressed as a contribution coming from v^s plus corrections due to the presence of $\gamma^3 v^L$. Such a theory has been elaborated by P. Résibois, Y. Pomeau and J. Piasecki[3,8] using the Prigogine–Balescu graphs (see for example Prigogine[4]). It rests on a new classification of these graphs and leads to the derivation of the transport properties of van der Waals fluids.

B. In spite of the fact that we only discussed graphs for *classical* equilibrium or non-equilibrium statistical mechanics, we want to report some new results[5] related to *quantum* non-equilibrium statistical mechanics.

This author has shown that the density response function may be expressed as a binary expansion[6,7] or, alternatively, as a linked cluster expansion. Both approaches make use of graph theory and provide an extension of previous work[2] to quantum non-equilibrium statistical mechanics.

C. The use of graphs for intermolecular forces rests on the additivity of the potential energy. It is supposed to be expressible as a sum of binary intermolecular potentials (see equation II.3). In the following expression

$$H = \frac{1}{2m} \sum_i p_i^2 + \sum_{i>j=1}^{N} v_{ij} + \sum_{i>j>k=1}^{N} w_{ijk}$$

such an additivity is no longer assumed. However it is still possible to derive[1] the non-equilibrium properties of a system where the Hamiltonian is of the above form but, in this case, graph theoretical expansions are useless. We wanted to cite this example because it shows an important limitation to the use of graphs in statistical mechanics.

REFERENCES

1. Braun, E. (1972). *Phys. Rev. A*, **5**, (1941).
2. Cohen, E. G. D. (1962), *Physica*, **28**, 1025.
3. Piasecki, J. and Résibois, P. (1973). *J. Math. Phys.*, **14**, 1984.
4. Prigogine, I. (1962). "Non-Equilibrium Statistical Mechanics", Interscience, New York.
5. Reichl, L. E. (1973). *Physica*, **64**, 433, 450.
6. Résibois, P. (1963). *J. Math. Phys.*, **4**, 166.
7. Résibois, P. (1965), *Physica*, **31**, 645.
8. Résibois, P., Pomeau, Y. and Piasecki, J. (1974). *J. Math. Phys.*, **15**, 1238.

CHAPTER 9

Applications of Graph Theory to Organometallic Chemistry

by M. Gielen

Vrije Universiteit Brussel
Fakulteit der Toegepaste Wetenschappen
Dienst: Algemene en Organische Scheikunde
and
Université Libre de Bruxelles
Faculté des Sciences
Collectif de Chimie Organique Physique

1   INTRODUCTION

If one tries to draw a parallel between the behavior of tetrahedral molecules having either carbon or a metal atom as central element, one notices that, for carbon as a chiral center, enantiomers may almost not be converted into one another by nonbond-rupture mechanisms:[63] carbon may thus be characterized

TABLE I

Set of permutations for the tetrahedral system

σ symbols for R	Permutation[14,70,78] transforming R into 1,243 (S)	Permutation type
1,234	(34)	(..)
1,342	(23)	(..)
1,423	(24)	(..)
2,143	(12)	(..)
2,431	(1243)	(....)
2,314	(1234)	(....)
3,412	(1324)	(....)
3,124	(1342)	(....)
3,241	(13)	(..)
4,132	(1432)	(....)
4,321	(1423)	(....)
4,213	(14)	(..)

by its stereochemical stability.[8] For a metal as chiral center, such an isomerization is possible.

The number of isomers **I** obtained for a polytopal complex built up with a central element surrounded by n different ligands is equal to $n!/\sigma$, if σ is the symmetry number, i.e. the order h_r of the rotational subgroup of the point group to which the polytopal complex belongs. For tetrahedral molecules, $n = 4$ and $\sigma = 12$ (for the T_d point group, the proper axes are $E + 8C_3 + 3C_2$);[17] $\mathbf{I} = 4!/12 = 2$: 1,234* (R) and 1,234 (S).[38]

One may find the answer to the question 'how could one isomerize R into S?' by looking at all the possible permutations transforming one of the isomers ML_4 (written under the form of σ different symbols) into the other one (Table I).[37]

One sees that it is possible to go from R to S using two different permutation types.

(a) a transposition[78] or digonal twist D, which gives the same chemical result as
(b) a Hamiltonian circuit H[45] (which is a 'Berry-type' pseudo-rotation)

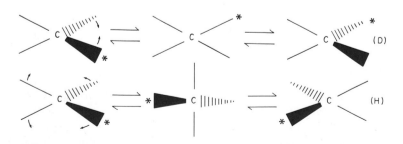

*The ligands are numbered according to Cahn, Ingold and Prelog's chirality rule.[13]

 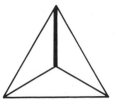

FIG. 1   Possible permutations on tetrahedra shown on typical Schlegel diagrams.

These permutations may be represented on typical Schlegel diagrams[45] by using thick lines to show which ligands are permuted (Fig. 1):

The intermediate obtained for both isomerization mechanisms has a planar structure (and hence a dsp^2 hybridization, which is energetically not feasible for C as central element) with D_{4h} symmetry ($\sigma = 8$). The number of such square planar (SP) isomers is $4!/8 = 3$, i.e.

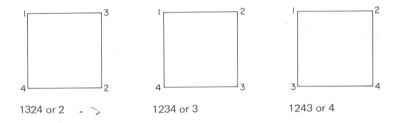

1324 or 2   - ⟩   1234 or 3   1243 or 4

Different representations have been proposed for this rearrangement. Topological graphs[38,65] may be used to describe kinetically[86] how the two interconverting tetrahedral isomers may be transformed into one another (Fig. 2): the vertices represent the elements of the set (isomers) and the edges represent the operations (pseudo-rotations). Analogous topological matrices[79] may be used instead.[38]

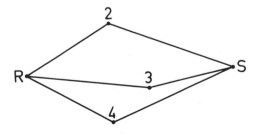

FIG. 2   Topological representation for the transformation $R \rightleftharpoons SP \rightleftharpoons S$. The connectivity δ is equal to 3 for the Td set and to 2 for the D_{4h} set; hence, $I\delta = 6$ for both sets; furthermore, D and H represent odd permutations:[78] the class is closed.[66]

Each coefficient g_{ij} of this general matrix **G** represents the rate constant describing the transformation of isomer i into isomer j. If all the ligands are equivalent but distinguishable, then all the rate constants are identical;[11] g_{ij} may then be set equal to 1 if isomer i may be transformed into isomer j in one step and $g_{ij} = 0$ otherwise

$$
\text{G}(1{,}234) = k \times
\begin{array}{c|ccccc}
 & R & S & 2 & 3 & 4 \\
\hline
R & 0 & 0 & 1 & 1 & 1 \\
S & 0 & 0 & 1 & 1 & 1 \\
2 & 1 & 1 & 0 & 0 & 0 \\
3 & 1 & 1 & 0 & 0 & 0 \\
4 & 1 & 1 & 0 & 0 & 0 \\
\end{array}
$$

If the central element is carbon, the square planar structures cannot be reached easily[8,49,59,63] and one may write $2 = 3 = 4 = 0$. This introduces a chemical constraint which can be put in the form of a constraint matrix **C**, which describes the relationship existing between the isomers of a specific case (here R and S) and the whole set of isomers of the general case

$$
\text{C} =
\begin{array}{c|ccccc}
 & R & S & 2 & 3 & 4 \\
\hline
R & 1 & 0 & 0 & 0 & 0 \\
S & 0 & 1 & 0 & 0 & 0 \\
\end{array}
$$

$\mathbf{C} \times \tilde{\mathbf{C}} = \mathbf{D}$ and is called the degenerating matrix ($\tilde{\mathbf{C}}$ is the transpose of $\mathbf{C}$). Here **D** is a 2 by 2 matrix; $\mathbf{D}^{-1}$ is thus also a 2 by 2 unit matrix.

It may be shown[32,91] that

$$\mathbf{C} \times \mathbf{G}\,(1{,}234) \times \tilde{\mathbf{C}} \times \mathbf{D}^{-1} = \mathbf{G}' \text{ valid, for the specific case.}$$

Here, one has

$$
\begin{array}{c|cc}
 & R & S \\
\hline
R & 0 & 0 \\
S & 0 & 0 \\
\end{array}
$$

This means that R cannot be transformed into S when carbon is the central element because the square planar structures 2,3 and 4 are normally too difficult to reach.[49]

Another type of topological matrix description may be given which takes only the starting and resulting isomers (R and S) into account (reaction matrix **R**)

$$\text{R}(1{,}234) = \begin{array}{c} R \\ S \end{array} \begin{array}{|cc|} R & S \\ 0 & 3 \\ 3 & 0 \end{array}$$

$r_{RS} = 3$ shows that there are three distinct ways to isomerize R into S.

In fact, the four-atom family is so simple that one does not really need such topological descriptions. This is not the case for higher coordination numbers.

2   FIVE-COORDINATE COMPLEXES

2.1   Description of the 12,345 Case

The most stable arrangement taken by five ligands surrounding a central element seems to be the $5,9,6$-D_{3h}-trigonal-bipyramidal one.[54,60] If one defines first the two apiaxial substituents (a) and afterwards the three apiequatorial ones (e)[18] going clockwise when viewed from the first cited apiaxial ligand[12] then it is possible to write $\sigma = 6$ different five-digit symbols to describe one given isomer.[37]

2.1.1   Intramolecular rearrangements

If one gets interested in intramolecular rearrangements of such stereochemically nonrigid ML_n complexes, one may describe the $n!$ permutations which transform a given isomer (represented by the σ possible symbols) into each of the $n!/\sigma$ isomers. One notices then that there is only a limited number of permutation-types (six in the case of the D_{3h} complexes) which define the possible isomerizations. It is thus easy to find among the different isomers those which result from a given chemical rearrangement and to define by the way the connectivity of the process.[37,84]

The results obtained for the five-atom family may be represented on Schlegel-diagrams (Table II)[45] using thick lines to define the permuted ligands or by using the symbol $(L_1 \ldots L_n)$.[37,78,84] The isomere may be symbolized by the two apiaxial substituents;[55,56] $12(12{,}354)$ is the enantiomer of $12(12{,}345)$; the enantiomer which has the highest five-digit symbol is barred.

It is possible to associate an intramolecular movement to a given permutation[37] and to show that the Berry process is energetically the apparently most favorable of these five pseudo-rotations.[46,82]*

---

*A $(2 + 3)$ – turnstile mechanism [(ae)(aee)] has been shown[37,84] to be equivalent to the $(1 + 4)$ – pseudo-rotation mechanism [(aeae)]. Both are odd permutations (the same is true for P3 and for P5); P2 and P4 are even permutations.

TABLE II
Possible permutations of trigonal bipyramids shown on typical Schlegel diagrams

Process	P Permutation used (Schlegel diagrams)	Connectivity	Isomers obtained from 12	Process	Permutation used (Schlegel diagrams)	Connectivity	Isomers obtained from 12
E		1	12	P3	(ae)	6	$\overline{13}$ $\overline{23}$ / 14 $\overline{24}$ / 15 25
P1	(aeae) Berry mechanism	3	$\overline{34}$ / $\overline{35}$ / 45	P4	(aeaee) Hamiltonian circuit	3	34 / 35 / 45
P2	(aee)	6	$\overline{13}$ $\overline{23}$ / $\overline{14}$ 24 / $\overline{15}$ 25	P5	(ee) Digonal twist	1	$\overline{12}$

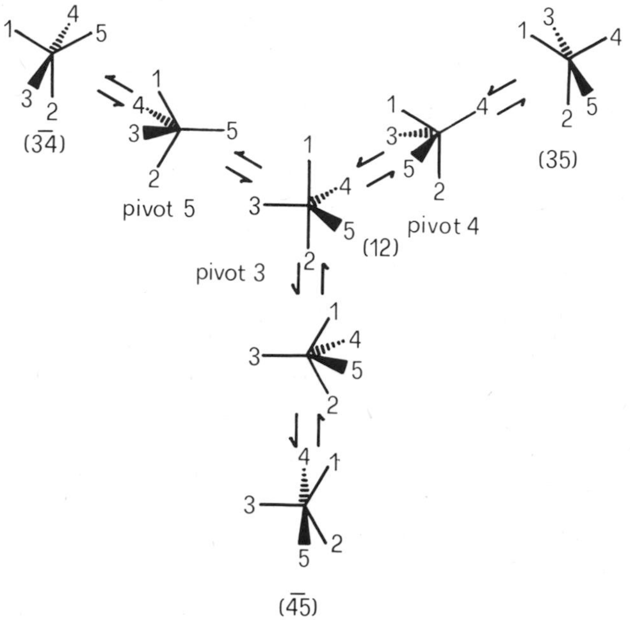

FIG. 3 The Berry-mechanism or (*aeae*) odd cyclic permutation. The class is closed (*Iδ* products in each set are equivalent).[66]

2.1.2 The Berry-process

The intramolecular rearrangement mechanism defined by Berry (Fig. 3) for the five-atom family may thus be described by a cyclic (*aeae*) permutation *P*1 of four ligands: one ligand (named 'pivot'[19,55,56]) remains unchanged during this process. It may be visualized as two synchronized bending motions through which the two apiaxial substituents become apiequatorial and two of the three apiequatorial ones (the pivot excepted) become apiaxial.[22,50] This *P*1 mechanism describing the isomerization of the D_{3h} trigonal-bipyramidal isomers goes through a 5,8,5-C_{4v}-tetragonal pyramidal intermediate (or transition) state.[47,48,53,57,83]

These reactions may be symbolized by

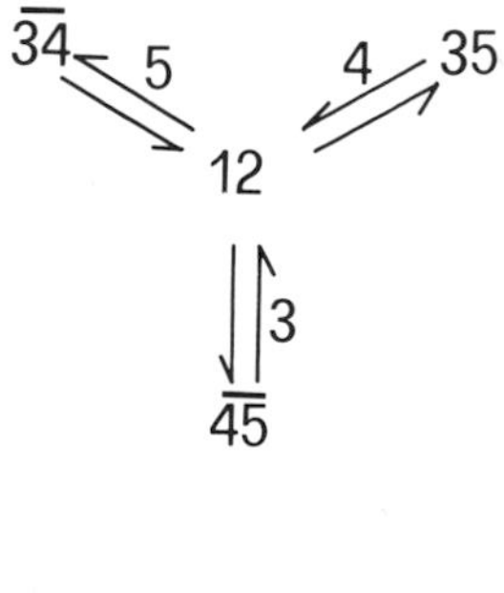

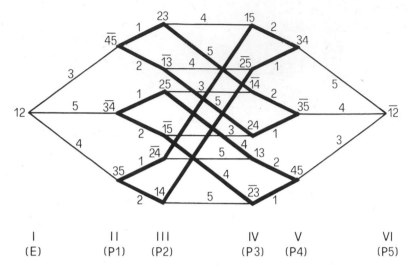

I	II	III	IV	V	VI
(E)	(P1)	(P2)	(P3)	(P4)	(P5)

FIG. 4   Topological representation of the five-atom family: the Desargues–Levi graph projection, the index of the pivot ligand for each pseudorotation being shown alongside the line.[56]

The pivots which may be used to transform a given isomer are the apiequatorial ligands, and thus are the ligands which do not appear in the two-digit symbol; the isomer formed by a Berry pseudo-rotation has, as apiaxial substituents, those which do not appear in the equation (neither the apiaxial substituents of the initial chiromer, nor the apiequatorial pivot). The absolute configurations of two isomers connected through one Berry process are the same if the pivot is even and opposite when the pivot is odd.[40]

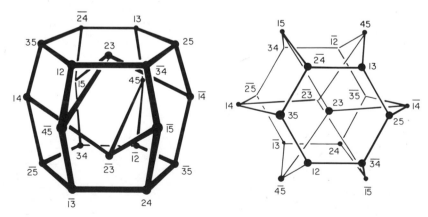

FIG. 5   Three-dimensional topological representation of the five-atom family describing the Berry mechanism $P1$.[28,35,56]

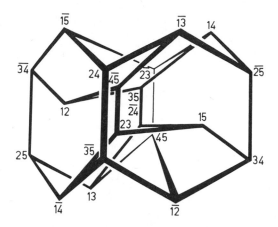

FIG. 6   Three dimensional topological representation of the five-atom family describing the Berry mechanism $P1$.[44]

It is possible to describe the whole set of isomer(ization)s[2,35] by a Desargues–Levi graph (Fig. 4)[4,5,56] and to notice that six families are then obtained. The first one (12) is the starting isomer (E), which may of course be chosen arbitrarily since the chemical system is symmetrical.[56] The second one gives the three isomers obtained by one Berry pseudorotation. The third one defines six isomers which are obtained by two successive $P1$ mechanisms or by one $P2$ type isomerization.[40,41] The fourth family is obtained by three $P1$ steps or by one $P3$. The fifth one defines three isomers which may be obtained either by four $P1$ steps (by two $P2$ steps) or by one $P4$. In order to transform a given isomer into its enantiomer, one needs at least five Berry pseudorotations (or one digonal twist $P5$).

2.1.3   Topological graphical representations

It is interesting to represent all the possible isomerizations on a symmetrical[56] graph for which each *point* stands for an isomer and each *line* connecting two given points stands for an isomerization path. Several isomorphous graphical topological representations of the Berry pseudo-rotations have been described[2,21,55,61] among which are three-dimensional cubic[3] ones[18,19,35] of girth[4] six [6,6,6][26] which belong to the D_{3d} symmetry point group.[28,46,56]*

Two other much more symmetrical graphical representations have been described by Balaban (Fig. 7).[4] On the first one, which contains one evident Hamiltonian circuit, two families of points may however be distinguished, which

---

*It must be noticed that parallel lines describe reaction paths characterized by the same pivot.[19]

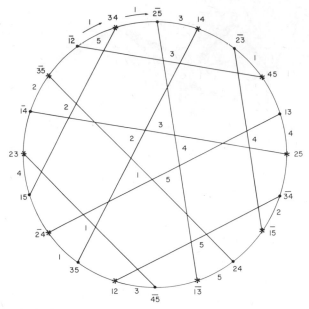

FIG. 7   Topological representation of the Berry pseudorotation describing the Berry-mechanism $P1$.[4]

are defined by two connected ten membered cycles:[65] the first one (*I*) uses alternatingly even and odd pivots to go from x to $\bar{x}$, the other one (II) uses first the odd ones, then the even ones.

The other symmetric graph (Fig. 9) contains two evident connected ten-membered cycles defining the two same families (the outer connected points and the inner ones).

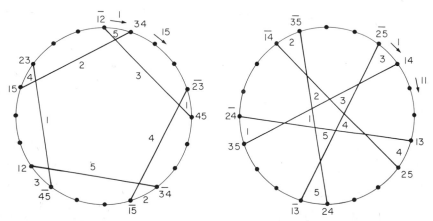

FIG. 8   The two families of isomers found on the Balaban graph; the connected points are resp. 1 and 15, 1 and 11 apart on the Hamiltonian circuit.

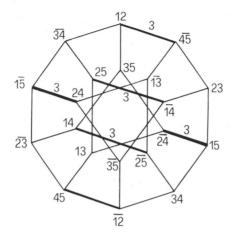

FIG. 9   Topological representation of the Berry pseudo-rotation $P1$.[4]

The midpoint of each line of all these topological graphs represent the C_{4v} intermediate formed in the Berry pseudo-rotation (Fig. 9).

It may also be seen that these topological graphs are alternating ones: a starred isomer is never connected to another starred one: the chromatic number γ is thus equal to 2.[7]

2.1.4   Operator-description
The different isomers of the five-atom family may also be generated by operators,[10] $Q = (12345)^*$ (or any other Hamiltonian circuit on the polytope)[11] and the inversion I ($I[12] = [\overline{12}]$). It may be shown that IQ acting on 13 generates one of the families found on Balaban's graph ($IQ[13]$ or IQ acting on any isomer belonging to the same family generates all the isomers of this family through the Berry mechanism Pl). IQ acting on one of the isomers of the other family generates this family through $P3$ operations. Furthermore, Q acting on an isomer of the same family is equivalent to $P2$ (and only generates one half of this family: for $P2$, the class is open):[67] analogously, Q acting on an isomer of the second family is equivalent to $P4$. I, acting on any isomer, is of course identical to $P5$.

2.1.5   Topological matrix representations
Instead of a graphical representation, an *analogous* matrix representation may be defined.[32,35,67]

---

*Q is an even operator, $I(\equiv P5)$ is an odd one; hence, IQ is odd. Therefore, $P1$ and $P3$ are odd permutations (for these mechanisms, the bipyramidal set is closed); $P2$ and $P4$ are even permutations (for these two mechanisms, the bipyramidal set is open: it is not possible to tranverse *all* isomers using these processes).[67]

R (12,345) describes thus the possible isomerizations of one of the ten starred (non starred) isomers into one of the ten non starred (starred) ones. The coefficients r_{yz} of this reaction matrix **R** (12,345) are set equal to the rate constant for the one step transformation of y into z. In the case of a ML_5 molecule in which all the ligands are equivalent but distinguished by labels, r_{yz} may be set equal to 1 if isomer y may be converted into isomer z by only one pseudo-rotation, and to zero otherwise.[11]

$$
\mathrm{R}(Px)(12,345) = \begin{array}{|c|c|} \hline \mathrm{A}(Px)(12,345) & \mathrm{B}(Px)(12,345) \\ \hline \hat{\mathrm{B}}(Px)(12,345) & \hat{\mathrm{A}}(Px)(12,345) \\ \hline \end{array}
$$

$\hat{\mathrm{B}}$ is the transpose of B; $\mathrm{A}(P1)(12,345)$ is a zero matrix and

$$
\mathrm{B}(P1)(12,345) =
$$

	$\overline{12}$	13	$\overline{14}$	$\overline{23}$	24	$\overline{34}$	15	$\overline{25}$	35	$\overline{45}$
12	·	·	·	·	·	1	·	·	1	1
$\overline{13}$	·	·	·	·	1	·	·	1	·	1
14	·	·	·	1	·	·	·	1	1	·
23	·	·	1	·	·	·	1	·	·	1
$\overline{24}$	·	1	·	·	·	·	1	·	1	·
34	1	·	·	·	·	·	1	1	·	·
$\overline{15}$	·	·	·	1	1	1	·	·	·	·
25	·	1	1	·	·	1	·	·	·	·
$\overline{35}$	1	·	1	·	1	·	·	·	·	·
45	1	1	·	1	·	·	·	·	·	·

The non-zero elements of this submatrix $\mathrm{B}(P1)(12,345)$ may be considered as the turning points (C_{4v} transition states or intermediate complexes) to go from one D_{3h} isomer to another.

The Berry process $P1$ is also represented by a generalized matrix $\mathrm{G}(12,345)$ which corresponds to a semi-regular graph[4] and which describes the transformations of D_{3h} isomers j to C_{4v} intermediates i (pyramidal isomers are designated[38] by the apical substituent followed by the basal substituent which is in trans position with respect to the smaller basal one; the symbol thus obtained is underlined in order to avoid confusion between *pyramidal* and bipyramidal structures).

$$
\mathrm{G}(P1)(12,345) = \begin{array}{|c|c|} \hline \mathrm{C}(P1)(12,345) & \mathrm{D}(P1)(12,345) \\ \hline \mathrm{D}(P1)(12,345) & \mathrm{C}(P1)(12,345) \\ \hline \end{array}
$$

where $D(P1)(12,345)$ has all coefficients $d_{i,j} = 0$ except $d_{\overline{13},23}$, $d_{14,\overline{24}}$ and $d_{\overline{15},25}$, which are equal to 1 and where

$$C(P1)(12,345) =$$

	12	$\overline{13}$	14	$\overline{15}$	$\overline{23}$	24	$\overline{25}$	$\overline{34}$	35	$\overline{45}$
$\overline{13}$	·	·	·	·	·	·	·	·	·	1
$\overline{23}$	·	1	·	·	·	·	·	·	·	1
$\overline{32}$	1	·	·	·	·	·	·	·	·	1
$\overline{42}$	1	·	·	·	·	·	·	·	1	·
$\overline{52}$	1	·	·	·	·	·	·	1	·	·
14	·	·	·	·	·	·	·	·	1	·
24	·	·	1	·	·	·	·	·	1	·
$\overline{34}$	·	·	1	·	·	·	1	·	·	·
$\overline{43}$	·	1	·	·	·	·	1	·	·	·
$\overline{53}$	·	1	·	·	·	1	·	·	·	·
$\overline{15}$	·	·	·	·	·	·	·	1	·	·
25	·	·	·	1	·	·	·	1	·	·
$\overline{35}$	·	·	·	1	·	1	·	·	·	·
45	·	·	·	1	1	·	·	·	·	·
$\overline{54}$	·	·	1	·	1	·	·	·	·	·

The product $\widetilde{G} \times G$ gives $\mathbf{R}$ except for the diagonal coefficients (analogously, $G \times \widetilde{G}$ gives the adjacency matrix valid for the isomerizations of tetragonal pyramids occurring through the same $P1$ mechanism).

2.1.6   Non-topological graphical representations

A non-topological representation of the bipyramidal five-coordinate complexes has also been described[31] and consists of two enantiomeric four-dimensional tetrahedra: the first one contains the starred isomers, the other one, the non-starred ones (Fig. 10).

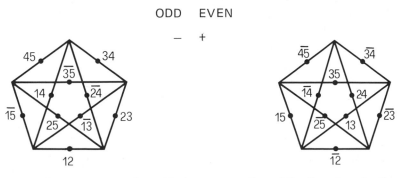

FIG. 10   Non-topological graphical representation of the five-atom family.

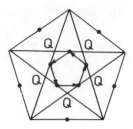

FIG. 11   Representation of the Q-operator on the non-topological graph *Left:* connections between points lying on connected lines *Right:* connections between points lying on unconnected lines.

If one assigns a given parity to the substituents (+ or even, − or odd) and if one takes into account the absolute configuration ($R \equiv +$ and $S \equiv -$) of the chiromers, one finds, in each four-dimensional tetrahedron, isomers of the same parity (if the parity of an isomer is arbitrarily defined as the product of the parities associated to the apiaxial ligands and to the absolute configuration).

The Q-operator is very simply expressed on this non-topological graph (Fig. 11) and defines the two families of Balaban's graph (the outer pentagons and the inner star pentagons).

The different isomerization mechanisms are also easily represented on this non-topological graph (Fig. 12).[35,38]

For $P1$, one finds the isomers formed from a given isomer x (E) in the other tetrahedron on the only triangle which has nothing in common with the edge bearing $\bar{x}$ ($P5$). Odd isomerization mechanisms indeed occur with a change in the isomer's parity: these are odd transformations.

$P2$ gives isomers which are located on the three triangles which have, as common edge, the line bearing E; thus, it occurs within a given tetrahedron, without parity-change: this is an even transformation.

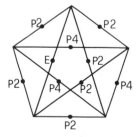
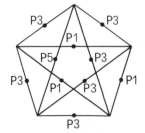

FIG. 12   Representation of the different isomerization processes on the non-topological graph (it must be noticed that obviously racemization cannot occur through E, $P2$ and $P4$).

2.1.7   The other processes

The other possible isomerization processes have also been described by reaction matrices or by topological graphs.

$$
A(P2)(12,345) =
\begin{array}{c}
\\
12\\
\overline{13}\\
14\\
\overline{23}\\
\overline{24}\\
34\\
\overline{15}\\
25\\
35\\
45
\end{array}
\begin{array}{cccccccccc}
12 & \overline{13} & 14 & 23 & \overline{24} & 34 & \overline{15} & 25 & \overline{35} & 45 \\
\cdot & 1 & 1 & 1 & 1 & \cdot & 1 & 1 & \cdot & \cdot \\
1 & \cdot & 1 & 1 & \cdot & 1 & 1 & \cdot & 1 & \cdot \\
1 & 1 & \cdot & \cdot & 1 & 1 & 1 & \cdot & \cdot & 1 \\
1 & 1 & \cdot & \cdot & 1 & 1 & \cdot & 1 & 1 & \cdot \\
1 & \cdot & 1 & 1 & \cdot & 1 & \cdot & 1 & \cdot & 1 \\
\cdot & 1 & 1 & 1 & 1 & \cdot & \cdot & \cdot & 1 & 1 \\
1 & 1 & 1 & \cdot & \cdot & \cdot & \cdot & 1 & 1 & 1 \\
1 & \cdot & \cdot & 1 & 1 & \cdot & 1 & \cdot & 1 & 1 \\
\cdot & 1 & \cdot & 1 & \cdot & 1 & 1 & 1 & \cdot & 1 \\
\cdot & \cdot & 1 & \cdot & 1 & 1 & 1 & 1 & 1 & \cdot
\end{array}
$$

and $B(P2)(12,345) = 0$

For this (aee) ligand permutation, one has thus in fact two distinct matrices or two unconnected chemical sets; the bipyramidal class is open. A topological graph has also been given for the $P2$ mechanism (Fig. 13).[38] Two enantiomeric Hamiltonian graphs may be constructed: each of them contains two five-membered subcycles (identical to those defined on the non topological graph; the connected points are resp. 2,2 and 4,4-apart on the Hamiltonian circuit of one of the unconnected sets), and two ten-membered subcycles (or Hamiltonian circuits on one of the unconnected chemical sets; the connected points are 1,1 and 5,3-apart on the Hamiltonian circuit of one of the unconnected chemical sets).

For $P3^*$, $A(P3)(12,345) = 0$ and $B(P3)(12,345) = A(P2)(12,345)$. An explicit topological illustration might be given on a Hamiltonian graph,[7] but would be much too complicated to have any practical use; the bipyramidal set is closed (odd permutation).[66]

$A(P4)(12,345) = B(P1)(12,345)$ and $B(P4)(12,345) = 0$: several isomorphous topological graphs (Fig. 14) have been given for this process (the bipyramidal set is open, but the class is closed).[67]

The matrix describing the $P5$ process has $A(P5)(12,345) = 0$ and $B(P5)(12,345)$ is a unit matrix. Graphically, $P5$ is thus represented by ten (parallel) lines connecting enantiometric pairs. The class is open (I8 products in each set are not equivalent).

---

*$P2$ and $P3$ are complementary; they go through the same C_S intermediate structure (twisted wedge);[37] each of them gives an open class (I8 products in each set are different) but if one uses the two mechanisms, one can reach all the C_S intermediates.

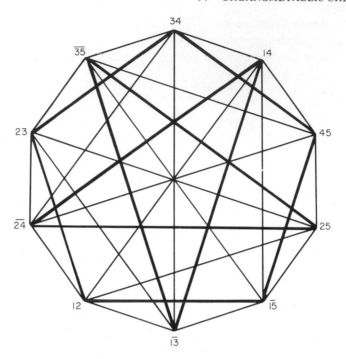

FIG. 13   Topological representation of the (*aee*) permutation: the *P*2-process (only half the isomers are given; the enantiomeric set is easily obtained from this one).

2.2   Description of Specific Cases

2.2.1   Chemical constraints

An experimental study of the general system formed by 20 different isomers should normally be much too complicated and simpler systems, with a restricted number of chiromers, are thus very interesting.

Three types of constraints may be used to reduce the number of isomers:

(a) *Bidentate ligands* suppress isomers having this ligand in *aa* position [if the formed chelate contains a small (for instance 4-membered) ring, the *ee* isomers do not exist either].

(b) *Ligands of quite different electronegativities* will force the most electro-negative one(s) in to the apiaxial (or 'polar') position(s) according to Bent's rule [or the less electronegative one(s) in apiequatorial position(s)].[6,27] These constraints may very easily be expressed: one needs only to suppress the isomers which cannot be formed owing to these geometrical or electronic constraints[56] either in the topological graph or in the reaction matrix.

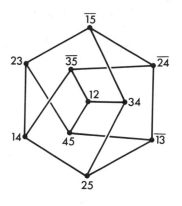

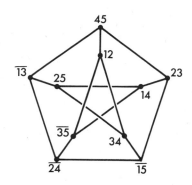

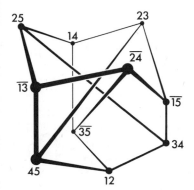

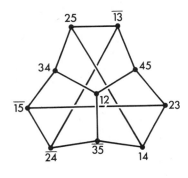

FIG. 14 Topological representations of the *P*4-process: the Petersen graphs[2,21,35] (only half the isomers are given; the enantiomeric set is easily obtained from this one).

(c) The introduction of *identical ligands* renders indistinguishable two or several isomers which were initially different.

This may be seen in an example: let us look at the 12,245 system for instance (ligand $2 \equiv$ ligand 3). Isomers which have ligand 2 *or* ligand 3 in apiaxial position collapse

12 = 13		$\overline{12} = \overline{13}$
24 = 34	and	$\overline{24} = \overline{34}$
25 = 35		$\overline{25} = \overline{35}$

Besides, the isomers which do not contain either ligand 2 *or* ligand 3 in apiaxial position contain a symmetry plane[76,77] defined by the two apiaxial ligands and the apiequatorial one different from 2. One has thus furthermore

$$14 = \overline{14} \qquad 15 = \overline{15} \qquad 45 = \overline{45}$$

The isomers having 2 and 3 *both* in apiaxial position become also optically inactive[76,77] when $2 \equiv 3$ (the apiequatorial plane is then a symmetry plane) or $23 = \overline{23}$ becomes 22).

2.2.2   Matrix transformations

It is possible to express these constraints in a **C** matrix,[32] which is a rectangular one containing 20 columns (the number of isomers of the general case) and a number of lines equal to the number of allowed isomers for a specific case (i.e. 10 for the 12,245 case); the C_{ij} coefficients of this constraint matrix are equal to *1* if two isomers are identical and zero if they are not. **C** $(2 \equiv 3)$ is for instance equal to

$$C(2 \equiv 3) =$$

	12	24	25	$\overline{12}$	$\overline{24}$	$\overline{25}$	14	15	22	45
$\overline{12}$	0	0	0	1	0	0	0	0	0	0
13	1	0	0	0	0	0	0	0	0	0
$\overline{14}$	0	0	0	0	0	0	1	0	0	0
$\overline{23}$	0	0	0	0	0	0	0	0	1	0
24	0	1	0	0	0	0	0	0	0	0
$\overline{34}$	0	0	0	0	1	0	0	0	0	0
15	0	0	0	0	0	0	0	1	0	0
$\overline{25}$	0	0	0	0	0	1	0	0	0	0
35	0	0	1	0	0	0	0	0	0	0
$\overline{45}$	0	0	0	0	0	0	0	0	0	1
12	1	0	0	0	0	0	0	0	0	0
$\overline{13}$	0	0	0	1	0	0	0	0	0	0
14	0	0	0	0	0	0	1	0	0	0
23	0	0	0	0	0	0	0	0	1	0
$\overline{24}$	0	0	0	0	1	0	0	0	0	0
34	0	1	0	0	0	0	0	0	0	0
$\overline{15}$	0	0	0	0	0	0	0	1	0	0
25	0	0	1	0	0	0	0	0	0	0
$\overline{35}$	0	0	0	0	0	1	0	0	0	0
45	0	0	0	0	0	0	0	0	0	1

If, furthermore, a $\overline{22}$ bidentate ligand is used which forms 4-membered chelates, $14 = 15 = 22 = 45 = 0$ and the constraint matrix describing this particular case will be a 6 by 20 one obtained from $\tilde{C}(2 \equiv 3)$ after suppression of the four last columns. $C(2 \equiv 3) \times \tilde{C}(2 \equiv 3) = D(2 \equiv 3) = 2\delta_{ij}$. $D^{-1}(2 \equiv 3)$ is thus, in this case, equal to $\tfrac{1}{2}\delta_{ij}$.

These constraint matrices have been used to determine the reaction matrix of simpler systems from $R(12,345)$ valid for the general case by the following matrix transformation $R'(Px)(\text{particular system}) = \check{C} \times R(Px)(12,345) \times C \times D^{-1}$ $(2 \equiv 3)$.

3 APPLICATION TO THE CASE OF DISPLACEMENTS AT TETRAHEDRALLY SUBSTITUTED METAL ATOMS

3.1 Addition-elimination displacements

3.1.1 Description of the general case

If one assumes that a tetrahedral complex $(\overset{\pm}{5},\overset{\pm}{4},\overset{\pm}{3},\overset{\pm}{2},\overset{\pm}{1})$† may be attacked by a fifth ligand (5,4,3,2,1) respectively to give a trigonal-bipyramidal intermediate,[16] then the five reactions in Fig. 15 will lead to the assumed pentacoordinate structure.[75]

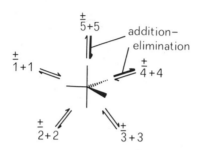

FIG. 15 Addition-elimination reactions at tetrahedrally substituted atoms.

3.1.2 Apiaxial and apiequatorial attack

The first question which arises when such a complicated system is studied is: how may a fifth ligand be added to a tetrahedral complex to form a trigonal bipyramid?[56]

Each of the ten possible tetrahedral molecules may undergo attack at any one of the four distereotopic faces (apiaxial attack) or of the six diastereotopic edges

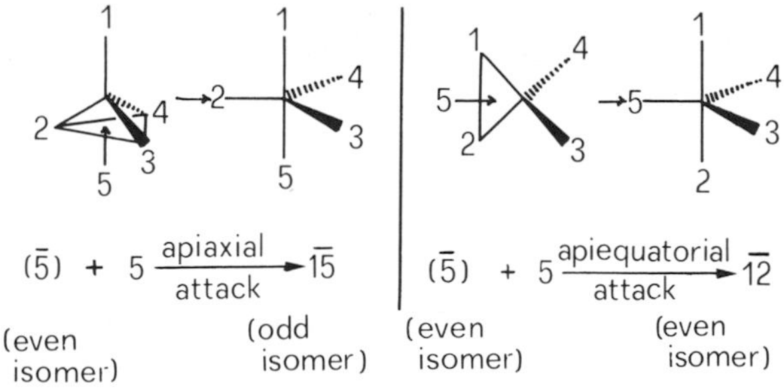

FIG. 16 Apiaxial and apiequatorial attack: parity rule $[\overset{+}{5} \equiv (R)1,234; \overline{5} = (S) 1,243]$.

†The tetrahedral species are considered as being formally derived from the pentacoordinate intermediate by the loss of a ligand[75] and may be named by giving the ligand which has been lost with a "+" for the R-isomer[13] and a "−" for the enantiomer having the S-configuration.

(apiequatorial attack) to give ten diastereoisomeric pentacoordinate inter-mediates (Fig. 16).[19]

The set of ten diastereomeric trigonal bipyramids produced from a (R) tetrahedral complex is of course enantiomeric to the set produced from its enantiomer. Furthermore, it may be seen that apiequatorial attack takes place without any parity-change, when apiaxial attack causes a parity-change.

3.1.3   Matrix formulation

These addition-elimination reactions are best described in matrix form (**AE**)

apiequatorial attack (upper-left) · · · · · · · · · · · · · · · · · · · **apiaxial attack** (upper-right)

		even isomers										odd isomers								
AE =		$\overline{12}$	13	$\overline{14}$	15	$\overline{23}$	24	$\overline{25}$	34	35	$\overline{45}$	12	$\overline{13}$	14	$\overline{15}$	23	$\overline{24}$	25	34	$\overline{35}$/45
even isomers	$\bar{1}$	·	·	·	·	1	1	1	1	1	1	1	1	1	1	·	·	·	·	·	·
	$\overset{+}{2}$	1	1	1	·	·	·	·	1	1	1	1	·	·	·	1	1	1	·	·	·
	$\bar{3}$	1	·	1	1	·	1	1	·	·	1	·	1	·	·	1	·	·	1	1	·
	$\overset{+}{4}$	1	1	·	1	1	·	1	·	1	·	·	·	1	·	·	1	·	1	·	1
	$\bar{5}$	1	1	1	·	1	1	·	1	·	·	·	·	·	1	·	·	1	·	1	1
odd isomers	$\bar{1}$	1	1	1	1	·	·	·	·	·	·	·	·	·	·	1	1	1	1	1	1
	$\bar{2}$	1	·	·	·	1	1	1	·	·	·	·	1	1	1	·	·	·	1	1	1
	$\overset{+}{3}$	·	1	·	·	1	·	·	1	1	·	1	·	1	1	·	1	1	·	·	1
	$\bar{4}$	·	·	1	·	·	1	·	1	·	1	1	1	·	1	1	·	1	·	1	·
	$\overset{+}{5}$	·	·	·	1	·	·	1	·	1	1	1	1	1	·	1	1	·	1	·	·

apiaxial attack (lower-left) · · · · · · · · · · · · · · · · · **apiequatorial attack** (lower-right)

and may be condensed as suggested by Robson[75]

	$\overline{12}$	13	$\overline{14}$	15	$\overline{23}$	24	$\overline{25}$	$\overline{34}$	35	$\overline{45}$
$\bar{1}$	−	−	−	−	+	+	+	+	+	+
$\overset{+}{2}$	−	+	+	+	−	−	−	+	+	+
$\bar{3}$	+	−	+	+	−	+	+	−	−	+
$\overset{+}{4}$	+	+	−	+	+	−	+	−	+	−
$\bar{5}$	+	+	+	−	+	+	−	+	−	−

3.1.4   Pseudo-rotations

A second question which has been discussed already is connected with the stereochemical lability of the trigonal-bipyramidal complexes, which may interconnect by the five isomerization processes $P1(aeae)$, $P2(aee)$ $P3(ae)$, $P4(aeeae)$ or $P5(ee)$.

3.1.5   Exchange matrix

It is thus possible to define exchange matrices $E(Px)$ which describe this addition-elimination mechanism with pseudo-rotating[56] trigonal-bipyramidal intermediate complexes.[40,41]

$$E(Px) = \begin{array}{c|c|c} & \bar{1}\ldots\bar{5} \quad \overset{+}{1}\ldots\overset{+}{5} & \overline{12}\ldots\overline{45} \quad 12\ldots45 \\ \hline \begin{array}{c} \bar{1} \\ \vdots \\ \bar{5} \\ \overset{+}{1} \\ \vdots \\ \overset{+}{5} \end{array} & O & AE \\ \hline \begin{array}{c} \overline{12} \\ \vdots \\ \overline{45} \\ 12 \\ \vdots \\ 45 \end{array} & \widetilde{AE} & R(Px) \end{array}$$

3.2   Chemical Constraints

3.2.1   The phosphetane case (retention of configuration): constraints

If one wants to look at some particular cases, for instance at those studied by Mislow.[15,19,20,56] one has to introduce specific constraints (Fig. 17):

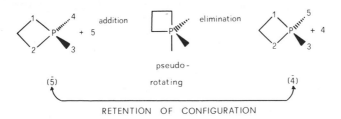

RETENTION OF CONFIGURATION

FIG. 17   Stereochemistry for the phosphetane case.

(a) if 5 (nucleophilic reagent) and 4 (leaving group) are the only groups which can be cleaved in the pentacoordinate structure, then[40,41]

$$\overset{\pm}{1} = \overset{\pm}{2} = \overset{\pm}{3} = 0$$

(b) a severe ring strain is introduced when the four-membered ring is forced to span the *aa* or the *ee* positions (the profound effect of ring strain on rearrangement barriers is discussed by Westheimer[87])

$$\overset{(-)}{12} = \overset{(-)}{34} = \overset{(-)}{35} = \overset{(-)}{45} = 0$$

(c) ligands 4 and 5 are electronegative groups and may not both be equatorial

$$\left[\overset{(-)}{12}\right] = \overset{(-)}{13} = \overset{(-)}{23} = 0$$

3.2.2   Topological graphical representations

This has been expressed in a constraint matrix[35] which may be used to transform $\mathbf{E}(Px)$ into new exchange matrices $\mathbf{E}'(Px)$ valid for this specific case. Each of these new matrices $\mathbf{E}'(Px)$ may finally be visualized in the form of a topological graph (Fig. 18).

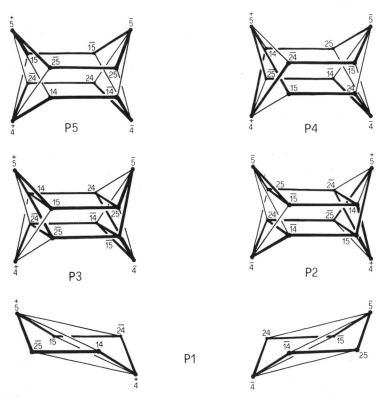

FIG. 18   Topological graphical representation of the displacement reactions at the phosphorus atom of phosphetanes for different intramolecular isomerization mechanisms of the trigonal bipyramidal intermediate complexes.[40,41]

In order to discuss the stereochemistry of these addition–elimination-type displacements, it is necessary to know the relative configurations, and thus the connection between the absolute and the relative configurations of the tetrahedral isomers. This has also been expressed by Robson under the form of a matrix **(RC)** which is equivalent to saying that any two of $\overset{+}{1}, \bar{2}, \overset{+}{3}, \bar{4}$ and $\overset{+}{5}$ have inverted configuration with respect to each other ('inverted relative configuration').

$$RC = \begin{array}{c} \\ \overset{+}{1} \\ \bar{2} \\ \overset{+}{3} \\ \bar{4} \\ \overset{+}{5} \end{array} \begin{array}{ccccc} \overset{+}{1} & \bar{2} & \overset{+}{3} & \bar{4} & \overset{+}{5} \\ + & - & - & - & - \\ - & + & - & - & - \\ - & - & + & - & - \\ - & - & - & + & - \\ - & - & - & - & + \end{array}$$

3.2.3 Discussion and conclusions

The study of the graphs describing $E'(P5)$, $E'(P4)$, $E'(P3)$, $E'(P2)$ and $E'(P1)$ may be done with two different hypotheses:[40,41]
(a) one has apiaxial attack and apiaxial loss only[56] (only the thick lines of the graphs are then to be considered);
(b) one has any apiaxial and/or apiequatorial attack or loss (all the lines must be considered).

It is now obvious from the P5 graph that $\bar{5}$ will never be transformed into $\bar{4}$ or $\overset{+}{4}$ if one considers only apiaxial attack and loss for this mechanism, but that $\bar{5}$ may be transformed into $\overset{+}{5}$; one has thus *no substitution reaction* but a *racemization* of the initial product. If one considers any attack or loss, the $\bar{5}$ isomer may give either $\bar{4}$ or $\overset{+}{4}$ leading to *racemization*.

For P4, $\bar{5}$ may be converted into $\overset{+}{4}$ (thus *inversion*) if one has apiaxial attack or loss only; otherwise, one has *racemization*. The same is true for the other even mechanism, P2.

For P3, one has *racemization* for both hypotheses.

For the Berry pseudo-rotation P1, one has *retention* of configuration both for apiaxial and for any attack or loss, as Mislow has pointed out before.[19,56]

From these results, two things may be concluded:

(a) the mechanism by addition–pseudo-rotations–elimination allows us to explain the experimental stereochemistry: the overall retention of configuration;[19]
(b) the only pseudo-rotation which is compatible with the experimental results is the Berry mechanism P1.[40,41]*

---

*It must be noticed that Whitesides[88] has also shown unequivocally that only the Berry mechanism is consistent with his experimental data (see also Holmes[50])

3.2.4   Minimum constraints for stereospecificity

It may be seen that one of the minimum sets of constraints for stereospecificity is that the two rows of matrix **AE** (given under Robson's condensed form) corresponding to the initial and final tetrahedral molecules are opposite in sign, i.e.

$$\overset{(-)}{12} = \overset{(-)}{13} = \overset{(-)}{23} = \overset{(-)}{45} = 0$$

for the case

$$\overset{\pm}{1} = \overset{\pm}{2} = \overset{\pm}{3} = 0.$$

This case may be described similarly to the one studied experimentally by Mislow and is characterized by the same stereochemical features although the topological graphs corresponding to the $E'(Px)$ matrices are a little more complicated.

Another set of constraints is worth mentioning: it is the complementary one:

$$\overset{(-)}{14} = \overset{(-)}{15} = \overset{(-)}{24} = \overset{(-)}{25} = \overset{(-)}{34} = \overset{(-)}{35} = 0$$

In this case, one gets inversion of configuration for apiaxial attack or loss (except for $P5$, which of course gives racemization). One also gets inversion for both apiaxial and apiequatorial attack in the case of the Berry mechanism $P1$. This case is represented by the graph in Fig. 19. For all the other mechanisms, apiaxial and apiequatorial attack or loss yield racemization.

The inversion of configuration obtained in the case of only apiaxial attack or loss accounts for the experimental inversion of configuration generally observed for S_N2 reactions on unstrained tetrahedral systems and occurring either by an addition-elimination-type complex mechanism or by a synchronous one-step reaction in which 45 is then a transition state (Walden inversion).[22,56]

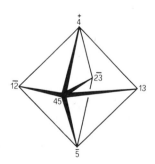

FIG. 19   Minimum constraints for the inversion: the $P1$ mechanism.

4   SIX-COORDINATE COMPLEXES

4.1   Description of the 123,456 case

4.1.1   Isomerization twist mechanisms

An octahedrally substituted complex may conventionally be numbered by naming first the ligands which define the face giving the lowest number when turning clockwise (and when defining the substituents according to Cahn, Ingold and Prelog[13]); the three other ligands are given afterwards, turning in the same direction and beginning with the ligand being found on the same triangular face as the two first cited ones.[29]

One may define an isomer univocally by giving the two last digits of this six-digit symbol and by barring the structure which has the higher number of those defining two enantiomers.

There are

$$\frac{6!}{\sigma} = \frac{6!}{24} = 30 \text{ isomers}$$

for the octahedral class and it is possible to find four different interconversion processes from the 6! permutations.[37]

A first isomerization process, $P6$, may be defined by two analogous[9,36] cyclic permutations: that of three ligands which are in cis (c) position (which define a face of the octahedron) – the Bailar[1] trigonal twist (ccc) – and a second (equivalent) one of four ligands forming two adjacent faces of the polytope – the Ray–Dutt[73] rhombic twist (Fig. 20).[47]

Bailar trigonal twist   Ray–Dutt rhombic twist

FIG. 20 $P6$ process shown on Schlegel diagrams; the Bailar twist is an even permutation; the equivalent Ray–Dutt twist is an odd permutation: this might be connected to the fact that exactly half of the permutations associated with rotations of the polyhedron are even, whereas they are all even for the trigonal bipyramid.[66]

This isomerization process transforms 56 into $\overline{25}$, $\overline{26}$, $\overline{34}$, $\overline{35}$, 43, 45, $\overline{63}$, 64 through D_{3h} intermediate trigonal prismatic structures.

Other processes are: $P7$ which is equivalent to the transposition of two adjacent ligands ($c\bar{c}$), and may be called digonal twist (Fig. 21).[30]

$P8$, which is equivalent either to a double digonal twist performed with ligands defining parallel edges [$(c\bar{c})(c\bar{c})$] (even permutation) or to a C_{3v}

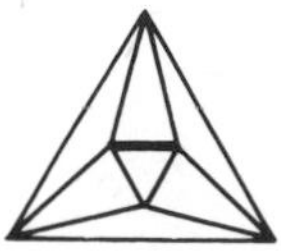

$$56 \xrightarrow{\;P7\;} \begin{array}{l} 24,\overline{24} \\ 36,\overline{36} \\ 46,\overline{46} \\ 53,\overline{53} \\ 54,\overline{54} \\ 65,\overline{65} \end{array}$$

FIG. 21 $P7$ process shown on a Schlegel diagram.

Hamiltonian circuit (odd permutation) which may be called the inversion (Fig. 22).[38]

$$(56 \xrightarrow{\;P8\;} \overline{56})$$

FIG. 22 $P8$ process shown on Schlegel diagrams.

$P9$, which is equivalent either to a triple concerted digonal twist $(c\bar{c})^3$ or to a C_{2v} Hamiltonian circuit. $P9$ gives the enantiomers of the chiromers obtained by $P6$ (Fig. 23).[38]

FIG. 23 $P9$ process shown on Schlegel diagrams.

4.1.2   Non-topological representation

A non-topological representation of these 30 octahedral isomers[29] consists of a five dimensional tetrahedron if one places two enantiomers on one edge (Fig. 24).

The trigonal or the rhombic twist may be represented on Fig. 25 if all the permitted one-step transformations starting from one isomer are given by C_3-rotations performed on each of the four triangular faces which contains the edge bearing that isomer.[36]

For the digonal twist, the formed isomers are located on the three-

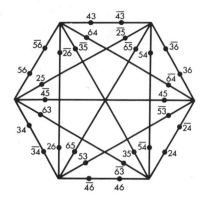

FIG. 24   Non-topological representation of the six-atom family.

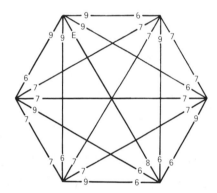

FIG. 25   Representation of the different isomerization processes on the non-topological graph.

dimensional tetrahedron which has no common point with the line bearing the starting isomer[30] (this shows why there is no parity rule for the isomerizations of octahedral structures: only *half* the permutations defining the elements of the rotational subgroup are even).

4.1.3   Topological representations
Matrix representations may be given for these processes[30,32,36] and are analogous to (but larger than) those described for the five-coordinate cases.

Reaction matrices have been defined for the four processes $P6$, $P7$, $P8$ and $P9$. $\mathbf{R}(P6)(123,456)$ consists of the following submatrices:[36]

$$\mathbf{A(P6)(123,456)} =$$

	$\overline{56}$	$\overline{45}$	$\overline{64}$	63	35	25	$\overline{43}$	34	26	$\overline{53}$	$\overline{24}$	36	46	$\overline{54}$	$\overline{65}$
56	0	0	0	0	0	0	0	0	0	0	0	0	0	0	0
45	0	0	0	1	0	1	0	1	0	0	0	0	0	0	0
64	0	0	0	0	1	0	1	0	1	1	1	1	0	0	0
$\overline{63}$	0	1	0	0	0	1	0	1	0	0	1	0	0	0	0
$\overline{35}$	0	0	1	0	0	0	1	0	1	0	0	0	1	1	0
$\overline{25}$	0	1	0	1	0	0	0	1	0	0	0	1	0	1	0
43	0	0	1	0	1	0	0	0	1	0	0	0	0	0	1
$\overline{34}$	0	1	0	1	0	1	0	0	0	1	0	0	1	0	1
$\overline{26}$	0	0	1	0	1	0	1	0	0	0	0	0	0	0	0
53	0	0	1	0	0	0	0	1	0	0	1	1	1	0	1
24	0	0	1	1	0	0	0	0	0	1	0	1	0	0	0
$\overline{36}$	0	0	1	0	0	1	0	0	0	1	1	0	0	1	0
$\overline{46}$	0	0	0	0	1	0	0	1	0	1	0	0	0	1	1
54	0	0	0	0	1	1	0	0	0	0	0	1	1	0	0
65	0	0	0	0	0	0	1	1	0	1	0	0	1	0	0

$$\mathbf{B(P6)(123,456)} =$$

	56	45	64	$\overline{63}$	$\overline{35}$	$\overline{25}$	43	$\overline{34}$	$\overline{26}$	53	24	$\overline{36}$	$\overline{46}$	54	65
56	0	1	1	1	1	1	1	1	1	0	0	0	0	0	0
45	1	0	1	0	0	0	0	0	0	1	1	1	0	0	0
64	1	1	0	0	0	0	0	0	0	0	0	0	0	0	0
$\overline{63}$	1	0	0	0	1	0	0	0	0	0	0	0	1	1	0
$\overline{35}$	1	0	0	1	0	0	0	0	0	0	1	0	0	0	0
$\overline{25}$	1	0	0	0	0	0	1	0	0	0	0	0	0	0	1
43	1	0	0	0	0	1	0	0	0	0	0	1	0	1	0
$\overline{34}$	1	0	0	0	0	0	0	0	1	0	0	0	0	0	0
$\overline{26}$	1	0	0	0	0	0	0	1	0	1	0	0	1	0	1
53	0	1	0	0	0	0	0	0	1	0	0	0	0	0	0
24	0	1	0	0	1	0	0	0	0	0	0	0	1	1	0
$\overline{36}$	0	1	0	0	0	0	1	0	0	0	0	0	0	0	1
$\overline{46}$	0	0	0	1	0	0	0	0	1	0	1	0	0	0	0
54	0	0	0	1	0	0	1	0	0	0	1	0	0	0	1
65	0	0	0	0	0	1	0	0	1	0	0	1	0	1	0

Furthermore[30]

$$A(P7)(123,456) = \\ B(P7)(123,456) =$$

	56	45	64	$\overline{63}$	$\overline{35}$	$\overline{25}$	43	$\overline{34}$	26	53	24	$\overline{36}$	$\overline{46}$	54	65
56	0	0	0	0	0	0	0	0	0	1	1	1	1	1	1
45	0	0	0	0	1	0	1	0	1	0	0	0	1	1	1
64	0	0	0	1	0	1	0	1	0	0	0	0	1	1	1
$\overline{63}$	0	0	1	0	0	0	1	0	1	1	0	1	0	0	1
$\overline{35}$	0	1	0	0	0	1	0	1	0	1	0	1	0	0	1
25	0	0	1	0	1	0	0	0	1	1	1	0	1	0	0
43	0	1	0	1	0	0	0	1	0	1	1	0	1	0	0
$\overline{34}$	0	0	1	0	1	0	1	0	0	0	1	1	0	1	0
26	0	1	0	1	0	1	0	0	0	0	1	1	0	1	0
53	1	0	0	1	1	1	1	0	0	0	0	0	0	1	0
24	1	0	0	0	0	1	1	1	1	0	0	0	0	0	1
$\overline{36}$	1	0	0	1	1	0	0	1	1	0	0	0	1	0	0
$\overline{46}$	1	1	1	0	0	1	1	0	0	0	0	1	0	0	0
54	1	1	1	0	0	0	0	1	1	1	0	0	0	0	0
65	1	1	1	1	1	0	0	0	0	0	1	0	0	0	0

$A(P8)(123,456) = 0$ and $B(P8)(123,456) = 1$ $A(P9)(123,456) = B(P6)(123,456)$ and $B(P9)(123,456) = A(P6)(123,456)$.[37]

4.2   Description of Specific Cases

4.2.1   Topological graphs

Topological graphs are generally much too complicated to be of practical use and only become interesting when the number of isomers has been reduced by chemical constraints. In the case of six-coordinate complexes, constraint matrices may be defined as for the five-atom family and may be used to transform $R(Px)(123,456)$ into smaller matrices which are then valid for the specific cases. When the number of isomers is reduced (to less than 20), then it is again reasonable to express the topological matrix as a graph where all pathways are immediately apparent. This has been done for a large number of specific cases.[29,30,34,36] If one looks for example at the 123,456 case, then all the chiromers having ligands 1 and 2, 3 and 4, 5 and 6 in trans position are excluded

$$[\text{i.e. } \overset{(-)}{24}, \overset{(-)}{25}, \overset{(-)}{26}; \overset{(-)}{(24)}, \overset{(-)}{56}, \overset{(-)}{65}; \overset{(-)}{(24)}, \overset{(-)}{34}, \overset{(-)}{43}].$$

$R(\overline{123},\overline{456})$ can be calculated.

$$B(P6)(\overline{123},\overline{456}) =
\begin{array}{c}
 \\
45 \\
\overline{64} \\
\overline{63} \\
35 \\
\overline{53} \\
36 \\
46 \\
\overline{54}
\end{array}
\begin{array}{c}
\overline{45}\ 64\ 63\ \overline{35}\ 53\ \overline{36}\ 46\ 54 \\
\left[
\begin{array}{cccccccc}
0 & 1 & 1 & 0 & 1 & 1 & 0 & 0 \\
1 & 0 & 0 & 1 & 1 & 1 & 0 & 0 \\
1 & 0 & 0 & 1 & 0 & 0 & 1 & 1 \\
0 & 1 & 1 & 0 & 0 & 0 & 1 & 1 \\
1 & 1 & 0 & 0 & 0 & 1 & 1 & 0 \\
1 & 1 & 0 & 0 & 1 & 0 & 0 & 1 \\
0 & 0 & 1 & 1 & 1 & 0 & 0 & 1 \\
0 & 0 & 1 & 1 & 0 & 1 & 1 & 0
\end{array}
\right]
\end{array}$$

and $A(P6)(\overline{123},\overline{456}) = 0$.

This matrix has been transformed into an alternating topological graph which takes the form of a four-dimensional cube (see Fig. 26).[63-65]

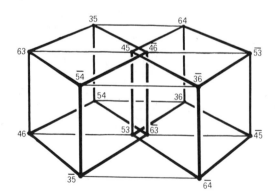

FIG. 26   Topological representation of the $\overline{123},\overline{456}$ case: the $P6$ process

If one wants furthermore to go on simplifying this system[24] by setting for instance $1 \equiv 3 \equiv 5$ and $2 \equiv 4 \equiv 6$, then for the $\overline{121},\overline{212}$ system,[63,64] $35 = 36 = 46 = \overline{53} = \overline{54} = \overline{64}$ becomes 12^{36} or $T\Delta$;[42] $45 = \overline{63}$ becomes $\overline{22}$ or $C\Delta$; $\overline{35} = \overline{36} = \overline{46} = 53 = 54 = 64$ becomes 12 or $T\Lambda$; $\overline{45} = 63$ becomes 22 or $C\Lambda$.

The constraint matrix $\mathbf{C}$ may be constructed

$$
\mathbf{C} = \frac{\overline{22}}{\overline{12}}
\begin{array}{c}
\overline{45}\ \ 64\ \ 63\ \ \overline{35}\ \ 53\ \ \overline{36}\ \ \overline{46}\ \ 54 \\
\begin{bmatrix}
1 & 0 & 1 & 0 & 0 & 0 & 0 & 0 \\
0 & 1 & 0 & 1 & 1 & 1 & 1 & 1
\end{bmatrix}
\end{array}
$$

$\mathbf{C} \times \widetilde{\mathbf{C}} = \mathbf{D} = \begin{vmatrix} 2 & 0 \\ 0 & 6 \end{vmatrix}$, thus $\mathbf{D}^{-1} = \begin{vmatrix} \tfrac{1}{2} & 0 \\ 0 & \tfrac{1}{6} \end{vmatrix}$ and

$\mathbf{C} \times \mathbf{B}(\overline{123},456) \times \widetilde{\mathbf{C}} \times \mathbf{D}^{-1}$ gives $\mathbf{B}(\overline{121},212)$. One gets

$$
\mathbf{B}(P6)(\overline{121},212) = \frac{\overline{22}}{\overline{12}}
\begin{array}{c}
22\ \ 12 \\
\begin{bmatrix}
1 & 1 \\
3 & 3
\end{bmatrix}
\end{array}
$$

which may be represented by a 'two-dimensional cube' (a square).[36,42]

4.2.2   The digonal twist

The same sort of calculations may be done for the other processes. For the digonal twist, one obtains

$$\mathbf{A}(P7)(\overline{123},456) = \mathbf{B}(P7)(\overline{123},456)$$

	$\overline{45}$	64	$\overline{63}$	$\overline{35}$	53	36	$\overline{46}$	54
45	0	0	0	1	0	0	1	1
64	0	0	1	0	0	0	1	1
$\overline{63}$	0	1	0	0	1	1	0	0
$\overline{35}$	1	0	0	0	1	1	0	0
53	0	0	1	1	0	0	0	1
$\overline{36}$	0	0	1	1	0	0	1	0
$\overline{46}$	1	1	0	0	0	1	0	0
54	1	1	0	0	1	0	0	0

$\mathbf{R}(P7)(\overline{123},456)$ may be represented by a three dimensional cube if each point stands for two enantiomers (Fig. 27).

For the $\overline{121},212$ case (Fig. 28), which may be obtained from $(\mathbf{A})(P7)(\overline{123},456)$

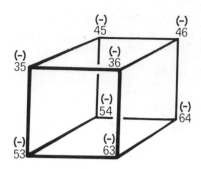

FIG. 27   Topological representation of the $\overline{123},\overline{456}$ case: the $P7$ **process.**

using the already described **C** constraint matrix one has

$$
\mathbf{R}(P7)(\overline{121},\overline{212})=
\begin{array}{c|cccc}
 & 22 & 12 & \overline{22} & \overline{12} \\
\hline
22 & 0 & 1 & 0 & 1 \\
12 & 1 & 2 & 1 & 2 \\
\overline{22} & 0 & 1 & 0 & 1 \\
\overline{12} & 1 & 2 & 1 & 2
\end{array}
$$

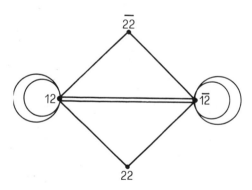

FIG. 28   Topological representation of the digonal twists in the $\overline{121},\overline{212}$-case.

4.2.3   Generalized matrix description

In the case of the apparently most favorable process, the trigonal (or rhombic) twist, a generalized matrix description has also been given[39] which expresses how octahedral and trigonal prismatic stereoisomers are converted into one another. From this 30 by 120 matrix, the 30 by 30 **R**(P6)(123,456) matrix may be defined (if one excepts the diagonal elements) and also another 120 by 120 matrix, which represents the rearrangements of trigonal prisms occurring by the same isomerization process. Graphical topological representations of some specific cases have also been described,[41],[65] and have been found starting from the general $\mathbf{G}(P6)(123,456)$ matrix, using constraint matrices. These more sophisticated topological representations are very useful if one wants to analyze in more detail the isomerization reactions of specific octahedral complexes:[42] the degeneracies of the possible intramolecular rearrangement processes become immediately apparent.

5   APPLICATION TO THE CASE OF INTRAMOLECULAR REARRANGEMENT REACTIONS OF TRIS-CHELATE COMPLEXES

Topological correlation diagrams have been derived for the $\overline{121},\overline{212}$ case for three different reaction pathways:

(1) A twist mechanism involving no bond rupture.[29],[36]
(2) A bond rupture mechanism going through trigonal-bipyramidal *TBP*

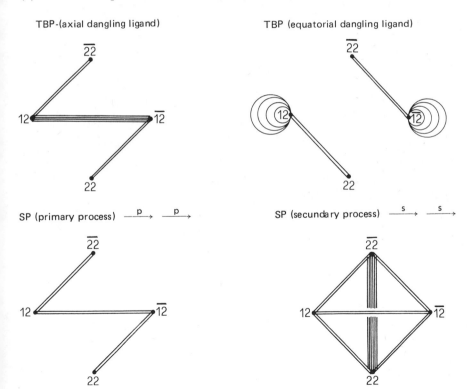

FIG. 29   Topological representations of different reaction pathways on $\overline{121},\overline{212}$.

transition states with dangling ligand in e or a positions.[25,63,64] It may be shown that this bond rupture mechanism gives in fact the same products as the digonal twist and that the same topological representation may thus be used.

(3) A bond rupture mechanism going through square pyramidal (SP) transition states[25] with the dangling ligand in basal (this case is equivalent with the TBP-a case) or in apical positions.

For this last case, two processes may be visualized: a primary one (p), where the SP formed after bond-rupture is isomerized into another a-SP by one 'Berry-type' pseudorotation, and a secondary one(s), where more than one 'Berry-type' pseudorotation is needed.

Topological graphs have been described for these different mechanisms (Fig. 29).

Experimental isomerization and racemization rates have been compared with predicted values for the three mechanisms; this has led to the exclusion of the twist mechanism. The mechanisms consistent with the kinetic results are of the bond-rupture type going essentially through TBP-axial transition states.

ACKNOWLEDGEMENTS

In the preparation of this chapter, I have been much endebted to various colleagues and friends. In particular, I have received much helpful advice and constructive criticism from Prof. Kurt Mislow and from his coworkers, from Prof. Jacques Nasielski and from Dr. Jean Brocas.

I wish also to express my grateful thanks to MM. Rudi Willem, Michel De Clercq, Jean Topart, Guy Mayence, Christiane Dehouck and Hassan Mokhtar-Jamai for their invaluable assistance. Deprived of their kind cooperation, this treatise would not have been possible.

REFERENCES

1. Bailar, J. C. (1958). *J. Inorg. Nucl. Chem.* **18**, 165
2. Balaban, A. T., Farcasiu, D. and Banica, R. (1966). *Rev. Roum. Chim.* **11**, 1205
3. Balaban, A. T., Davies, R. O., Harary, F., Hill, A. and Westwick, R. (1970). *J. Aust. Math. Soc.* **11**, 207
4. **Balaban, A. T. (1972). *Rev. Roum. Math. Pures et Appl.* 17, 3**
5. **Balaban, A. T. (1971). *J. Combinatorial Theory* (B) 12, 1**
6. Bent, H. A. (1961). *Chem. Rev.* **61**, 275
7. Berge, C. (1970). "Graphes et Hypergraphes", Dunod
8. Berry. R. S. (1960). *Inorg. Chem.* **32**, 933
9. Brady, J. (1969). *Inorg. Chem.* **8**, 1208

10. Brocas, J. and Gielen, M. (1971). *Bull. Soc. Chim. Belges,* **80**, 207
11. Brocas, J. (1971). *Theoret. Chim. Acta (Berl.),* **16**, 79
12. Buekenhout, F. (1971), private communication
13. Cahn, R. S., Ingold, C. and Prelog, V. (1966). *Angew. Chem. Intern. Ed. Engl.* **5**, 385
14. Carmichael, R. D. (1956). "Introduction to the Theory of Finite Groups". Dover
15. Casey, J. P. and Mislow, K. (1970). *Chem. Comm.,* 1410
16. Corfield, J. R., Death, N. J. and Trippett, S. (1970). *Chem. Comm.,* 1502
17. Cotton, F. A. (1967). "Chemical Applications of Group Theory", Interscience (sixth printing)
18. Cram, D. J., Day, J., Rayner, D. R., Von Schriltz, D. M. and Garwood, D. C. (1970). *J. Amer. Chem. Soc.* **92**, 7369
19. Debruin, K. E., Naumann, K., Zon, G. and Mislow, K. (1969a). *J. Amer. Chem. Soc.* **91**, 7031
20. Debruin, K. E. and Mislow, K. (1969)b). *J. Amer. Chem. Soc.* **91**, 7393
21. Dunitz, J. D. and Prelog, V. (1968). *Angew. Chem. Intern. Ed. Engl.* **7**, 726
22. Eaton, D. R. (1968). *J. Amer. Chem. Soc.* **90**, 4272
23. Ege, G. (1971), private communication
24. Farago, M. E., Page, B. A. and Mason, C. F. V. (1969). *Inorg. Chem.* **8**, 2270
25. Fortman, J. I. and Sievers, R. E. (1967). *Inorg. Chem.* **6**, 2022
26. Frucht, R. (1949). *Can. J. Math.* **1**, 365
27. Furtsch, R. A., Dierdorf, D. S. and Cowley, A. H. (1970). *J. Amer. Chem. Soc.* **92**, 5759
28. Gielen, M. and Nasielski, J. (1969a), *Bull. Soc. Chim. Belges* **78**, 339
29. Gielen, M. and Depasse-Delit, C. (1969b). *Theoret. Chim. Acta (Berl.),* **14**, 212
30. Gielen, M. (1969c). *Bull. Soc. Chim. Belges* **78**, 351
31. Gielen, M., Depasse-Delit, C. and Nasielski, J. (1969d), *Bull. Soc. Chim. Belges* **78**, 357
32. Gielen, M., De Clercq, M. and Nasielski, J. (1969e), *J. Organometal. Chem.* **18**, 217
33. Gielen, M. Mayence, G. and Topart, J. (1969f). *J. Organometal. Chem.* **18**, 1
34. Gielen, M. and Topart, J. (1969g). *J. Organometal. Chem.* **18**, 7
35. Gielen, M. (1969h). *Mededelingen Vlaamse Chem. Veren.* **31**, 185
36. Gielen, M. (1969i). *Mededelingen Vlaamse Chem. Veren.* **31**, 201
37. Gielen, M. and Vanlautem, N. (1970a). *Bull. Soc; Chim. Belges.* **79**, 679
38. Gielen, M., Brocas, J., De Clercq, M., Mayence, G. and Topart, J. (1970b). "Proceedings 3rd Symposium on Coordination Chemistry" Akademiai Kiado, Budapest, Vol. 1, 495
39. Gielen, M. (1971a). *Bull. Soc. Chim. Belges.* **80**, 9
40. Gielen, M., Brocas, J., De Clercq, M., Mayence, G. and Topart, J. (1971b). "Proceedings 3rd Symposium on Coordination Chemistry", Akademiai Kiado, Budapest, Vol. 2
41. Gielen, M. (1971c). *Ind. Chim. Belges.* **36**, 815
42. Gordon, J. G. and Holm, R. H. (1970). *J. Amer. Chem. Soc.* **92**, 5319
43. Gorenstein, D. G. and Westheimer, F. H. (1967). *J. Amer. Chem. Soc.* **89**, 2762
44. Gorenstein, D. G. and Westheimer, F. H. (1970). *J. Amer. Chem. Soc.* **92**, 634
45. Grunbaum, B. (1967). "Convex Polytopes", Interscience

46. Heller, J. (1969), *Chimia.* **23**, 351
47. Hellwinkel, D. (1966a) *Ber.* **99**, 3628, 3642, 3660
48. Hellwinkel, D. (1966b). *Angew. Chem. Internat. Ed. Engl.* **5**, 725
49. Hoffmann, R., Alder, R. W. and Wilcox, C. F. (1970). *J. Amer. Chem. Soc.* **92**, 4992
50. Holmes, R. R. and Deiters, R. M. (1968). *J. Amer. Chem. Soc.* **90**, 5021
51. Hudson, R. F. (1965). "Structures and Mechanisms in organophosphorus Chemistry", Academic Press, New York
52. Hudson, R. F. (1967), *Angew. Chem. Intern. Ed. Engl.* **6**, 749
53. King, R. B. (1969), *J. Amer. Chem. Soc.* **91**, 7211
54. King, R. B. (1970). *J. Amer. Chem. Soc.* **92**, 6455, 6460
55. Lauterbur, P. C. and Ramirez, F. (1968). *J. Amer. Chem. Soc.* **90**, 6722
56. Mislow, K. (1970). *Accounts Chem. Res.* **3**, 321
57. Muetterties, E. L., Mahler, W. and Schmutzler, R. (1963). *Inorg. Chem.* **2**, 613
58. Muetterties, E. L., Mahler, W., Packer, K. J. and Schmutzler, R. (1964). *Inorg. Chem.* **3**, 1298
59. Muetterties, E. L. (1965). *Inorg. Chem.* **4**, 769
60. Muetterties, E. L. and Schunn, R. A. (1966). *Quart. Rev.* **20**, 245
61. Muetterties, E. L. (1967a). *Inorg. Chem.* **6**, 635
62. Muetterties, E. L. and Wright, C. M. (1967b). *Quart. Rev.* **21**, 179
63. Muetterties, E. L. and Knoth, W. H. (1968a). "Polyhedral Boranes" Marcel Dekker, New York
64. Muetterties, E. L. (1968b). *J. Amer. Chem. Soc.* **90**, 5097
65. Muetterties, E. L. (1969a). *J. Amer. Chem. Soc.* **91**, 1636
66. Muetterties, E. L. and Storr, A. T. (1969b). *J. Amer. Chem. Soc.* **91**, 3098
67. Muetterties, E. L. (1969c). *J. Amer. Chem. Soc.* **91**, 4115
68. Muetterties, E. L. (1976). "General Structural Stereochemical Principles and Topological Analysis", to be published
69. Muetterties, E. L. (1970b). *Accounts Chem. Res.* **3**, 266
70. Murnaghan, F. D. (1938), "The Theory of Group Representations" John Hopkins Press, Baltimore
71. Porter, C. W. (1928), "Molecular Rearrangements", The Chemical Catalog Company
72. Ramirez, F. Pfohl, S., Tsolis, E. A., Ugi, I., Marquarding, D., Gillespie, P. and Hoffman, P. (1976). *J. Amer. Chem. Soc.*, to be published
73. Ray, P. and Dutt, N. K. (1943). *J. Indian Chem. Soc.* **20**, 81
74. Riordan, J. (1958). "An Introduction to Combinational Analysis", John Wiley
75. Robson, A. J., private communication
76. Ruch, E. and Schönhofer, A. (1970a). *Theoret. Chim. Acta (Berl.),* **19**, 225
77. Ruch, E., Hässelbarth, W. and Richter, B. (1970b). *Theoret. Chim. Acta (Berl.),* **19**, 288
78. Rutherford, E. E. (1948). "Substitutional Analysis" Edinburgh University Press
79. Schmidtke, H. H. (1967). *Coord. Chem. Rev.* **2**, 3
80. Schmutzler, R. (1965), *Advan. Fluorine Chem.* **5**, 31
81. Springer, C. S. and Sievers, R. E. (1967). *Inorg. Chem.* **6**, 852
82. Tee, O. S. (1969). *J. Amer. Chem. Soc.* **91**, 7144
83. Udovich, C. A., Clark, R. J. and Haas, H. (1969). *Inorg. Chem.* **8**, 1066
84. Ugi, I., Marquarding, D., Klusacek, H., Gokel, G. and Gillespie, P. (1970). *Angew. Chem. Intern. Ed. Engl.* **9**, 703

85. Ugi, I., Marquarding, D., Klusacek, H., Gillespie, P. and Ramirez, F. (1971). *Accounts Chem. Res.* **4**, 288
86. Wajc, S. J., Mansour, A. H. and Jottrand, R. (1965). *Rev. Institut francais Pétrole*, **20**, 849
87. Westheimer, F. H. (1968). *Accounts Chem. Res.* **1**, 70
88. Whitesides, G. M. and Bunting, W. M. (1967). *J. Amer. Chem. Soc.* **89**, 6801
89. Whitesides, G. M. and Mitchell, H. L. (1969). *J. Amer. Chem. Soc.* **91**, 5384
90. Willem, R., unpublished results
91. Brocas, J. and Willem, R. (1973). *Bull. Soc. Chim. Belges.* **82**, 479

NOTE ADDED IN PROOF

Recently, some new properties of graphs related to organometallic chemistry have been examined.

The multiplication properties of the topological matrices associated to the graphs described above have been written down for the ligand partition (12345) for the ·modes* corresponding to trigonal bipyramidal symmetry[2,3] and for the ligand partition (123456) for the modes corresponding to octahedral symmetry.[4] These results based on the definition of the modes proposed by Ruch,[5] have been established independently using the double coset formalism.[5] Alternative definitions have however been proposed.[6,7]

The eigenvalues of the matrices associated to the graphs for any mode and any ligand partition of the trigonal bipyramid and the octahedron as well as the corresponding time evolution of the concentrations of the isomers may be found in references (2), (3), (4) and (8).

Balaban[9] has represented mode P7 of the octahedron by a 15-vertex regular graph of degree 6 and a 30-vertex regular graph of degree twelve. Mode P6 corresponds to a 30-vertex regular graph of degree 8. For pentacoordinate stereoisomerizations, the consideration of the regular combination graph for P1 leads the same author[10] to a notation of the isomers which is different from that adopted in this work.

The introduction of a generator of order ten classifies the 30 isomers of octahedral symmetry into three families and gives rise[11] to new and simple graphical topological representations of the isomerization modes of these complexes.

Mislow[12] has defined the concept of stereochemical correspondence, i.e. the isomorphic behaviour of different chemical systems (such as bidentate spiro-phosphoranes and substituted diaryl derivatives).

REFERENCES

1. Musher, J. I. (1972). *J. Am. Chem. Soc.* **94**, 5662; Also *Inorg. Chem.* **11** (1972), 2335
2. Brocas, J. and Willem, R. (1973). *Bull. Soc. Chim. Belges.* **82**, 469
3. Brocas, J. (1972). *Topics in Current Chemistry*, **32**, 43

*This word has been introduced by Musher(1).

4. Brocas, J., Fastenakel, D., Hicquebrand, J. and Willem, R. (1973). *Bull. Soc. Chim. Belges.* **82**, 629
5. Hässelbarth, W. and Ruch, E. (1973). *Theoret. Chim. Acta.* **29**, 259
6. Meakin, P., Muetterties, E. L., Tebbe, F. N. and Jesson, J. P. (1971). *J. Am. Chem. Soc.* **93**, 4701
7. Klemperer, W. G. (1972). *J. Chem. Phys.* **56**, 5478; Also *Inorg. Chem.* **11**, (1972), 2668; *J. Am. Chem. Soc.* **94**, (1972), 6940, 8360; **95**, (1973), 380
8. Brocas, J. and Willem, R. (1973). *Bull. Soc. Chim. Belges.* **82**, 479
9. Balaban, A. T. (1973). *Rev. Roum. Chim.* **18** 841
10. Balaban, A. T. (1973). *Rev. Roum. Chim.* **18**, 855
11. Gielen, M., Willem, R. and Brocas, J. (1973). *Bull. Soc. Chim. Belges.* **82**, 617
12. Gust, D. and Mislow, K. (1973). *J. Am. Chem. Soc.* **95**, 1535; Also Gust, D., Finocchiaro, P. and Mislow, K. (1973). *Proc. Nat. Acad. Sci, USA,* **70**, 3445

CHAPTER 10

The Graph-like State of Matter and Polymer Science

Manfred Gordon and William B. Temple

*Institute of Polymer Science and Department of Chemistry,
University of Essex, Colchester, Essex, England*

1   INTRODUCTION

The following five propositions usually seem far from self-evident to chemists and physicists:

(1) The basic mathematical tool of chemical kinetics and statistical mechanics is graph theory.
(2) Graphs are one-dimensional objects
(3) A molecular graph, once it is embedded in 2D or 3D (two-dimensional or three-dimensional) space, is not just a combinatorial shorthand device. It must be taken seriously as a mechanical model for a structured particle.
(4) The difficult and important aspects of molecular graphs lie in their symmetry properties.
(5) The model known as f-functional random polycondensation is the basic paradigm of chemistry.

The following sections (1.1–1.5) provide arguments in favour of these propositions, and serve to introduce the notion of a graph-like state of matter, which forms the subject of the remainder of this chapter.

1.1   Graph theory as a basic tool

The most fundamental model of physics and chemistry is the perfect gas. It no longer raises interesting problems, but as we proceed to generalize it, the immediate problems posed are graph-theoretical. Figure 1 shows schematically the two directions in which the perfect gas model can conceivably be generalized: through introducing forces between the 0D (zero-dimensional or point-) particles, or through structuring the particles by raising their dimensionality from zero to unity. The first route, involving intermolecular forces, led Mayer and Mayer[48] to the theory of cluster integrals, and the central graph-theoretical notion of the reducibility of cluster graphs (cf. Isihara[40]). The second route involves 1D particles which – by definition (see Section 1.2) – are graphs. Here the important graph-theoretical notion is that of an ordered rooted graph. Figure 1 shows the interesting analogy between the rejection of reducible clusters and of terminal-rooted trees in the graph-theoretical simplification of the two routes of generalizing the perfect gas. The analogy continues far beyond the rejection of irrelevant graphs. The graphs retained as relevant in both routes to generalization are enumerated by the coefficients of suitable generating functions, and the mathematical manipulation of the generating functions forms the pathway to the results concerning physical or chemical properties of the models. Along both routes, these generating functions are closely related to partition functions: the cumulants in non-ideal fluids[40] or the weight–fraction generating functions (see Section 7.2) for matter in the graph-like state treated here.

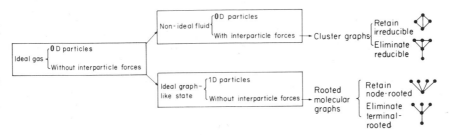

FIG. 1   Hierarchy of the simplest physico-chemical theories. 0D and 1D mean zero dimensional and one dimensional respectively.[31]

Apart from the central role graphs play in serving as models for the structure of particles, they also serve in the analysis at the next higher level, i.e. as models of networks of chemical reactions, which change the structure of particles. The representation of reaction networks by graphs is illustrated elsewhere in this book. Important results on the stability of solutions to systems of differential kinetic rate equations, in terms of network graphs, have been published by Horn.[37] He and his co-workers have shown that qualitative features (like the existence of oscillatory solutions) are deducible from network graphs, are important in cell biochemistry, and in the theory of microscopic reversibility.

In summary: the first serious problems posed by the models of physics and chemistry are graph-theoretical in nature.

1.2   Graphs are one dimensional

Every mathematician, but few physicists or chemists, knows that a graph is a 1D object, even though it may be *embedded* (or realized) in spaces of higher dimensionality (for instance, all the graphs shown in this book are embedded in the plane of the printed page). The dimensionality of a space (e.g. a line, a surface or a cube) is defined as the number of independent coordinate parameters required to locate a point within the space, as is familiar from vector algebra. Of course, a single coordinate will locate every point on a line. At this stage, physicists often object that a branched line, or a graph containing loops, must surely be two- or three-dimensional. This is not so, because a single coordinate still locates every point — all we require is a rule for putting the different branches or loops into a defined sequence, (a task accomplished by certain coding systems).

Graph theory has two divisions: combinatorial and topological graph theory. The combinatorial division is illustrated by problems of counting graphs: the topological one is entirely concerned with problems arising from embedding graphs in spaces of higher dimensionality, e.g. in the plane. The history of graph theory exhibits a trend: problems thought at first to be topological in nature, on analysis move into the combinatorial arena. This happened to the notion of planarity of graphs,[34],[45] and to the famous four-colour problem of maps.[58] Sections of this chapter deal with the elementary combinatorics necessary in statistical mechanics and kinetics.

1.3   The graph-like particle in a box

Much of physics, and all of chemistry, deals with assemblies of more than one particle in some enclosure or 'box', i.e. with many-body problems. We have seen that the first step in introducing internal structure into the particles turns them into 1D graphs (see Section 1.1). A *molecular graph* is typically a set of points, labelled (e.g. ^{12}C) to represent atoms, plus a set of lines representing the inter-atomic bonds (for more details see Section 6). It is tempting to attribute to molecular graphs the role often played by graphs in mathematics, viz. what H. Sachs, in his excellent introduction to the theory of finite graphs,[52] called a shorthand for combinatorial information. This is, indeed, *one* role played by molecular graphs, and the main part fulfilled by Feynman graphs (see Mattuck[47]) in the quantum-theoretical treatment of many-body theories.

However, the laws of statistical mechanics assign to molecular graphs the much more fundamental role of real particles. This is because neither statistical mechanics, nor molecular kinetic theory, can be practised without a properly defined (classical or quantum) dynamic model, typically consisting of particles with defined generalized coordinates, moving under defined (zero or finite) force fields. Thus the molecular graph has to be embedded in some higher-dimensional space, typically 3D Euclidian space contained in a box. We shall show in Section 1.4 that the interesting and useful properties of molecular graphs stem from their symmetry properties. The embedding of a molecular graph in 3D space should, therefore, preserve its symmetry properties, i.e. it should remain *graph-like*. In general this is achieved by maintaining the particle, not in its 'real state' such as in Fig. 2a, but in a completely flexible, floppy condition, as illustrated in Fig. 2b. Then any symmetry operation which permutes sets of labelled points of the molecular graph can still be performed on the graph-like particle embedded in three dimensions. If, however, we 'freeze' the 3D particles into the more familiar kind of molecular model (Fig. 2a), some symmetry operations are lost, and many theories become hopelessly difficult. The ideal graph-like state (floppy particles without interparticle forces) is based on molecular models which are repulsive to chemists (Fig. 2b), but it furnishes the only rigorous foundation for the familiar elementary approximate theories of many-body systems. In Section 1.5, we shall recall some of the controversies which for many years tried to call in question the theory of f-functional random poly-condensation, because *ad hoc* combinatorial arguments were used without reference to defined particle structure. As a second example of a graph-like state theory we mention the Flory–Huggins model of solutions, a fundamental model

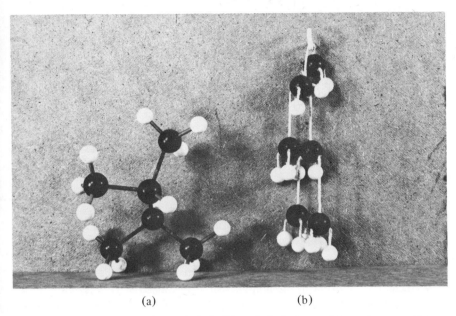

(a)   (b)

FIG. 2   Molecular models of 2,3 Dimethyl butane. (a) customary three-dimensional model (b) graph-like model.

for polymer physics. Although both Huggins[38] and Flory[13] originally embedded the particles in a rigid 3D ('pseudo') lattice structure, Kasteleyn[43] stressed that the problem, as formulated, relates purely to the lattice graph. Silberberg[54] recently rederived the result of the theory much more simply, without the lattice graph, by embedding graph-like polymer chains in an otherwise empty box. In fact, he described the chemical bonds as freely extensible, so that each polymer repeat unit could be anywhere in the box, irrespective of the positions of the other repeat units in the same polymer chain. Indeed, the whole stock-in-trade of elementary pre-quantum theories applies most simply to 1D particles illustrating Schmidtke's dictum that 'graph theory anticipates molecular orbital theory'.[53]

1.4   The source of difficulty in graph-like theories: combinatorial weights due to symmetry

Partition functions constitute the basic tool of statistical mechanics, because from them all thermodynamic quantities are immediately accessible. The general form of the partition function is

$$Z = \sum_i g_i e^{-\epsilon_i/kt} \tag{1}$$

For any model under study, this presents only two sources of difficulty: the energy levels ϵ_i and the statistical or combinatorial weights (degeneracies) g_i. The ϵ_i are in general found from quantum theory, but in graph-like state models they follow from simple classical considerations. In the classical theory of rubber elasticity, for instance, only a single energy level ϵ_0 is occupied! Again, in the Flory–Huggins theory (see 1.3), the energy levels are replaced by their maximum term, which is immediately written down as a quadratic function of concentration. The extension of this model by Gibbs and DiMarzio, towards a model for the glass transition, still has an exceedingly simple energy contribution to the partition function.[16] These three statistical–mechanical theories are quite typical, and by no means trivial. Their real content must come from the weights g_i, and so it does: in each of the three cases only very rough, approximate g_i's are known.

The weights g_i which cause difficulty are combinatorial rather than statistical weights. In classical theory they are deterministic, no information is condensed, and so probability theory, statistics or 'randomness' does not come in. (We shall retain the word 'random mixture' for its intuitive value, but define it exactly). The difficulties inherent in the combinatorial weights arise from symmetries in the internal structure of molecular graphs, or in their configuration structure after embedding them in 1D, 2D or 3D spaces. Quantum theory is difficult, because the differential equations in which we clothe it are hard to solve. Pre-quantum or graph-like-state theories are difficult, because it is confusing to count symmetrical objects.

1.5   The basic paradigm of chemistry

Chemistry is about the making and breaking of bonds. Chemists agree that the most elementary instance is the reaction

$$A + A \rightleftharpoons A_2 \tag{2}$$

This formation of a diatomic element still poses important problems even for hydrogen, but they are problems to be tackled by quantum theory. A search for more basic (pre-quantum) problems, but problems of wide generality not devoid of intellectual content, immediately directs us towards Flory's f-functional random polycondensation model (for details see Section 7), as the first significant generalization of (2), which merely constitutes the special case $f = 1$.[14]. (Each monomer has a single reactive site with bond forming capability.)

The statistical groundwork for the model was correctly derived by Flory, using ad hoc combinatorial arguments. Stockmayer[56] made a move towards obtaining the relevant results with the rigorous tools of statistical mechanics, and to generalize them in various ways. He used the principle of maximization of entropy, which is a useful route in statistical mechanics, but equally characteristic of many purely statistical (non-*mechanical*) theories (see e.g. Good[21]). In fact, the combinatorics were still of an *ad hoc* nature. Plates, nuts, bolts and washers were introduced, not of course as mechanical models for actual molecules, but as a combinatorial shorthand. Later Fukui and Yamabe[15] applied Stockmayer's approach, based on Lagrangian multiplier techniques, to more complex models. Good[21,22] used the method of Lagrange expansion of weight fraction generating functions for treating the basic paradigm (cf. Gordon[24]), and Whittle founded a new approach to it on the theory of stochastic processes.[59,60] Whittle's work deserves special study because of its rigour and because it points to a gap in the bridge between kinetic and statistical mechanical derivations of the theoretical results (see Section 2). Meanwhile, the whole theory of *f*-functional random polycondensation had become the target of a number of attacks.[3,46]. Though few were convinced by claims that the theory was downright erroneous, the diverse methods of derivation did leave something to be desired.

Two weaknesses may be discerned. First, to attribute symmetry numbers of one-dimensional molecular graphs to the customary 3D models for molecules would violate the requirement of a properly defined dynamical model, without which statistical *mechanics* cannot be practiced. Yet such an attribution was made implicitly in all past treatments of *f*-functional condensation, and then explicitly by Gordon and Judd.[61] The implicit attribution is easily exposed (see Section 7.2), if we look at the partition function derived and, from its algebraic structure, deduce the symmetry numbers attributed to the particles in the system. The relevant combinatorial invariance principle, which regulates the relation between graph-like and 3D modes, is derived in Section 5.

Secondly, there was a gap between kinetic (or stochastic) derivations on the one hand, and statistical mechanical ones on the other. Whittle remarked that his stochastic equation 10 'has no basis in statistical mechanics'.[59] The missing link is combinatorial, and concerns the relation between statistical rate factors in chemical kinetics and symmetry numbers in statistical mechanics. This issue is raised as soon as we generalize eq. (2) merely to *f*-functional random polycondensation, but it is a perfectly general problem for a unified theory of chemical kinetics, thermodynamics and statistical mechanics. In Section 2 we review the theorem which solves this problem.

2   A THEOREM CONNECTING STATISTICAL RATE FACTORS WITH SYMMETRY NUMBERS

The following simple equations embody the relations between (a) *chemical kinetics*, through rate constants k_1 in the forward direction and k_2 in the backward direction of a reaction; (b) *thermodynamics*, through equilibrium constant K and standard free energy change ΔF° (an adequate approximation to the standard free enthalpy change ΔG° for a condensed system); and

(c) *statistical mechanics*, through the partition functions Z_i and stoichiometric coefficients ν_i, negative for reactants and positive for products:

$$\frac{k_1}{k_2} = K = e^{-\Delta F^\circ / RT} = \prod_i Z_i^{\nu_i} \tag{3}$$

We justify omitting the customary exponential factor multiplying $\Pi Z_i^{\nu_i}$ by choosing the ground state of the monatomic elements as energy zero. In almost every detail, these equations have long been fully understood. For instance, the activation energies arising when k_1 and k_2 are considered as functions of temperature, appear in ΔF° ($= \Delta U^\circ - T\Delta S^\circ$) as the difference $E_2^* - E_1^* = \Delta U^\circ$; and in $\Pi Z_i^{\nu_i}$, this energy ΔU° becomes distributed in some way between the energy levels of reactants and products. However, if we turn to the statistical rate factors (of k_1 and k_2) which often appear, due to symmetry of reactants or products, the situation has seemed less clear. In ΔF° they must become converted to some contribution to ΔS° alone, because they are temperature-invariant. But what happens to this 'combinatorial entropy', contained in ΔS°, when we pass to the partition functions? The following theorem shows that in all possible reactions, the (temperature-invariant) rotational symmetry numbers in the partition functions Z_i exactly take care of the situation.

We first recall that all chemical reactions can be obtained as linear combinations of a set of 'canonical' reactions, those assembling any species from its constituent atoms. It suffices to consider the synthesis of an arbitrary species from its atoms. For a species containing y chemical links, i.e. counting each chemical bond as a link, irrespective of its multiplicity, the synthesis will be conducted in $y + 1$ steps.

In the first y steps the links are inserted one by one, between the relevant pairs of atoms in arbitrary sequence, while all molecules successively produced in this way are taken in their graph-like floppy states (Fig. 2b). In the final, $(y + 1)$th, step, the graph-like product species is converted unimolecularly to its 3D state (Fig. 2a), which may be rigid or contain freely rotating bonds. The first y steps can be made in $y!$ sequences, according to the order of inserting the links – though some of these sequences may in fact be indistinguishable because of symmetry. For each of the y steps of a given sequence, we can form the ratio κ_1/κ_2 of the statistical rate factors in the forward and backward directions. Forming the product of their ratios over the y steps of the sequence, leads to a temperature-invariant factor in the overall equilibrium constant K of the synthesis. Thermodynamics demands that this factor must be invariant also to the sequence chosen for the y synthetic steps. Of course it is so, but it is not obvious. Two sequences of forward statistical rate factors alone are not necessarily constant. For instance forming A–B–B–A in the sequences A–B, A–B–B, A–B–B–A gives the set of statistical rate factors 1, 1, 1 (each step being asymmetric). But in the sequence A–B, B–A, A–B–B–A the third step is symmetric, and the set of rate factors is 1, 1, ½. But the ratio of the forward set to the backward set is indeed constant (= ½). The (purely graph-theoretical) result (see Gordon and Temple[31,32]) is quoted without proof:

$$\prod_{i=1}^{y} \frac{\kappa_{1i}}{\kappa_{2i}} = 1/|G| \tag{4}$$

|G| is the order of the automorphism group of the molecular graph of the final product (e.g. for A–B–B–A, |G| = 2). This |G| is clearly independent of the sequence of steps, and of the properties of all the intermediates.

In the $(y+1)$th step, the graph-like particle becomes 'frozen', or at least partially frozen, up to the point where free bond rotation is still allowed, into its 3D structure. It then has a new symmetry group, viz, its rotational symmetry group R, consisting in general of the external symmetry operations on the molecule as a whole, plus any internal symmetry operations like those arising from the three-fold axis of a rotating methyl group. This group R is, of course, a subgroup of the automorphism group G of the graph-like molecule.[51] By Lagrange's theorem, the ratio of the orders $|G|/|R|$ is an integer g, and here g tells us in how many equivalent ways the graph-like particle can become converted to the 3D form; the 3D form can 'relax' to the graph-like state in only one way. In other words, g is the statistical equilibrium factor of the $(y+1)$th step in the synthesis of the 3D molecule. Accordingly, the equilibrium constant of the formation of the real 3D molecule from its atoms thus has an overall factor.

$$g \prod_{i=1}^{y} \kappa_{1i}/\kappa_{2i} = g/|G| = 1/|R| \tag{5}$$

In words: the product of the statistical equilibrium factors in the phenomenological equilibrium constants along any route for synthesizing a molecule from its constituent atoms equals the reciprocal symmetry number of the molecule.

The theorem of eq. 5 shows that the phenomenological equilibrium constant of a canonical reaction (assembling a species from its atoms) has a factor 1/|R|. If this factor is removed, the *fundamental equilibrium constant* remains, which is a measure of chemical driving forces, suitable for comparing such forces without accidental effects due to symmetries. It follows that the fundamental equilibrium constant of a non-canonical reaction, e.g. a metathesis, is found by removing the reciprocal symmetry numbers of the products, and the symmetry numbers of the reactants, which occur as factors in the phenomenological equilibrium constant. This procedure is the rationale behind the calculation of 'relative rate factors' in organic chemistry. The theorem also disposes of the controversies concerning conflicting uses of statistical factors or symmetry numbers in the theory of activated complexes (see Murrell and Laidler[50]), whose very purpose is a bridging between kinetic and thermodynamic theories.

Once a defined set of reagents and a defined transition complex are postulated, their symmetry numbers contain all the information required, and statistical factors can be dispensed with. This follows because the synthesis equilibria covered by eq. (5) form a canonical set from which all possible equilibria of reactants and complexes can be constructed.

Turning to the product $\Pi Z_i^{\nu i}$ over powers of the partition functions in eq. 3, we see as the result of our purely kinetically based argument that the partition function of a 3D molecule must have a factor $1/|R|$, its reciprocal 3D symmetry number. We already know of the presence of this factor from statistical mechanics, where it prevents the overcounting — as a result of symmetry — of the states in the phase integral. The proof of eq. (4) (and hence eq. 5) thus forms a bridge between chemistry and physics. This graph-theoretical theorem explains why statistical kinetic and statistical mechanical treatments of equilibria, especially in polymer science, must actually agree. It should also dispose of the paradox inherent in the following situation: the properties of a complex equilibrium state (say a cross-linked polystyrene gel) can be found from the statistical rate factors of polymerization and depolymerization, even though the actual rates of these processes may be immeasurably slow. The reason why the physical properties of a system follow entirely from a set of non-occurring processes, lies in the graph-theoretical theorem (5), which guarantees that we weight each species present correctly by its reciprocal symmetry number in the phase integral (partition function).

3 PARTITION FUNCTIONS AND MOLECULAR DISTRIBUTIONS

Next, we develop the role of graph-like models of molecules in simplifying the calculation of molecular distributions, and the invariance principle which regulates such simplifications. The basic mechanical principles are put in suitable form for our purpose as follows.

Consider the typical canonical reaction which assembles the ith species S_i, from its constituent atoms, say ν_{i1} atoms of type 1, ν_{i2} atoms of type 2, etc.:

$$\nu_{i1}A_1 + \nu_{i2}A_2 + \ldots \rightleftarrows S_i \qquad (6)$$

In terms of the partition functions Z_i of the species, and Z_j of an atom of type j, we have for any equilibrium mixture containing S_i

$$n_i = \mathcal{N} Z_i / \Pi_j Z_j^{\nu_{ij}} \qquad (7)$$

Here n_i is the mole fraction of species i in the mixture, and $\mathcal{N}$ will denote throughout the appropriate normaliser. We may choose the reference state (energy zero) of each atom such that

$$Z_j = 1 \quad (j = 1, 2, \ldots)$$

This means the reference state for Z_i represents separated atoms at absolute zero provided the ground state of the atom is non-degenerate. From Z_i we may factor out the reciprocal symmetry number $1/|G_i|$ for a graph-like molecule, or $1/R_i$ for a 3D type molecule, etc. We shall use the general symbol σ_i for the symmetry number when we do not wish to specify the geometrical form of species i:

$$Z_i = a_i \sigma_i^{-1} \tag{8}$$

We propose to concentrate on the lower route in Fig. 1, i.e. on intraparticle effects. Then a_i is the internal partition function of one molecule of species i, unaffected by the presence of other particles. In other words, in an ideal-gas-like system, the partition function of the overall system factorizes into the partition functions of the individual particles. In real systems free from ionic charges e.g. a liquid polycondensate, this is often still a good approximation. If not, the usual theory of correction for non-ideality — e.g. those outlined in the top route in Fig. 1 — are applicable to the treatment we are about to develop.

From equations six to eight, the molecular distribution is given by

$$n_i = a_i \sigma_i^{-1} / \Sigma_j a_j \sigma_j^{-1} \equiv \mathcal{N} a_i \sigma_i^{-1} \tag{9}$$

where the summation is over all species included in the normalization.

Quantum theory provides in principle a pathway to the *a priori* calculation of the internal partition function a_i of any species i. In practice, such calculation is rarely feasible. Graph-theory provides the makeshift which is very general, and practically universally employed, for the calculation from empirical data of standard thermodynamic data, and hence implicitly of partition functions. As yet, few chemists are aware of this graph-theoretical formulation though they would generally recognize individual instances, usually termed 'additivity schemes'. Recently, Gordon and Kennedy systematized the underlying theory in a form addressed to chemists.[2,5] In the following section, which the reader may omit because we shall not be concerned with it again, we summarize their work in formal graph-theoretical language, and therefore very economically. Here it suffices to say that additivity of free energies inevitably means the development of the partition function as a multifactor product, and we shall require only the two terms which are used to start such an expansion. These two first terms constitute the so-called *bond additivity* assumption.

$$a_i = \prod_{j_0, j_1} c_{j_0}^{s_{ij_0}} c_{j_1}^{s_{ij_1}} \tag{10}$$

The contribution to the partition function a_i by an *atom* of type j_0 (e.g. carbon) is denoted by c_{j_0}, while s_{ij_0} signifies the number of such atoms in a molecule of species i. Similarly c_{j_1} denotes the contribution from a *bond* of type j_1 (e.g., a C–H bond) and s_{ij_1} the number of such bonds in species i. (The bond additivity model is not generally successful unless different contributions c_{j_0} are assigned to different valency states of the same atom. Then the set of bonds fully defines the set of atoms present.) It is useful here, to retain the factor due to the c_{j_0}, i.e. contributions from atoms, even though it is usually absorbed in the bond contribution factor of the c_{j_1} by virtue of the choice of reference states for the atoms. Substituting eq. 10 in 9, we obtain our working equation for equilibrium molecular distributions;

$$n_1 = \sigma_i^{-1} \prod_{j_0, j_1} c_{j_0}^{S_{ij_0}} c_{j_1}^{S_{ij_1}} \Big/ \sum_k \prod_{j_0, j_1} c_{j_0}^{S_{kj_0}} c_{j_1}^{S_{kj_1}} \qquad (11)$$

$$= \mathscr{N} \sigma_i^{-1} \prod_{j_0, j_1} c_{j_0}^{S_{ij_0}} c_{j_1}^{S_{ij_1}} \quad \text{(bond-additivity)}$$

Under the bond-additivity regime, no potential exists which can cause the bonds within a particle to interact with each other, just as in the ideal gas no potential causes interaction between particles. Anticipating the detailed discussion in Section 6, we note that stereoisomers and structural isomers have identical a_i according to (10), and by (11) stereoisomers have mole fractions proportional to that of their reciprocal symmetry numbers. Such a mixture of stereoisomers is termed *random*; and such randomness is a feature of *f*-functional random polycondensates, or generally systems obeying bond-additivity. The special case of bond-additivity and graph-like particles (Fig. 2b), i.e. with σ_i being denoted by |G| is called the *ideal graph-like state*. Before we extend our analysis of this state, we interpose the next section which deals with the generalization to non-ideal graph-like systems.

3.1   The non-ideal graph-like state

Equation 10 constitutes the case $n = 0$ of the general equation

$$a_i = \prod_j \prod_{k=0}^{n+1} c_{j_k}^{[G_{j_k}, G_i]} \qquad (12)$$

which follows from the formulation by Gordon and Kennedy of a canonical hierarchy of additivity schemes.[25] Here G_i is the molecular graph of species i; G_{j_k} is the jth connected section graph of diameter k (in arbitrary fixed enumeration) belonging to G_i, while $[G_{j_k}, G_i]$ denotes the strong lattice constant of G_{j_k} in G_i. Moreover, c_{j_k} is the contribution to the partition function of one section graph of type j_k, and n denotes the range of the substituent effects considered (*n*th shell substitution effect). A multitude of additivity schemes in the literature have been shown to be special cases of (12), which is itself a special case of LCGI (linear combination of graph theoretical invariants), a kind of scheme widely used in chemical physics. The invariants concerned are then not, in general, restricted to strong lattice constants.

The choice of an optimum scheme conforming to eq. 12 to fit given thermodynamic data has been discussed in terms of statistical principles.

Attention has been drawn to graph-theoretical constraints on some schemes proposed in the literature.[2 5]

4   DEFINITIONS

4.1   3D-rot state and (contracted) molecular graphs

We propose to call *3D-rot state* the state of a molecule, in three-dimensional space, possessing freely rotating links (chemical bonds). As always, the *molecular graph* of such a molecule consists of a set of points (representing its atoms) plus a set of lines (representing its bonds), together with colours or atomic labels for the points to identify their chemical nature to the degree of detail required (e.g. [12]C). If a 3D-rot molecule has rigid (non-rotating) links (e.g. double bonds) as well as freely rotating ones, we shall simplify its molecular graph by submitting all the lines representing rigid links to *elementary contraction* (i.e. we contract them away and identify the end-points of each line contracted away). What remains could be called the *contracted molecular graph*. Clearly each point of such a contracted graph represents a rigid portion of the 3D-rot molecule. In chemistry such a rigid set of atoms (or a single atom) is conveniently called a *centre*, by generalization of the notion of a centre in inorganic chemistry (e.g. a rigid octahedral centre to which ligands are attached). The notion of contracted molecular graphs is illustrated in Fig. 3. Note that the centres are given labels to identify their chemical nature; and links bearing a univalent atom — which cannot be said to 'rotate' at all — are left uncontracted. Henceforth we shall

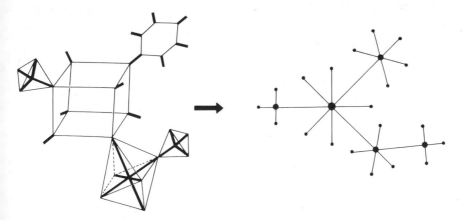

FIG. 3   Illustration of the contraction of a 3D molecular structure to a molecular graph. Thick bonds in the chemical structure indicate those left uncontracted. Chemically distinct elements and centres retain their labels in the contracted molecular graph.

TABLE I

3D-rot molecule		molecular graph
structural unit { multivalent centre ⟶	node	} point
univalent centre ⟶	terminal	
link ⟶		line

refer to contracted molecular graphs merely as molecular graphs. The Table I establishes the nomenclature of pairs of objects which correspond under the mapping relating a 3D-rot molecule and its molecular graph.

4.2   Isomers

Two or more distinguishable (i.e. non-isomorphic) 3D molecules which have indistinguishable (isomorphic) molecular graphs are called *stereoisomers* (we include 'geometric' isomerism in stereo-isomerism). Molecules whose molecular graphs are non-isomorphic, but share the same specification for their set of labelled centres and set of bonds, are called structural isomers.

4.3   Mapping (embedding) models of molecular graphs in 3D-rot forms

A given molecular graph can be mapped (or embedded) into one of its corresponding stereoisomeric forms; often, because of symmetry, this can be achieved in more than one way. The mapping process, illustrated in Figs. 3-5, typically sends a node into an interior point of an (eight-cornered) cube. For convenience the interior point is chosen as the geometric centre of the polyhedron, and it is implied further that the lines which in the molecular graph radiate from the node, become embedded as lines radiating from the centre of the polyhedron to its corners. Then the term node of a graph corresponds to the word centre in both the chemical and the geometrical sense.

In Figs. 3—5, we merely *imply* the familiar notion, that the locations and directions of the (freely rotating) links issuing from a polytope (polygon or polyhedron) are completely defined by the positions of its corners and its centre. The same information can be explicitly given by inserting pins (to represent the links), and this is done at the bottom of Fig. 5, where the polyhedra are replaced by spheres.

4.4   The number of distinct rooted ordered trees

A molecular graph may be embedded in 2D space, specifically in the half-plane, $(y \geqslant 0)$. Let one of the points (atoms) of the embedded graph be chosen as *root*,

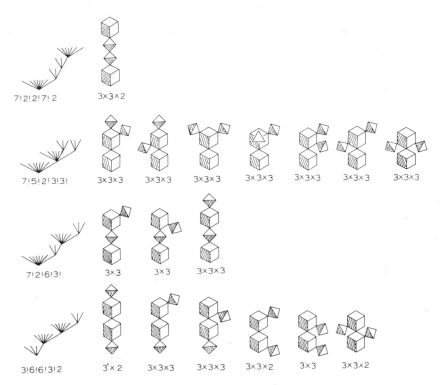

FIG. 4   Four of the six structural isomers and their seventeen stereoisomers from the system composed of two cubes and two tetrahedra free to rotate about their linkage points. The differing graph-like and 3D symmetry numbers and the differing number of stereoisomers are reconciled under the invariance relation furnished by eqn. (15).

and placed on the x-axis. The number of distinct rooted planar trees, which can be obtained by choosing various points as root and by ordering the branches in the half-plane in various ways, is found to be a very convenient concept. It is illustrated for ethane in Fig. 6. We see there that of the 7 distinct rooted planar trees of ethane, three are terminal-rooted, and $T = 4$ are node-rooted. As hinted in Fig. 1, the terminal-rooted forms may be discarded. The number T of distinct node-rooted planar trees plays the important role of a kind of reciprocal 2D symmetry number in the sequel.

5   THE INVARIANCE PRINCIPLE

The principle to be derived shows that important quantities connected with molecular symmetry remain proportional to each other as we pass from one dimension to two or three. The principle will be cast in the form of two simple equations, one relating symmetry numbers of 1D molecular graphs to the number of distinct rooted ordered 2D (i.e. planar) trees, and the other to the symmetry numbers of 3D molecules. The principle is a combinatorial one, derived by the principle of labelling and delabelling (cf. De Bruijn[2]), which is natural, because space coordinates are labels given to points.

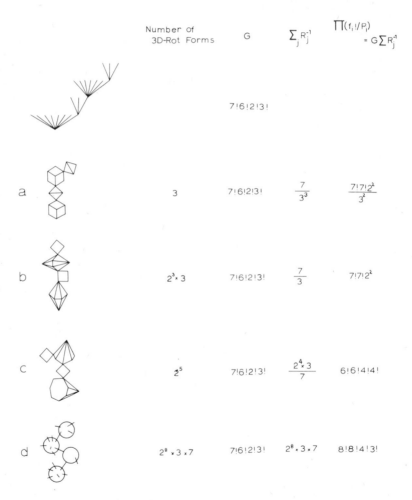

Number of 3D-Rot Forms	G	$\sum_j R_j^{-1}$	$\prod(f_i!/P_i)$ $= G\sum_j R_j^{-1}$
		$7!6!2!3!$	
	3	$7!6!2!3!$	$\dfrac{7}{3^3}$	$\dfrac{7!7!2^2}{3^2}$
	$2^3 \times 3$	$7!6!2!3!$	$\dfrac{7}{3}$	$7!7!2^2$
	2^5	$7!6!2!3!$	$\dfrac{2^4 \times 3}{7}$	$6!6!4!4!$
	$2^8 \times 3 \times 7$	$7!6!2!3!$	$2^8 \times 3 \times 7$	$8!8!4!3!$

FIG. 5　The invariance law of eqn. (15) applied to specific structural isomers of four concatenated centres. The centres become increasingly asymmetrical as we descend the series, although all the 3D structures share the same graph-like representation. The cubes and tetrahedra (a) form the system of highest symmetry, while the four spheres with totally asymmetric linkage points represented by pins (d) is completely devoid of symmetry.

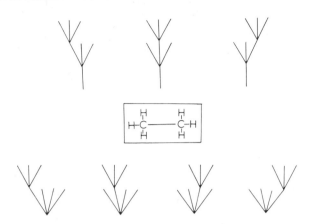

FIG. 6   The seven distinct rooted ordered trees derived from ethane. The top three are the *terminal*-rooted trees, and the bottom four are the *node*-rooted trees, which form the basis of our interest. The roots in each line lie on the x-axis which has been omitted.

We turn first to the invariance relation between 1D and 2D objects. Let the symmetry number of the molecular graph be $|G|$ and let its ith vertex bear f_i links. We then have the following theorem:[29]

$$T_M = (\mathbf{f} - 1)! \sum_i f_i / |G| \tag{13}$$

where T_M denotes the number of distinct rooted ordered trees furnished by a molecular tree, a concept illustrated in Fig. 6. We have used the notation $(\mathbf{f} - 1)! = \prod_i (f_i - 1)!$. The subscript i runs over all the vertices in the tree.

PROOF: Assume first that every edge in the tree is distinctly labelled (which — for trees of > 2 vertices — is equivalent to labelling each vertex). Then the number of distinct rooted ordered trees is given by the numerator on the right hand of eq. (13), since the edges sprouting upwards from each vertex may be permuted in $(\mathbf{f} - 1)!$ ways except for the root vertex i, which has an extra link which contributes an extra factor f_i. T_M is then given by the summation over i. Where the tree possesses elements of symmetry, however, we will have seriously overcounted the number of configurations— because they are no longer distinct after delabelling the edges. This is simply corrected by dividing our answer by the order of the symmetry group of the molecular graph (i.e. delabelling). Note that if $|G| = 1$ the graph has no symmetry and every vertex will be distinct.

This already simple expression may be further simplified if we enumerate only T, the number of distinct *node*-rooted ordered trees, i.e. we discount the number of trees rooted on terminals (Fig. 6). This changes eq. (13) merely by restricting the summation on the right to the degrees of the *nodes*:

$$T = (f - 1)! \sum_{\text{nodes}} f_j / |G| = \frac{\prod_i f_i! \sum_{\text{nodes}} f_i}{|G| \prod_k f_k} \tag{14}$$

We now consider an invariance theorem spanning the step from one to three dimensions, for molecules consisting of arbitrary centres, pairs of them being connected by freely rotating links. Let R_j be the symmetry number of the jth stereoisomer of a given structural isomer, whose molecular graph has symmetry number $|G|$. Let its ith centre bear f_i links and have symmetry number P_i in its completely isolated and unlabelled ('bare') state, i.e. 12 for a tetrahedron or benzene ring, and 60 for an icosahedron such as $B_{12}H_{12}^{2-}$. Then

$$|G| \sum_j R_j^{-1} = \prod_i (f_i!/P_i) \tag{15}$$

PROOF: The number of distinct 3D-rot forms obtainable by labelling distinctly all the y, say, links of each stereoisomer belonging to the given structural isomer is $y! \Sigma R_j^{-1}$. This must equal the number $y!/|G|$ of distinct ways of line-labelling the molecular graph, times the number $\prod_i (f_i!/P_i)$ of ways of mapping one line-labelled molecular graph into all possible completely unlabelled 3D-rot forms. This proof was given by Gordon and Temple.[31]

5.1 Stereochemical significance

Because of the generality of the arguments, the simplicity of the result of eq. (15) is all one is entitled to expect. However, the simplicity is rather of the kind pertaining to number theory, and is perhaps a little deceptive. The example in Fig. 4 has been chosen, not for its chemical topicality (though it is perfectly realistic), but because it is just sufficiently complex to reveal adequately the surprise element. The example surveys four structural isomers and seventeen stereoisomers, consisting of two identical cubical and two identical tetrahedral centres, with all their monovalent substituents identical (e.g., hydrogen).

In the first column, the molecular graphs of four out of the six possible structural isomers are shown. Each of the four molecular graphs can be embedded into a different number of stereoisomeric forms, which are those shown in the same row.

Each object has its symmetry number indicated below it. It is sufficient to explain this for the simplest case, that in the top row. Here the molecular graph has symmetry number $|G| = 7!2!2!7!2!$ because it bears seven free links on the bottom 'floor' which can be permuted in 7! ways; and two free links on the next two floors above, which can each be permuted in 2! ways, and then again seven free links on the top floor, which give rise to 7! permutations. Finally, the two ends of the molecular graph, its bottom and top floors, can be interchanged which accounts for a further factor of 2! in the symmetry number. Turning to the unique stereoisomer on the right in the bottom row, its symmetry number R is given as $3 \times 3 \times 2$: each cube has a freely rotating 3-fold axis, and the two ends of the molecule can again be interchanged.

The most interesting feature of eq. (15) is its implied assertion that $|G|\Sigma R_j^{-1}$ should have the same value for each structural isomer. This follows because $\Pi_i(f_i!/P_i)$ is invariant not only to stereoisomerism, but even to structural isomerism. In verifying the assertion for Fig. 4, i.e. that the product of the number on the left in a given row times the sum of the reciprocals of the remaining numbers on the same row is a constant (here $(7!)^2 2^2/3^2$), one touches on an interesting feature of the structure of space.

5.2   The surprise element

It is seen, when we examine the top row of Fig. 4, that of the two prime factors of 7 in the product just referred to, both have their source in the molecular graph. In the second row, however, only one 7 is supplied by the graph, while the other comes from the fact that there are seven different stereoforms; each of these conveniently contributes the same symmetry number (3^3), so that the sum of their reciprocal symmetry numbers factorizes into $7/3^3$. In the third row, again only one of the factors 7 comes from the graph, but now there are not seven, but only three, stereoisomers. Fortunately, the sum of their reciprocal symmetry numbers is $(3+3+1)/3^3$, so we recover the missing factor seven as the sum $3+3+1$ in the process of converting to a common denominator. Finally, in the bottom row, the molecular graph does not even contribute a single 7; but on converting the sum of the reciprocal symmetry numbers of the six stereo-forms to a common denominator, the numerator emerges as $1+6+6+9+18+9=7^2$ (q.e.d.). Although the term $|G|\Sigma R_j^{-1}$, being invariant for all four structural isomers, always contains the square of the magic number 7 as a factor, yet for different structural isomers this factor is variously encoded wholly, partly or not at all in the 1D graph structure (i.e., in $|G|$). By changing the structure of the graph (by passing from one structural isomer to another), we may squeeze one or both sevens out from $|G|$, but then such factors inevitably appear, in reciprocal form and distributed over the stereoisomers, in the third dimension, when we embed the graphs in 3D space. The effect is not a peculiar feature of cubes and tetrahedra, but truly a property of euclidian space. This follows because eq. (15) leaves us with complete freedom in choosing the shapes and topologies of the four 3D centres, which in the example of Fig. 4 make up the 3D-rot forms into which we imagine each of the four molecular graphs to be embedded. The solid bodies need not have corners; Möbius strips or Klein bottles can replace cubes and tetrahedra. There is one reservation: the points of attachments and directions of the links are *implied* by the corners when they are present. In solid bodies devoid of corners, we must, for the purpose of computing P_i indicate the missing information by embedding unlabelled pins in the body representing the attachments and directions of the links (see Fig. 5 later). In selecting cubes and tetrahedra, we have merely opted for the highest attainable symmetry of the centres. We briefly discuss the other extreme case of maximal asymmetry, and two intermediate cases of appreciable but non-maximal symmetry. To this end we pass from Fig. 4 to Fig. 5, in which, for the same graph of Fig. 4, the nature of the centres is allowed to vary, viz: (a) cubes and tetrahedra as before, (b) squares and hexagonal bipyramids, (c) squares and heptagonal pyramids, (d) spheres. Only a single typical 3D-rot stereoisomer is shown for each case by way of illustration. The completely asymmetric case (bottom row) may be visualized as having centres which are spheres, with pins inserted in completely asymmetrical random positions. Two of the spheres bear eight pins each and the other two bear four each. The pins then serve as the links between pairs of centres (which allow free rotation) or between a centre and a terminal. In this asymmetric case, eq. (15) simplifies because $P_i = 1$ ($i = 1,2,3,4$), and its meaning is more readily visualized. In fact one now finds very large numbers of stereoisomeric forms, since each way of mapping the line-labelled graph into a 3D-rot form is *sui generis*. Moreover, $R_j = 1$ for all j. Thus eq. (15) reduces to

$$N = \Pi f_i! |G| \tag{16}$$

where N is the number of distinct stereoisomers belonging to a given molecular graph. The truth of eq. (16) is easy to understand directly by labelling and delabelling.

If we proceed through the spectrum of symmetries (P_i) of centres, from the most symmetric case of cubes and tetrahedra (Fig. 4 and case a, Fig. 5) to the completely asymmetric one (d Fig. 5) we observe the trends, which are to be expected from eq. (15): the number of stereoisomers belonging to a given structural isomer (i.e., to a given molecular graph) increases, as does the sum of their reciprocal symmetry numbers ΣR_j^{-1} of the two intermediate cases chosen by way of illustration, that of hexagonal bipyramids and squares solves the 'problem' of satisfying eq. (15) in the same way as the corresponding case of cubes and tetrahedra ΣR_j^{-1} has seven as a factor in the numerator. However, the case of heptagonal pyramids and squares solves the problem by featuring 7 as factor in the denominator! This illustrates again the basic meaning of the invariance law, in that the symmetry element reflecting the number 7 can be made manifest in three-dimensional euclidian space (the heptagonal pyramid has a sevenfold rotation axis), or else it must appear (in reciprocal form) in the one-dimensional subspace furnished by the molecular graph.

6   STATISTICAL–MECHANICAL SIGNIFICANCE

Equation (11) allows the molecular distribution at equilibrium in any system with bond-additivity to be calculated. If two species j and k are stereoisomers of the same structural isomer (molecular graph), it follows from eq. (10) that

$$a_j = a_k \quad \text{(stereoisomers)} \tag{17}$$

Summing the mole fractions of the complete set $n_t (t = 1,2, \ldots)$ of stereoisomers of a given structural isomer (eq. 11), using (17):

$$n = \Sigma n_t = \mathcal{N} \left(\prod_{j_0, j_1} c_{j_0}^{s_{j_0}} c_{j_1}^{s_{j_1}} \right) \Sigma R_t^{-1} \tag{18}$$

For systems of molecules with freely rotating links the invariance law of eq. (15) is now used to substitute for ΣR_t^{-1}, and to eliminate the need to consider 3D-rot stereoisomers individually. To this end we group together in the product on the right of (15) all factors $(f_i!/P_i)$ which refer to the same type j_0 of atom, and substitute in (18):

$$n = \mathcal{N} \left(\prod_{j_0, j_1} \left(\frac{c_{j_0} f_{j_0}!}{P_{j_0}} \right)^{s_{j_0}} c_{j_1}^{s_{j_1}} \right) |G|^{-1} \tag{19}$$

The equilibrium distribution of molecular graphs generated by (19) is exactly of the ideal graph-like-state type. Comparison of (19) and (18) shows that, in systems which obey bond-additivity (eq. 11), we may calculate the molecular distribution by treating the particles as graph-like, which means that we lump any set of stereoisomers together into a single species and use the reciprocal symmetry number $|G|^{-1}$ of the floppy graph in the partition function of this species. The comparison also shows that the contribution of a given type of atom to the partition function must be changed by the (trivial) factor $f!/P$ for that atom when we pass from the 3D description (eq. 18) to 1D (eq. 19), precisely the change in the ground state degeneracy of the atom to be expected. Indeed, in the atoms which participate in a floppy graph, the points of attachment of f bonds can be permuted in $f!$ ways, while in the rigid-centre form of the 3D-rot atom, they can be permuted in P ways, where P is the symmetry number of the rigid centre (e.g. 24 for a cube).

Our purpose to clarify and justify the use of graph-like particles within the established framework of statistical mechanics is now achieved. The assumption of bond-additivity (eq. 10) is not conditional on the choice of a special energy zero: we may now simplify by putting all the atom contributions equal to unity in (19) (by re-adjusting the c_j):

$$n_i = \mathcal{N} \left(\prod_j c_j^{s_{ij}} \right) \; |\, G_i \,|^{-1} \tag{20}$$

With slight change of notation, this says that the mole fraction of species i is proportional to contributions c_1 from s_{i_1} bonds of type 1, c_2 from s_{i2} bonds of type 2, etc. and to the reciprocal symmetry number of its molecular graph.

7   THE f-FUNCTIONAL RANDOM POLYCONDENSATION MODEL

This basic model of chemistry concerns the equilibrium distribution obtained by linking a set of identical graph-like molecules with a star-shaped molecular graph into *tree-like* structures, as exemplified in Fig. 7. (*Cyclic* structures do arise in real chemical systems, and the model under discussion has been generalized to take these into account to an excellent approximation, the so-called spanning-tree approximation;[28,30] but even the classical tree-like model is usually realistic for bulk (undiluted) systems, and we restrict the derivation to this case.) Thus all molecules in the system will have molecular graphs which are $(1,f)$-trees, i.e. trees with nodes of degree f and terminals of degree 1. The molecular graph of

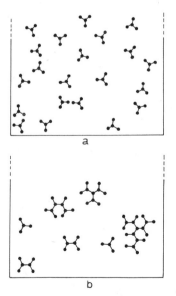

FIG. 7   Graph-like model for 3-functional condensation. (a) part of an infinite molecular system consisting entirely of 3-functional monomer graphs ($\alpha = 0$, see text). (b) Molecular system of 3-functional polycondensation formed from (a) by raising the conversion α(cf. Fig. 8).

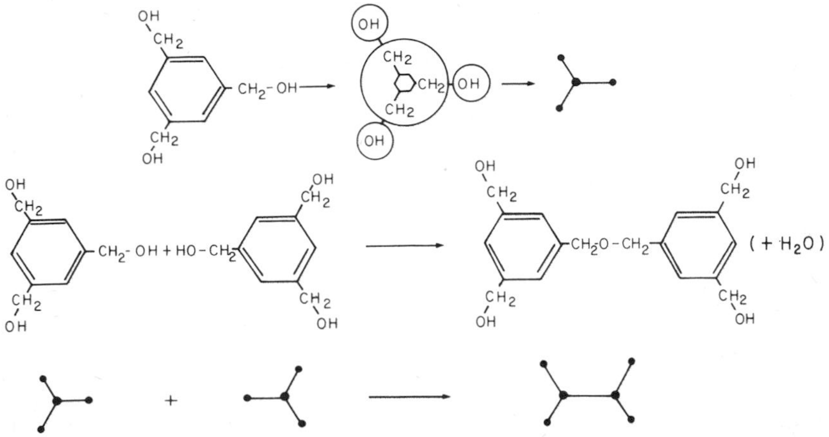

FIG. 8   The polycondensation of trimethylol benzene, and its reduction to graph-like form. The top illustration shows how the monomer molecule transforms to its molecular graph. The actual chemical bond forming (condensation) process is depicted in the middle line, with its graph-like analogue shown below. Reaction may continue at any of the remaining OH-functionalities (cf. Fig. 7).

ethane (Fig. 6) is a (1,4)-tree). Bond additivity is assumed (eq. 10). The model provides the first approximation to the standard thermodynamic data for the acyclic alkanes which may be regarded as polycondensates[25] of the star-like methane ($f = 4$) with elimination of hydrogen. A specific example of a practically irreversible type, with $f = 3$, which has been studied kinetically[32] rather than thermodynamically, is shown in Fig. 8; the top line presents the transformation of 1,3,5 trimethylol benzene into the civilized form of a contracted molecular graph. The next line shows the chemical linking process concerned, and the bottom line translates this into the graph-theoretical picture (cf. Fig. 7).

Apart from the trivial valency parameter f, this is a one-parameter model. The single parameter may be taken as the degree of advancement (or conversion) α of the equilibrium distribution, i.e. the fraction of all terminals present which have been eliminated by link formation.

We now derive the classical distribution law from eq. (20).[12,56] Already, this equation lumps together the set of stereoisomers corresponding to a given molecular graph. We further coarsen the level of description, by lumping together the mole fractions n_{xi} of all structural isomers with a given number x of nodes, which is feasible because they will share a constant factor $\Pi_j c_j^{s_{ij}}$. The overall mole fraction n_x of x-mer (i.e. all structural and stereoisomers with x nodes) is given by summation over structural isomers:

$$n_x \equiv \Sigma n_{xi} = \mathcal{N}(\Pi c_j^{s_{ij}})\Sigma |G_{ix}|^{-1} \tag{21}$$

We transform this to the classical distribution law by first attending to the symmetry numbers, which contribute to the entropic driving force. Subsequently we deal with the free-enthalpic driving force inherent in the bond energy terms c_j.

It seems difficult to obtain a generating function (g.f.) for the *reciprocal* symmetry numbers $|G_i|^{-1}$. Transformation from a 1D to a 2D picture, via eq. 14, exploits the fact that the *reciprocal* symmetry number is *directly* proportional to the number T of distinct node-rooted ordered trees. Let $T_{x,f}$ be the number of distinct node-rooted ordered $(1,f)$-trees with exactly x nodes of degree f (and, as is easy to verify, $fx - 2x + 2$ terminals). Then summing (14) over all such trees, and substituting in (21), yields

$$n_x = \mathcal{N}(\Pi c_j^{s_{ij}}) [(f-1)!]^{-x}(fx)^{-1} T_{x,f} \tag{22}$$

The factor x^{-1} on the right can be removed, by transforming from mole fraction n_x to weight fraction $w_x \propto n_x x$ and absorbing the renormalization adjustment $(\Sigma n_x x)^{-1}$ and the constant f^{-1} in $\mathcal{N}$ on the right:

$$w_x = \mathcal{N}(\Pi c_j^{s_{ij}}) [(f-1)!]^{-x} T_{x,f} \tag{23}$$

The constant factor $(f-1)!$ for each of the x nodes presents once again a degeneracy in the ground state of the node 'atom', and may be removed by a change in energy zero (i.e. by adjusting the c_j)

$$w_x = \mathcal{N}(\Pi c_j^{s_{ij}}) T_{x,f} \tag{24}$$

Since any tree of $x(=1,2,\ldots)$ nodes has $x-1$ links of type $(N-N)$ linking two nodes, and $fx - 2x + 2$ links of type $N-T$, where T is a terminal, we have

$$w_x = \mathcal{N} c_{N-N}^{x-1} c_{N-T}^{fx-2x+2} T_{x,f} \tag{25}$$

Now the fundamental equilibrium constant K of bond formation (after factoring out all symmetry factors, see Section 2), under the bond-additivity scheme, is given by

$$K/V = \alpha/(1-\alpha)^2 = c_{N-N}/c_{N-T}^2$$

where V is the volume. This shows that c_{N-N} and c_{N-T} may be taken as proportional to α and $(1-\alpha)$ respectively. As a result, eq. (25) may be written

$$w_x = \alpha^{x-1}(1-\alpha)^{fx-2x+2} T_{x,f} \tag{27}$$

That the normalization constant is now unity can be verified later by algebra, when the summation on the right has been evaluated (eq. 36). We could easily verify it, using the argument that $T_{1,f} = 1$ and that the weight fraction of monomer should be $(1-\alpha)^f$, but we wish to avoid all arguments based on probability theory in favour of a rigorous statistical mechanical treatment.

7.1 Mapping functions into trees

The combinatorial structures of the physical world are prefigured in those belonging to algebra. We calculate the number $T_{x,f}$ of distinct node-rooted ordered trees of x nodes of degree f by constructing the generating function (g.f.)

$$T_f(\theta) = \sum_{x=1} T_{x,f}\theta^x \qquad (28)$$

and extracting its coefficient $T_{x,f}$. Here θ is the auxiliary variable, required for any g.f.

The function $T_f(\theta)$ can be written down essentially without any calculation. It is

$$T_f(\theta) = \theta(1 + u)^f \qquad (29)$$

where u is itself defined as the following implicit function of θ:

$$u = \theta(1 + u)^{f-1} \qquad (30)$$

To prove this, we first substitute (30) iteratively in (29):

$$T_f(\theta) = \theta(1 + \theta(1 + \theta(1 + \theta(\ldots)^{f-1})^{f-1})^{f-1})^f \qquad (31)$$

To clarify the structure of $T_f(\theta)$, we propose the merely typographical convention[41] of writing any product of f factors, say, thus

$$a_1 a_2 \ldots a_f = a_1 \left\{ \begin{array}{l} a_2 \\ a_3 \\ a_4 \\ \vdots \\ a_f \end{array} \right. \qquad (32)$$

The first factor is singled out, and the others are arranged vertically, in Japanese fashion, in the new way of writing a product on the right of (32). Applying this to eq. (31), with $f = 3$ for the simplest illustration, we obtain Fig. 9a.

The function $T_f(\theta)$, written in this novel typographical way, is mapped into the infinite tree (a so-called Bethe lattice[11]) of Fig. 9b by letting each θ-symbol correspond to one of the nodes. Since θ occurs everywhere in $T_f(\theta)$ only as a first power, any contribution to the term in θ^x must contain x individual θ-symbols in Fig. 9a. In general there are several ways of selecting x θ-symbols forming a set which contributes to the term in θ^x. It is easily shown that the number of distinct such sets is $T_{x,f}$. Indeed, the structure of the function $T(\theta)$ in Fig. 9b ensures that if any θ-symbol (on generation $g_1, g_2, \ldots$) is picked as contributing to a term in the expansion of the function, then the θ symbol in front of the brace immediately on its left must also contribute to the same term (because it pre-multiplies the bracket in question). Therefore, each θ symbol

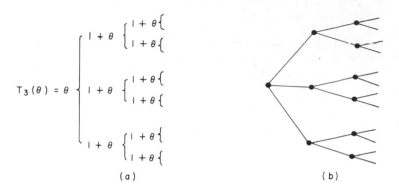

$$T_3(\theta) = \theta \left\{ \begin{array}{l} 1 + \theta \left\{ \begin{array}{l} 1 + \theta \left\{ \begin{array}{l} 1 + \theta \left\{ \\ 1 + \theta \left\{ \end{array} \right. \\ 1 + \theta \left\{ \begin{array}{l} 1 + \theta \left\{ \\ 1 + \theta \left\{ \end{array} \right. \\ \\ 1 + \theta \left\{ \begin{array}{l} 1 + \theta \left\{ \\ 1 + \theta \left\{ \end{array} \right. \end{array} \right.$$

(a)   (b)

FIG. 9   Using the typographical device explained in the text, the relation between the infinitely nested generating function ($f = 3$) of eqn. (29) and the rooted ordered tree is apparent.

picked entrains the appearance of all the θ-symbols which lie on the path from the corresponding point to the root in Fig. 9. Thus each contribution to θ^x maps into a node-rooted ordered tree in Fig. 9b. Also each such contribution has the form $1.\theta^x$ and there are $T_{x,f}$ such contributions, so that the total coefficient of θ^x is $T_{x,f}$.

7.2   Explicit evaluation of $T_{x,f}$

The explicit evaluation, by a routine procedure, of $T_f(\theta)$ as the power series in θ yields $T_{x,f}$ as the coefficient of θ^x. Using the accepted notation, this coefficient is denoted by $C(\theta^x)\{T_f(\theta)\}$. The routine is Lagrange's expansion (see, for a *simple* proof, Bromwich[1]), which for any suitable functions F, N, and R, are related thus:

$$F(\theta) = \theta N(u) \tag{33}$$

with

$$u = \theta R(u) \tag{34}$$

reads

$$C(\theta^x)\{F(\theta)\} = (x-1)^{-1} C(u^{x-2}) \{N'(u)(R(u))^{x-1}\} \tag{35}$$

(cf. Good[22]). For our case of eqs. (29) and (30), by comparison with (33) and (34), we must put $F(\theta) = T_f(\theta)$, $N(u) = (1+u)^f$ and $R(u) = (1+u)^{f-1}$. Then (35) gives easily:

$$T_{x,f} = (fx - x)!f / [(fx - 2x + 2)!(x - 1)!] \tag{36}$$

This is the desired combinatorial coefficient in the classical Flory–Stockmayer weight fraction distribution, which results from (27) and (36):

$$w_x = \frac{(fx - x)! \, f}{(fx - 2x + 2)!(x - 1)!} \, \alpha^{x-1}(1 - \alpha)^{fx-2x+2} \tag{37}$$

Our rigorous derivation from statistical mechanics has shown clearly why the number of node-rooted x-meric $(1,f)$-trees provides the combinatorial factor for molecules in the 3D-rot state. It has been shown by Gordon and Temple[31] that it is not in general applicable to rigid 3D molecules.

We have dealt in detail and rigorously with the basic model. The means of generalizing this in all the desired directions are well documented by various authors in the literature. We now sketch the course of such generalizations, giving references where appropriate. The broad directions of generalizations are:

(i) copolymerization of various monomer units with different valences (see e.g. Butler, Gordon and Malcolm[6]). This route essentially depends on replacing ordinary (scalar) g.f.'s by vectorial ones.

(ii) Introducing correlations (substitution effects) beyond the bond additivity scheme, i.e. the step from random to non-random systems. The graph-theoretical aspects were summarized in Section 3.1; concrete calculations are given by Gordon and Parker.[26] Gels (see Section 8.1 below) always harbour correlations of unlimited range. No exact treatment of any model for a gel is, therefore, known.

(iii) The calculation of various parameters of the molecular distributions, beside the mere weight fractions. Here the most diverse physical properties (e.g. Young's modulus,[8] the particle scattering factor[41] or the weight average reciprocal Stokes radius[42] of polymer systems emerge as averages over the molecular distribution concerned, and they are generally obtained by manipulation of the g.f.'s more readily than the weight fractions themselves – Lagrange expansion methods are rarely needed.

(iv) The introduction of molecular graphs with circuits (rings) has already been mentioned with some references above.

All four types of generalization are greatly facilitated by the fact that the problem has led us to a g.f. which deals with the only difficult aspect – molecular symmetries. The rest is essentially a set of tricks for getting the g.f. to throw out additional weights. Such a process can immediately be exemplified. It is easy to modify the generating function for the number of node-rooted ordered trees defined in eqs. (29) and (30), so as to produce the additional weights depending on α, i.e. to turn it into the weight fraction generating function

$$W(\theta) = \sum_x w_x \theta^x = \sum \alpha^{x-1}(1-\alpha)^{fx-2x+2} \, T_{x,f} \theta^x$$

$$= \sum \alpha^{x-1}(1-\alpha)^{fx-2x+2} f(fx-x)! \, \theta^x / (x-1)! (fx-2x+2)! \tag{38}$$

(cf. eq. 37). Let (29) be turned into

$$W(\theta) = \theta(1 - \alpha + \alpha u)^f \tag{39}$$

and (30) into

$$u = \theta(1 - \alpha + \alpha u)^{f-1} \tag{40}$$

This will change (30) into

$$W(\theta) = \theta(1 - \alpha + \alpha\theta(1 - \alpha + \alpha\theta(1 - \alpha + \alpha\theta(\dots)^{f-1})^{f-1})^{f-1})^f \tag{41}$$

It is immediately verified that this modification attaches the required factor α^{x-1} (see 38) to the term in θ^x, because each factor θ on the right, *except the first* which represents the tree root, is pre-multiplied by α in eq. (41). The origin of the additional factor $(1 - \alpha)^{fx - 2x + 2}$ in (38) is also readily traced in (41): each time a terminal of the tree occurs, a factor $1 - \alpha$ has to be chosen from the appropriate bracket in (41).

8   THE ROLE OF CASCADE OR MULTITYPE BRANCHING THEORY

Equations (39) and (40) are familiar examples of a class well known in the branch of applied mathematics dealing with processes variously described as Galton–Watson, or cascade, or multi-type branching processes. The importance of this field of Markov processes is underlined by the fact that among the first 34 volumes of Bellman's series on 'Modern analytic and computational methods in Science and Mathematics', two are entirely devoted to this subject.[49,55] The reader is referred to these works as regards the applications to genetics, particle physics, etc. The common feature of such applications lies in the generation of family trees of some kind. Much of the mathematical interest centres on a subtle singularity in the relevant generating functions such as eq. (38). This was first examined in connection with the problem of the indefinite survival of a family name in human genetics. The corresponding critical phenomenon in poly-condensation is called the gel point; it occurs in eqs. (37) or (41) at the critical value of α, viz. $\alpha_c = 1/(f - 1)$.

8.1   The gel point and the critically branched state

This point coincides with the divergence as α is increased, of the mean size of the molecular tree of which a node in the system forms part (as the root). The existence of an 'infinite' molecule in the chemical system causes the cessation of liquid flow, and the substance gels. The model yields this condition very simply, since the mean size $E(x)$ of tree attached to a root is found from the g.f. $W(\theta)$ by the standard operation for a generating function:

$$E(x) = W'(1) \tag{42}$$

This is evaluated from eq. (38) by a formal procedure (cf. Gordon[24]). Differentiating both (39) and (40):

$$W'(\theta) = (1 - \alpha + \alpha u)^f + f\alpha\theta(1 - \alpha + \alpha u)^{f-1} \, du/d\theta \tag{43}$$

$$du/d\theta = (1 - \alpha + \alpha u)^{f-1} + (f - 1)\alpha\theta(1 - \alpha + \alpha u)^{f-2} du/d\theta \tag{44}$$

Solving (44) for $du/d\theta$ and substituting in (43):

$$W'(\theta) = (1 - \alpha + \alpha u)^f + \frac{f\alpha\theta(1 - \alpha + \alpha u)^{2f-2}}{1 - (f-1)\alpha\theta(1 - \alpha + \alpha u)^{f-2}} \tag{45}$$

It remains to convert $W'(\theta)$ to $W'(1)$. Inspection of (39) and (40) shows that when $\theta = 1$, then $u = 1$ is a solution. Thus putting both θ and u equal to unity in (45), gives Stockmayer's classical formula for $E(x)$, here called the weight-average degree of poly-condensation (cf. Flory[14]):

$$E(x) = \frac{1 + \alpha}{1 - (f-1)\alpha} \tag{46}$$

which diverges at the gel point, as first explained by Flory:

$$\alpha_c = 1/(f - 1) \tag{47}$$

Matter close to (and on either side of) a gel point is said to be in a critically branched state,[5] and this is the state of matter — a special case of the graph-like state — in which life processes are generally observed.

Indeed, the gel point is closely connected with the biological problem of the origin of life on earth. Ironically, the critical point in a neutron cascade, which forms the basis of a nuclear bomb, is mathematically strictly analogous, though it is more relevant to the converse biological problem.

9   THE STRICT RELATION BETWEEN CASCADE THEORY AND THE GRAPH-LIKE STATE

The stochastic models treated in textbooks on cascade theory share with the statistical mechanics and chemical kinetics of the graph-like state a part of their graph-theoretical basis (rooted trees); they also manipulate similar sets of g.f.'s in similar ways. Nevertheless, the models are different in that the molecular graphs handled need not, and generally do not, arise from stochastic processes contemplated in cascade theory: the links between members of succeeding generations of a rooted tree-molecule were not actually formed in chronological order (but rather in more or less random order). As a result, the basic graph theory looms large in the chemical theory, while the two books cited on cascade theory do not consider graphs at all.* The relation between molecular graphs and stochastic cascade theory has been analysed by Gordon and Parker.[26] The chemical models may be translated into equivalent stochastic cascades, and the power tools of cascade theory may then be harnessed, provided a specific graph-theoretical restriction is imposed. Every tree generated by indefinitely repeated applications of the generating functions of the stochastic cascade process must occur rooted with equal frequency on each of its nodes. A forest of trees obeying this restriction has been called a clone. Sufficient conditions for a set of cascade g.f.'s to generate a clone have been given in the *basic theorem* of the subject, by Gordon, Parker and Temple.[32] Though not easy to prove, the theorem is easy to apply, and it legalizes as proper statistical mechanical treatments numerous applications of cascade processes to polymer systems in the literature, which have been made without rigorous justification. Once the basic conditions of the theorem have been shown to apply (as they often do), the temporal sequence in which the chemical bonds were formed become irrelevant, and the cascade formalism is rigorously applicable. Recent examples of direct applications of the cascade formalism to polymer science (which would be covered by the basic theorem) are exemplified by the works of Hasa,[35] Irshak and Kuzov,[39] Kajiwara,[42] Konstein and Pis'men,[44] Burchard *et al.*,[4] and Dušek.[9]

10   CONCLUSIONS

Ad hoc and intuitive treatments of statistical mechanics can be replaced by rigorous theory. This move leads naturally to the use of the appropriate mathematical tools — here well-known techniques for manipulating generating functions. In this way, a substantial part of physical science can be integrated within the notion of a specific state of matter: the graph-like state. There is a parallel between the graph-like state, based on floppy particles, and the solid state, based on the geometrical notion of a 3D space lattice. Each has its powerful but specific mathematical techniques. Both often constitute merely a first approximation to 'reality'. Yet physics is often all but identified with the solid state alone.

*A special tribute is due to I. J. Good whose pioneering work brought out clearly the connection between cascade processes and graph theory, and who contributed many important techniques.[18-20]

POSTSCRIPT (MAY 1975)

Two recent papers by Gibbs and co-workers illustrate the programme, sketched in Fig. 1 for advancing statistical mechanics. The relationship between Mayer cluster graphs and graph-like particles is clarified by Cohen, Gibbs and Fleming[7] using the bridging theory introduced by Hill,[36] through his criterion of a 'bond'. They are thus able to write a partition function for an f-functional random polycondensate which embodies both Stockmayer's[56] original combinatorial analysis for such a condensate and the Mayer cluster expansion. This led Gibbs et al.[17] to apply the same polycondensate paradigm as a model for liquid water (as it had been for water dissolved in benzene by Gordon et al.[27]), but now allowing for non-ideality by virtue of repulsions between particles which are not directly bonded. In this way, these authors combine the two routes outlined in Section 1.1 above for advancing from the ideal gas model in both directions at once.

The partition function used by Cohen et al.[7] at first sight appears very different from that written in eq. (8), though they are equivalent in the absence of interactions between unbonded particles. For instance, Cohen et al.[7] explicitly include a translational partition function for each *monomer* unit, and do not explicitly include a rotational symmetry number of the *polymer*, while eq. (8) takes the opposite course, but nevertheless the same phase integral is being evaluated: the translational partition function for monomer units can, of course, be considered included in our 'internal' partition function a_i, while for the rest we can group together the elements of phase space in various ways when arriving at the partition function (by implicit or explicit processes of integration). In particular, we can make use of combinatorial identities. The translation of the Stockmayer formalism into graph-theory is aided by the following combinatorial identity:

$$w_{x,f} = f^{x-1}(x-1)! \, T_{x,f}$$

Here $w_{x,f}$ is Stockmayer's combinatorial term, viz. the number of ways in which x labelled monomers each bearing f labelled functionalities can be connected into an x-mer. The reason for putting forward our method of formulating partition functions lies in its simplicity and generality; the avoidance of Lagrangian multipliers, whose meaning has to be found from physical arguments *a posteriori*, is also a benefit. Recourse to analytical theories (involving limits of continuous functions), such as the Lagrange expansion in Section 7.1 for determining $T_{x,f}$, is to be regarded as a blemish in the combinatorial theory of discrete structure. Indeed, Tutte[57] has now given a purely combinatorial proof, free from limits, of the multivariate Lagrange expansion formula of Good.[20] Better still for the present purpose, eq. (36) can now be written down as a special case of the following simple combinatorial result (Gordon and Leonis, to be published):

$$T_\mathbf{y} = 2(m!)/\mathbf{y}!$$

Here $T_\mathbf{y}$ is the total number of distinct rooted ordered trees of all isomers of a y-mer, i.e. a tree-like compound with y_1 units of type 1, y_2 units of type 2, ... etc., characterized by the vector $\mathbf{y} = (y_1, y_2, \ldots)$. Also $\mathbf{y}!$ denotes as usual $y_1! y_2! \ldots$; and $m = \frac{1}{2}\Sigma y_i i$, which equals the total number of lines in any of the isomer trees. To restrict $T_\mathbf{y}$ to the number $T_{n,\mathbf{y}}$ of distinct *node*-rooted ordered trees, we eliminate the trees rooted on terminals, using the general formula

$$T_{n,\mathbf{y}} = T_\mathbf{y}(2m - y_1)/2m$$

where y_1 is the number of terminals.

This immediately reduces to eq. (36) on putting $y_1 = fx - 2x + 2$ and $y_f = x$, and makes (37) accessible without recourse to analysis. Finally, the exploitation of graph theory for formulating the partition function of a perfect network, as a model for rubber elasticity by Eichinger[10] presents a recent example of the constant growth of graph-like-state theory.

REFERENCES

1. Bromwich, T. A. I'A. (1908). "An Introduction to the Theory of Infinite Series". Macmillan, London.
2. De Bruijn, N. G. (1964). *In* "Applied Combinatorial Mathematics" (E. F. Beckenbach, ed.) p. 144. Wiley, New York and London.
3. Bruneau, C.-M. (1967). *CR Acad. Sci. Paris* 264, Series C, 1168.
4. Burchard, W., Ullisch, B. and Wolf, Ch., (1974), *Faraday Disc.* 57, 56.
5. Burchard, W., Kajiwara, K., Gordon, M., Kálal, J. and Kennedy, J. W. (1973). *Macromolecules*, (in press).
6. Butler, D. S., Gordon, M. and Malcolm, G. N. (1966). *Proc. Roy. Soc.* A295, 29.
7. Cohen, C., Gibbs, J. H. and Fleming III, P. D. (1973), *J. Chem. Phys.*, 59, 5511.
8. Dobson, G. R. and Gordon, M. (1965). *J. Chem. Phys.*, 43, 705.
9. Dušek, K., (1974), *Faraday Disc.*, 57, 101.
10. Eichinger, B. E. (1972). *Macromolecules*, 5, 496.
11. Essam, J. W. and Fisher, M. E. (1970). *Rev. Mod. Phys.* 42, 272.
12. Flory, P. J. (1941). *J. Am. Chem. Soc.*, 63, 3091.
13. Flory, P. J. (1942). *J. Chem. Phys.*, 10, 51.
14. Flory, P. J. (1953). "Principles of Polymer Chemistry". Cornell University Press, Ithaca.
15. Fukui, K. and Yamabe, T. (1964). *J. Polymer Sci.*, A2, 3743.
16. Gibbs, J. H. and DiMarzio, E. A. (1958). *J. Chem. Phys.*, 28, 373.
17. Gibbs, J. H., Cohen, C., Fleming III, P. D. and Porosoff, H. (1973). *J. Solution Chem.*, 2, 277.
18. Good, I. J. (1948). *Proc. Camb. Phil. Soc.*, 45, 360.

19. Good, I. J. (1955). *Proc. Camb. Phil. Soc.*, **51**, 240.
20. Good, I. J. (1960). *Proc. Camb. Phil. Soc.*, **56**, 367.
21. Good, I. J. (1963a). *Ann. Math. Statist.*, **34**, 322.
22. Good, I. J. (1963b). *Proc. Roy. Soc.*, **A272**, 240.
23. Good, I. J. (1960). *Proc. Camb. Phil. Soc.*, **56**, 361.
24. Gordon, M. (1962). *Proc. Roy. Soc.*, **A268**, 240.
25. Gordon, M. and Kennedy, J. W. (1973). *J. Chem. Soc. Faraday II*, **69**, 484.
26. Gordon, M. and Parker, T. G. (1970/71). *Proc. Roy. Soc. (Edin.)*, **A69**, 181.
27. Gordon, M., Hope, C. S., Loan, L. D. and Roe, R.-J. (1960). *Proc. Roy. Soc. (London)*, **A258**, 215.
28. Gordon, M. and Scantlebury, G. R. (1966). *Proc. Roy. Soc.*, **A292**, 380.
29. Gordon, M. and Temple, W. B. (1970). *J. Chem. Soc.*, A, 729.
30. Gordon, M. and Temple, W. B. (1972). *Makromol. Chem.*, **152**, 277.
31. Gordon, M. and Temple, W. B. (1973). *J. Chem. Soc. (Faraday II)*, **69**, 1.
32. Gordon, M., Parker, T. G. and Temple, W. B. (1971). *J. Comb. Theory*, **11**, 142.
33. Gordon, M., Love, J. A., Parker, T. G. and Temple, W. B. (1971). *J. prakt. Chem.* **313**, 411.
34. Harary, F. (1969). "Graph Theory". Addison Wesley, London.
35. Hasa, J. (1971). *Coll. Czech. Chem. Comm.*, **36**, 1807.
36. Hill, T. L. (1956). "Statistical Mechanisms," Chs. 7 and 8 McGraw-Hill, New York.
37. Horn, F. (1972). *Arch. Rat. Mech. Anal.*, **49**, 172, and papers in *Proc. Roy. Soc.* in press.
38. Huggins, M. L. (1942). *J. Chem. Phys.*, **46**, 151.
39. Irshak, V. I. and Kuzov, L. I. (1971). *Dokl. Acad. Nauk. SSSR*, **6**, 1382.
40. Isihara, A. (1971). "Statistical Physics". Academic Press, New York, London and San Francisco.
41. Kajiwara, K., Burchard, W. and Gordon, M. (1970). *British Polymer J.*, **2**, 110.
42. Kajiwara, K. (1971). *J. Chem. Phys.*, **54**, 296.
43. Kasteleyn, P. W. (1963). *Physica*, **29**, 1329.
44. Konstein, S. E. and Pis'men, L. M. (1971). *Dokl. Acad. Nauk. SSSR*, **196**, 858.
45. Kuratowski, C. (1930). *Fund. Math.*, **15**, 271.
46. Masson, C. R., Smith, I. B. and Whiteway, S. G. (1970). *Canad. J. Chem.*, **48**, 33, 201, 1456.
47. Mattuck, R. D. (1967). "A Guide to Feynmann Diagrams in the Many-Body Problem". McGraw-Hill, London.
48. Mayer, J. E. and Mayer, M. G. (1940). "Statistical Mechanics". Wiley, New York.
49. Mode, C. J. (1971). "Multitype Branching Processes". (Modern Analytic and Computational Methods in Science and Mathematics — R. Bellman, ed.). Elsevier, New York.
50. Murrell, J. N. and Laidler, K. J. (1968), *Trans. Faraday Soc.*, **64**, 371.
51. Pólya, G. (1937). *Acta Math.* **68**, 145.
52. Sachs, H. (1970). "Einführung in die Theorie der endlichen Graphen", Teil I. B. G. Teubner, Leipzig.
53. Schmidtke, H.-H. (1966). *J. Chem. Phys.*, **45**, 3920.
54. Silberberg, A. (1970). *Disc. Faraday Soc.*, **49**, 162.

55. Srinivasan, S. K. (1969). "Stochastic Theory and Cascade Processes". (Modern Analytic and Computational Methods in Science and Mathematics — R. Bellman, ed.). Elsevier, New York.
56. Stockmayer, W. H. (1943). *J. Chem. Phys.,* **11**, 45.
57. Tutte, W. T. (1975). *J. Combinatorial Theory, Series B,* **18**.
58. Whitney, H. and Tutte, W. T., (1972), Utilitas Mathematica, 2, 241.
59. Whittle, P. (1965a). *Proc. Camb. Phis. Soc.,* **61**, 475.
60. Whittle, P. (1965b). *Proc. Roy. Soc.,* **A285**, 501.
61. Gordon, M. and Judd, M. (1971), *Nature (Lond.)* **96**, 234.

CHAPTER 11

Ordered Chromatic Graph and Limited Environment Concept

J. E. Dubois

Laboratory of Physical Organic Chemistry
University of Paris VII, Paris, France

The graph concept, written mathematically $G(X,U)$ where X is the set of nodes and U the set of edges linking these nodes,[1] plays a natural role in the field of chemical design. If the chemical formula or structural diagram taken as a whole is a natural representation of an entity, certain linguistic descriptions, systematic nomenclature or various codified descriptors are always associated with it in an attempt to arrive at a linear descriptor which can be easily manipulated for diverse operations.[2] However, although these descriptions are constantly being improved, none of them is satisfying unless its field of application is limited.

In reality, a very broad, flexible, coherent language, a real metalanguage derived by a logical process, can only be realized by identifying a chemical formula with an appropriate graph. The latter, either because of its associated polynomial or even more its connectivity or adjacency matrix[3,4] can theoretically lead to certain numerical or alphanumerical descriptions of the formula.

Thus, the graph contributes to the building of a metalanguage by the powerful tool of topological description and the vast potential of matricial calculus, useful for treating atom combinations in terms of entities (identifying), and entities in terms of populations (grouping).

In order to simultaneously retain all the advantages of the chemical formula and to convey these through the medium of an appropriate metalanguage a sophisticated use of the graph notion must be evolved. We plan to show that the usual notion of a mathematical graph must be broadened to take account of the diversity and wealth of localized data specific to chemistry.

The problem could be partly solved by a simple *matching* of the formula with its chromatic graph. In order to progress still further, we feel however that it is necessary to induce more order on this graph than exists in the usual chemical formula. Under these conditions, structural chemical information (bi- or tri-dimensional) can be expressed by using a general and coherent metalanguage in the form of descriptors which make possible not only manual treatment but above all computer analysis.

Correlation of properties can be used as mechanisms to introduce or justify the concept of induced order, but we shall prefer here a formalistic introduction of chromatic ordered graphs.

Identifying structures and grouping of entities

These are the essential operations in chemistry. We shall briefly define these terms, which refer to the multidirectional representation of structural formulas (diagrams or graphs in 'preorder').

Identifying

Identifying is a term which covers two basic steps: first that of assigning an identity to a chemical object, in other words, establishing the structure of such an object from structural and comportment data; second that of controlling the identity, that is, retrieving the structure of a chemical compound already registered, starting from partial knowledge of some of its component parts or of its comportment. Certain procedures set up to solve this important problem rely on results of graph theory combinatorics: counting and particularly listing compounds with specific restrictions, such as acyclic isomers with a specific functional group.[5]

Grouping

In design, whether in synthesis or correlation, one rarely reasons on the basis of isolated compounds. Instead, one seeks to situate a compound in a vaster set including dependence relationships (filiations, substitutions, permutations, transpositions) which give rise to analogies of preparation or comportment. This problem was defined long ago. Indeed as early as 1838, Laurent defined it as one of the principal objectives of chemistry and proposed, without further elucidation, to achieve it by considering affiliated chemical entities as being derived from each other by a formal mechanism of hydrogen substitutions. Today these hydrogens are often considered as transparent atoms within a molecular graph. This remarkable vision could not be exploited since, for many years, chemistry made only partial use of the potential inherent in the notions of isomerism and homology, considered as independent tools. We believe it possible to revive this train of thought by the idea of order imposed on a graph.

The identifying process is not at all hindered if the modelling of bonded atoms is *more ordered* than in the usual formulas. Besides, in certain specific cases, such as NMR, this order is already imposed.[6]

The *grouping of entities* organizes populations of isolated structures capable of being ordered and of constituting a set defined as a hyperstructure.[7] This notion of hyperstructure or necessary population hierarchization can also be approached by the graph theory. In addition, important simplifications can be achieved in the identifying and grouping operations by imposing on them a relationship of identical order called canonical. The ordered graph concept makes this possible.

The first part of this article analyses the chromatic graph concept, an extension of the topological graph with a wide impact in fields other than chemistry. The importance of the idea of order on a chromatic graph is demonstrated. The second part presents the main features comprising the methodology of ordering and description (DARC)[8] which makes it possible to induce on all or part a chromatic graph, a canonic reference order called 'Major Order' or derived 'finalized orders' oriented towards specific and definite goals. The identification of a structure by one of its parts is presented as an illustration of this methodology. The third part points out some problems which arise in the study of populations of chemical compounds. A general approach towards a solution, based on the hyperstructure concept, is proposed. The study of reaction pathways is given as an example.

1   CHROMATIC GRAPH AND ORDER

1.1   Chromatic graph concept

1.1.1   Definitions

A chromatic graph is one whose nodes (X) and edges (U) are symbolically differentiated by one or more colours.

Thus, a chemical compound can be assimilated to a chromatic graph whose nodes are atoms and whose edges are bonds, just as for a syntactic graph, the coloration refers to concepts and to the relationships which unite them.

Such a graph is written:

$$G_\chi(X, U, \chi_X, \chi_U)$$

where X is the set of the graph nodes, U is the set of edges linking the elements of X, χ_X (respectively χ_U) the function which, applied to an element of X (respectively U), expresses its colour.

Definition: The graph $G(X,U)$ is termed a topological graph associated with the chromatic graph $G_\chi(X, U, \chi_X, \chi_U)$. In chemistry, this topological graph represents the skeleton of the compound represented more completely by the chromatic graph G_χ.

Definition: Two chromatic graphs $G_\chi(X, U, \chi_X, \chi_U)$ and $G'_\chi(X', U', \chi_{X'}, \chi_U)$ are:
  − isotopologous if there is a bijection between X and X' compatible with a bijection between U and U' or, in other words, if their associated topological graphs are isomorphic (Fig. 1).
  − isochromatic on the nodes or on the edges if they are isotopological and if $\chi_X = \chi_{X'}$ or $\chi_U = \chi_{U'}$ (Fig. 1).
  − isomorphic if they are isochromatic on the nodes and on the edges (Fig. 1).

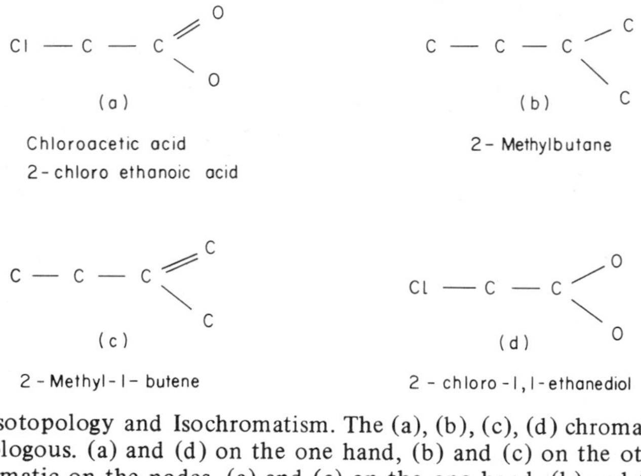

(a)

Chloroacetic acid
2-chloro ethanoic acid

(b)

2- Methylbutane

(c)

2 - Methyl-1- butene

(d)

2 - chloro -1,1-ethanediol

FIG. 1   Isotopology and Isochromatism. The (a), (b), (c), (d) chromatic graphs are isotopologous. (a) and (d) on the one hand, (b) and (c) on the other hand, are isochromatic on the nodes. (a) and (c) on the one hand, (b) and (d) on the other hand, are isochromatic on the links.

Definition: A partially chromatic graph is one where for certain elements of the X and U sets, functions χ_X and χ_U are undetermined (Fig. 2).

The use of the chromatic graph concept in a particular domain (D) requires a census-taking of all colours characteristic of this domain (χ_D). This involves

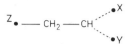

FIG. 2   Partially chromatic graph. The nature of the X, Y, Z nodes and the CX, CY links are undetermined.

defining two lists. The graph nodes take on one of the colours included in the chromatic list T_X, the edges, one of those in the chromatic list T_U. For example, during the representation of a document, T_X could group the concepts and T_U the possible relationships between these concepts.[9]

1.1.2   Chemistry and chromatism

In this paragraph we define primary chromatism as that chromatism associated with a chemical compound during a first approximation. We point out the close overlapping of topology and chromatism which underlines the specificity of the chromatic graph concept compared to that of the valuated graph. This rapprochement also enables us to link the adimensional model to a metric one.

1.1.2.1   Chromatic list

The χ_X function of the chromatic graph associated with a chemical compound theoretically takes its values from the pre-established and hierarchized (according to atomic numbers) list T_X of the Mendeleev table's 105 elements. The χ_U function takes its values from a far more limited set, the list T_U of the diverse types of bonds to be found in a compound (simple, double, triple, aromatic, tautomeric, dative). If the useful isotopes of a given atom are to be differentiated, the number of elements of T_X will in fact be multiplied by 3, 4 or even 5.

1.1.2.2   Chromatism and topology

In chemistry, chromatism and topology are interdependent. Thus, the colouring of a node corresponds to the nature of an atom. This, in turn, dictates the number of valences this atom exchanges with its neighbours. In topological terms, we say that the colouring imposes its maximum degree of connectivity on this node. The degree given a node indirectly determines certain types of colouring of its environment. In addition, all atoms, in as much as they are sites of a compound, have fixed degrees. Thus, the chromatic graph of a structure can be simplified by deciding, by convention, to exclude all hydrogen atoms (unitary degree); this results in lighter graphs called transparent graphs from which we can reconstruct the explicit graphs.

1.1.2.3   Chromatism, topology and metric

The chromatic graph in chemistry provides a basis for taking metric information into account and thereby representing a standard chemical form.

Thus, if we know the chromatism of two nodes and that of the bond between them, we can then deduce the length of the bond by simply consulting a table of reference. Similarly, if we know the chromatism of the nodes and the nature of two adjacent bonds, we can deduce their plane angle. If we dispose, therefore, of a chromatic graph and of metric tables containing the fundamental metric information, we can construct the standard chemical form (2D or 3D).[10]

1.2   Concept of order

The structural diagram displays a preordered structure (Fig. 3).[11] This rather peculiar type of order is general and intuitive. A very weak order composed of symmetrical local relationships, it is well adapted to the overall view of the structural diagram. As soon as a particular operation is linked to parts of the molecule, one must turn to stronger orders used either on the whole diagram or on certain of its parts.[12,13] Up to now, any customary imposed order can be characterized as a transitory expedient or an isolated description conceived for specific application.

The fact that an isolated description must associate linear representation with a structural diagram for purposes of computer processing makes it essential to search for more general methods for describing such a diagram. Instead of continuing to consider the description of molecular organization as trivial and temporary, a theory of this organization with well defined order functions must be elaborated.

Numbering of the sites of a structure, or of the chromatic graph which we associate with it, is a procedure which can be linked to the nature of the physicochemical problem studied (PMR, [13]CMR, ...), and it can thus be shown

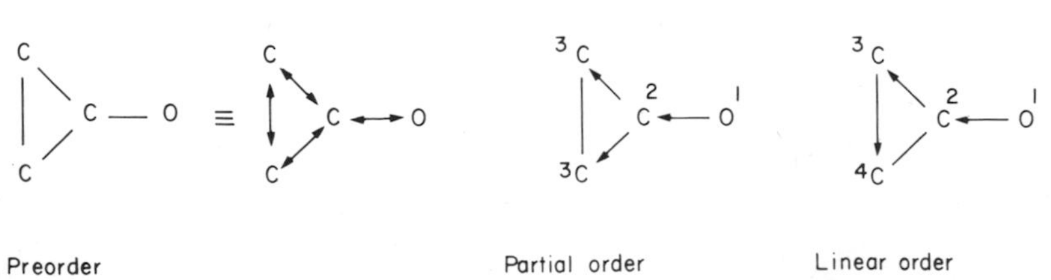

Cyclopropanol

Preorder   Partial order   Linear order

FIG. 3   Chromatic graph and ordering. *Linear order:* a father has only one son and vice versa. *Partial order:* a father may have several sons and a son may have several fathers. *Preorder:* a node may be its own father and its own grandfather.

that specific orders must be envisaged, oriented in their origin and their function from the standpoint of the aims pursued. The diversity of orders to be induced on a formula will be pointed out by transferring the analysis of the problem to the level of structural representation by graphs. The need to induce a certain order on a chromatic graph and the search for the nature of that order will be examined in terms of three important problems, useful for the operations of matching and grouping in chemistry. We shall thus limit ourselves to the search for isomorphism, homomorphism for a chromatic graph pattern.

1.2.1   Order and chromatic graph matching

To recognize whether or not two graphs are identical, a gradual progressive comparison operation called matching is used. In the case of simple graphs, the matching is easy, but global perception is no longer evident when dealing with complicated graphs. It can scarcely be rendered automatic by the use of an algorithm unless the matching operation can be transferred from the graph to its representation or alphanumerical description. Consequently, it is imperative to impose an order on graphs so as to identify the sites in the graph itself, in its representation and sometimes to render unique and unambiguous the passage from the graph to a representation of it.

To point out the importance of this order to be induced on the chromatic graph, we shall analyze, albeit briefly, the problem of isomorphism between two graphs and that of detecting the homomorphism of two graphs i.e. determining whether or not graph G'_χ is a subgraph of graph G_χ.

1.2.1.1   Chromatic graph isomorphism

The problem here is to verify if the three criteria (p. 340) defining the isomorphism of two graphs are satisfied.

For any two graphs G and G' an automatic solution will be cumbersome at best if not altogether impossible unless order is introduced on their sites. Indeed, one has to traverse both graphs trying out all the possibilities of node to node matching.† This is difficult if the structural diagrams or their linked chromatic graphs are in preorder, in other words, if a multidirectional matching is required. To solve this problem a linear order must be induced on the chromatic graph so as to reduce this multidirectional matching to a linear matching process. In fact, total order can be useful both on the chromatic graph level and on the level of its representation. The detection of isomorphism can be transferred from the comparison of the graphs to that of their alphanumerical representation (matricial descriptor) for which linear matching is possible.

This induction of order linking an *ordered chromatic graph* to a structural diagram replaces the difficulties of the matching process by the search for this ordered, unique form. Under these conditions, the detection of isomorphism is easier, more efficient and always possible.

---

†By applying the three isomorphic criteria each time.

Since 1965, two algorithms for inducing total order on a structure have been proposed and gradually extended to a vast group of chromatic graphs. One is Morgan's algorithm[14] used by the Chemical Abstract Service, and the other is the DARC/DUPEL†[15,16] algorithm worked out for the DARC system codes. The main features of this latter will be the object of the second part of this chapter. We shall now touch on the principles of the Morgan algorithm.

Whereas the DARC/DUPEL algorithm uses topology and chromatism to the utmost to introduce total order on a graph, the Morgan algorithm gives precedence to the topological graph. In numbering the nodes, which is a way of expressing the order induced, chromatism is used only to break the remaining ties (Fig. 4). From Morgan's total order an alphanumerical representation called

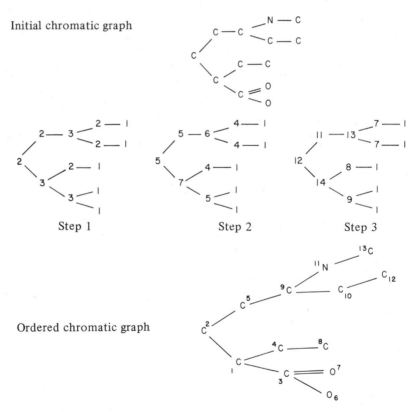

FIG. 4   Ordering by Morgan's algorithm. *Partial ordering by global topological valuation:* each node is valuated by its connectivity degree and is then valuated until stabilization by the sum of the connectivity values of its neighbours. *Linear ordering:* the final graph is renumbered according to the final values and by lexicographic chromatic decisions.

*DUPEL (graph Description in Unilinear form by Propagation of an Environment Limited to B).

a Unique Connection Table (UCT) is derived. This ordering can become cumbersome for very symmetrical compounds, as it is linked to the whole of the discoloured skeleton. Yet decisions on priorities would seem to be accessible on the level of limited topological modules, whose chromatism describes them most strongly. Moreover, Morgan's algorithm is not conceived for functional grouping. Indeed, the origin atom is often selected by purely topological considerations. Consequently, the origin evolves on the topological graph according to local hindrance and does not coincide with the characteristic functional focus. On the other hand, this order is well adapted for *isotopological grouping*, i.e. grouping of compounds with a same skeleton.

1.2.1.2 Chromatic graph homomorphism

Here the problem is to see if G'_χ is a subgraph of G_χ and consequently, to identify certain parts of the whole G_χ. This is a problem of pattern recognition.

At first glance, the induction of a linear order (intrinsic to G'_χ) does not present the same advantage here as in the search for isomorphism, since there is no reason for the order of the corresponding part of G_χ (impact of G'_χ on G_χ) to be the same as that of G'_χ (Fig. 5). Indeed the order induced on a part of a whole depends on its position in the whole. Save for rare exceptions, a chromatic graph G_χ and a subgraph G'_χ do not have the same origin of induced order. In more complex cases where G_χ or G'_χ are partly discoloured the correspondence of ordered sites is even more improbable. A study of the methods used in homomorphism searching can nevertheless serve to show the advantage of inducing total order on the level of certain privileged parts of G_χ and G'_χ.

This problem can be tackled today by a number of different procedures: node by node comparison, piece by piece comparison or frozen fragment approach[17] or again set reduction (Sussenguth algorithm[18]). These procedures can be applied either directly and this is called homomorphism searching by direct process, or after a preliminary screening stage and we call this homomorphism searching by ordered subset.

Homomorphism searching by direct process. Homomorphism searching procedures attempt to make subgraph G'_χ coincide with at least one of the parts of G_χ. Certain algorithms[18,19] attempt to lessen the strongly combinatoric aspect of this problem; nevertheless, their direct use on a large scale is time-consuming and prohibitive. For example, the direct search of a substructure in an index of 10,000 chemical compounds involves using the homomorphism procedure 10,000 times.

Homomorphism searching by ordered subset.[20] Standard subsets or subgraphs, not necessarily disjoint, are derived logically and according to the same rules from graph G_χ and from subgraph G'_χ simultaneously, Each subset, totally and identically ordered, on the level of G_χ and G'_χ, is considered as a whole. Then by a linear matching of their representation (cf. Isomorphism problem), one checks to see if all the standard subsets defined in subgraph G'_χ are also present in graph G_χ. If this condition is met, then one of the previously

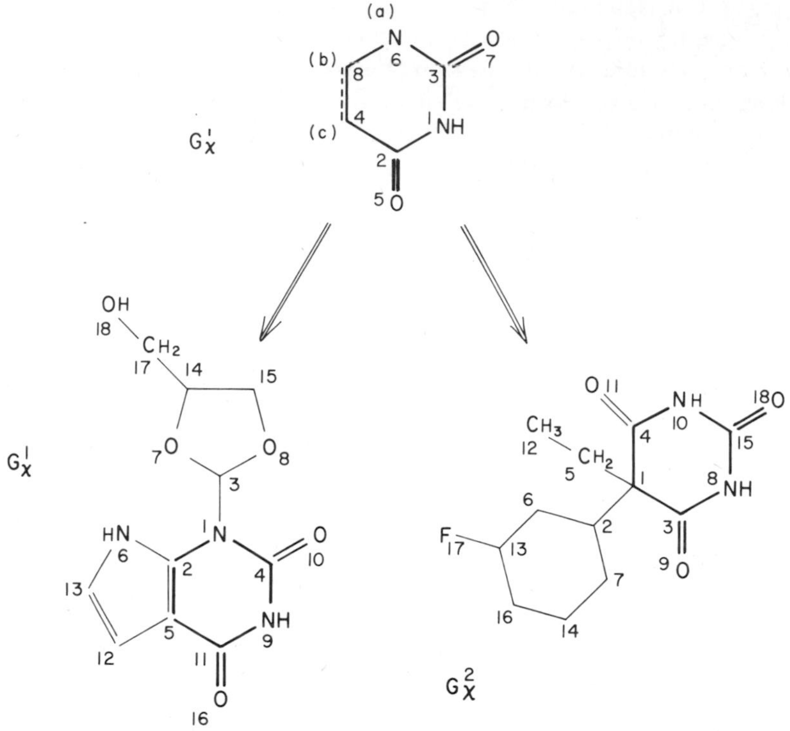

FIG. 5   Graph homomorphism problem. The combinatorial feature of sub-structure searching makes its implementation very cumbersome: the order on the nodes of a (G_X') isolated substructure is lost at the level of its impact within the (G_X^1), (G_X^2) structures.

mentioned procedures is employed. If this condition is not met, then one can already conclude that there is no homomorphism. Thus, by inducing total order on the level of standard subsets, the detection of homomorphism can be considerably optimized.

1.2.1.3   Order and pattern stereo-ordering

In chemistry, the passage from 2D plane representation to a 3D stereo-model assumes that one has incorporated in the 2D representations descriptors making it possible to differentiate between chiral and geometric stereoforms. If, therefore, one wants to retain all the information of each form (the object or its image) in its corresponding chromatic graph and, consequently, in its associated alphanumerical descriptor, a non ambiguous stereo-identification must be developed.

This is a very difficult general problem which we will approach here by referring to a simple case, that of forms with chiral motifs centred on a site, such as the asymmetrical carbon. As a focus with its immediate environment, this asymmetrical carbon (C*) constitutes a structural motif defined by its topology and its chromatism. Cahn, Ingold and Prelog (C.I.P.)[21] order this motif, regarded as a chiron† by imposing, primarily on first neighbour atoms of C*, a hierarchy introduced by an ordering law called a *sequence rule.*

Whereas, in Morgan's algorithm, non stereodescription is based mainly on the topological graph, the sequence rule deals with stereodescription in an opposite manner. Its priorities depend first of all on atom chromatism, but it ignores bond chromatism and thus sometimes engenders a fictional topology when assigning the stereochemistry of C* (Fig. 6).

This brief analysis shows that in order to account for the chiral feature of a stereoform, C.I.P. imposes a *local order* not necessarily coherent with the global order used when naming the molecule (Table I). With this in mind, we have shown[22] that the Major Order corresponding to the whole of the 2D non stereochemical description must be made coherent with the Finalized Local

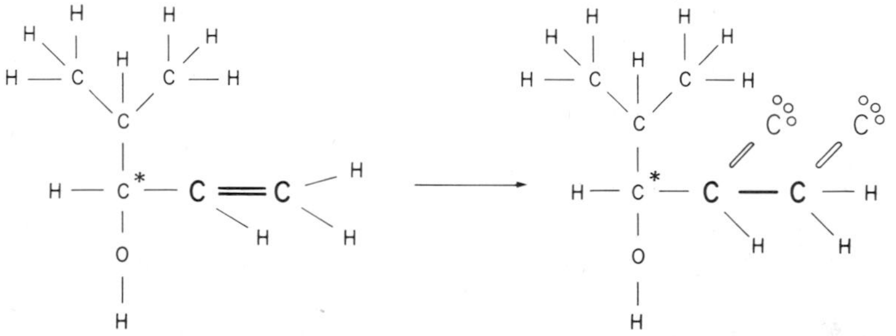

FIG. 6   Assigning stereochemistry by the Cahn–Ingold–Prelog sequence rule. When double bonds arise the sequence rule needs an artificial topology which results from creating phantom atoms.

Order oriented towards chiral description. In our opinion, this search for coherence goes further than the problem of indexing and must be based on a unitary theory of stereochemistry (focalization, ordering rule, observation rule and orientation procedures). Beginning with more general stereochemical concepts, the procedures based on modelling by an oriented chromatic graph which have been previously exposed[23,24] lead easily to a generative, homogeneous, unique and unambiguous 2D — 3D description.

The analysis of these several examples points out the difficulty encountered in defining *a priori* criteria for the choice of an order. Thus the search for isomorphism requires chromatic graphs to be totally ordered; the search for homomorphism can be undertaken between two graphs with no particular order,

†The chiron corresponds to the minimum structural organization which introduces chirality in an n-dimensional space.

TABLE I

Composite and homogeneous representations. Two types of representation are possible, depending on whether or not the local order is coherent with the general order.

Entities	Types of representation	Description	Encoding
	comp-osite	IUPAC-CIP IUPAC-DARC	(2S)-butan-2-ol (2T)-butan-2-ol
	homo-geneous	DARC	$(1000/0/8*) (1110/1/T:1*)$ $1(1000)11$
	comp-osite	IUPAC-CIP IUPAC-DARC	(5S)-2,2,5-trimethylcyclopentanone (5H)-2,2,5-trimethylcyclopentanone
	homo-geneous	DARC	$(1000*) (4210/2:11/8:11/H:12*)$ $1(1000)12[1:1.12, 1.21]$

i.e. in preorder (node by node searching) or locally ordered; the orientation of a stereoform assumes that a local order has been induced around chirality centers.

There are no absolute criteria for intrinsically defining the qualities of order to be induced on a chromatic graph. The choice of an order and, subsequently, the methods for establishing it always depend on a certain arbitrariness, and one cannot hope to find a universal solution perfectly adapted to all types of applications. A realistic solution consists of defining, within a given system, a *Major Order* economically arrived at by a rational method which leaves no room for empiricism. From such a Major Order it should be possible either to proceed to a *Finalized Order*, i.e. one optimized for a specific application (intraconvertibility), or to reach the order adopted by another system (interconvertibility). The interconvertibility of systems is concretized on the code level by transcoding. The ease with which this operation is realized depends both on the way in which the induced order is expressed on the ordered chromatic graph and on the logical level of the descriptive codes. Formulating the order must not be limited to a random numbering of the graph nodes but must, by a label assigned to each node, translate a specific organization of the graph and the local proximity relationships between atoms. The coded description in turn must make it possible to clearly explicit the order induced on the chromatic graph.

In the second part of this article, we shall describe the basic elements of the topological language used in the DARC System, conceived so as to deal with these different constraints. This is indeed a major topological language capable of engendering diverse finalized languages and from which we can proceed to descriptive languages in other systems or vice versa.

2   ORDER AND STRUCTURAL ENVIRONMENT

In the DARC system the need to work on the modelling of a structural diagram by an ordered chromatic graph was recognized. However, since the system originated from physicochemical and structural (correlation) considerations, it was necessary to base it on clear concepts capable of leading to ordered graphs in different ways. In particular, if the need for a principal or Major Order is essential, it seems clear that it must be surrounded by *Finalized Orders* or *Orders oriented* according to the nature of specific structural problems. This is shown when one turns to conceptual problems in chemistry, to oriented paths of synthesis or to site spectroscopy (PMR, ^{13}CMR, ...).

In fact, to develop a system for describing oriented structures by a total or linear order and to account for the constraints of chemical matching and grouping, we have just shown the advantage of having an ordered chromatic graph correspond to a structural diagram. However, the method used to achieve this type of graph must not make us forget the *possibilities contained in the diagram* and those *belonging to the graph.* To link properties to certain localities of the diagram, we have shown that the order imposed must be a function of this final aim. To this end it is often necessary to express mathematically the environment of a particular site in one case and of another site in a different case. To describe the state of an environment by the order imposed (partial or linear order) becomes possible if all the sites are indexed (linear order). Thus, a *quantitative description of the environment* is achieved by the passage: ordered graph → adjacency matrix → linear descriptor (GMD).

In the DARC system, our intention was to express the whole potential of the structural diagram which, in fact, makes it possible to visualize a large number of environments of sites. It is true that the idea of a site environment often remains qualitative and vague by its very nature. Many authors use this idea by assigning local, trivial, temporary numbers but do not propose a descriptive method or a general definition. Just such a definition was our aim in the DARC system when we proposed the concept of a finite or limited environment. This allows for a definition in terms of numbers of nodes and connectivity degrees through the associated graph of the environment and thereby for a general and quantitative description.† This limited environment concept serves as a base for the DARC Major Order which is oriented for computer mass storage as well as for focused physical properties (e.g. ^{13}CMR) and other derived Finalized Orders. Its use in inducing order on a chromatic graph, the generating of an ELCO (Environment/Limited, Concentric and Ordered) and the nature of the generating process will all be considered in more detail in the following section.

---

†The description of an environment by an ordered chromatic graph is indeed relevant since the standard metric, linked to the graph sites, assimilates the graph to an *ordered graph pattern* and not to a loose graph.

2.1   Order by generation or by fragmentation

To order a chromatic graph, the order may be linked to an ordered creative process of this graph, by generation. Formerly, and before a theory of order was elaborated, order was introduced on a graph by fragmentation of the latter into subgraphs. We will examine these two very different methods.

2.1.1   Ordering by fragmentation

The chromatic graph is perceived as the juxtaposition of disjoint chromatic subgraphs. To achieve total order by this method, one must, as a preliminary step, have taken a census and individually ordered each of the subgraphs used to construct the chromatic graph; one must moreover, have ordered the fragments as a whole. It is then possible to induce total order on all the atoms of the graph regarded as a set provided, of course, that the fragmentation process has entailed no deterioration of the information. Now it is extremely difficult, not to say impossible, to establish *a priori* an exhaustive dictionary of all fragments which, with no loss of information, can cover several million compounds. The combinations are almost infinite. In general, this fragmentary generation merely reflects the proximity of the atoms to the fragment. It is scarcely adapted to the definition of a finalized order convenient for conceptual applications.

2.1.2   Ordering by generation

To acquire a chromatic graph by generation,[7] as we have proposed, involves constructing it dynamically. One starts from an origin or Focus (FO) to progressively discover its environment. The creation of total order is merely a question of choosing a sequence which will generate the sites of the environment. This choice stems from an analysis of their intrinsic characteristics (topology, chromatism) and, exceptionally, of their vicinity.

Two essential features of this dynamic generation are brought out when the process is voluntarily stopped at any stage: the site corresponding to this stage is unambiguously located in the generation sequence and, furthermore, a specific, ordered, chromatic subgraph, ancestor of the chromatic graph in process of generation, is identified.

2.2   Order and labelling

The order induced by the generation process may be expressed in different ways. If the only aim is a hierarchization for the purpose of some particular computer processing, such as automatic documentation, a simple numbering, situating a node in an ordered sequence from 1 to N will suffice.

If, on the other hand, one wishes to have at one's disposal a form of indexing which reflects the local characteristics of the molecule, one must go beyond numbering to define a richer labelling which will take chemical constraints into account. For example, chemists have often evaluated the effects of sites equidistant from a focus, and it is therefore important that the labelling which translates this property become an element of the metalanguage. It should be noted, by the way, that chemists sometimes use a form of labelling to situate two functional groups with relation to each other (e.g., in the name diol-α, α can be considered as a label). In the following paragraphs, we shall show how the labelling becomes progressively more precise in the course of the different stages leading from preorder to total order.

2.3 Concentric partial order (A,B) and concentric environment

To introduce into our labelling the notion of 'intersite' topological distance or distance by valency bonds, we induce a partial order on the environment just before the generation leading to total order. *This concentric partial order is invariant no matter which linear order is sought.*

The focus environment ($\mathcal{E}$) is organized by concentric layers. Beginning from the focus, these layers are labelled alternately by the letters A and B. Two consecutive layers — the first, type A, the second, type B — constitute a limited environment E_B. An E_B is specified by an index 1 which denotes its rank with relation to the focus and is noted (E_B^1) (Fig. 7). Thus at this stage, one can already localize a site by its type (A or B) and the rank 1 of the E_B it belongs to. This site will be noted:

$(X)^1$ where $X = A$ or B

The partial order relationship is defined thus: given two sites X and Y, we say that X precedes Y and we note $X < Y$ if

(a) X and Y belong to the same E_B of rank 1

then $X = (A)^1$ $Y = (B)^1$

(b) X and Y do not belong to the same E_B

then $X \epsilon E_B^1$ $Y \epsilon E_B^k$ with $k > 1$

The concentric environment[15] thus defined is provided with a partial order structure and the $(X)^1$ type or *partial order labelling* makes it possible to localize sites by their distance from the focus and to evaluate intersite topological distances.

2.4 Linear order and ELCO

Having induced a partially ordered structure (A,B) on the environment, we must then order it totally; this means defining the generation sequence of sites belonging to a same layer and derived from the same node. To do this, we introduced the ELCO concept (Environment/Limited, Concentric, Ordered).

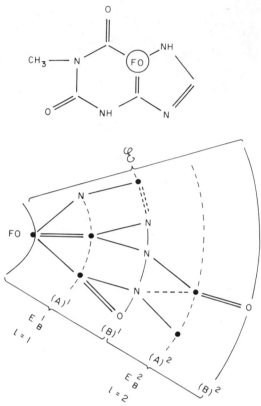

FIG. 7   Partial order and concentric environment. The partial order label $(A)^l$ or $(B)^l$ locates each node by its topological distance to the focus. Concentric organization brings forth: even rings – 2 nodes from the same layer linked to one node of the following layer; odd rings – 2 nodes from the same layer linked to each other.

2.4.1   Definition

The ELCO is the smallest environment module ordered as a whole, i.e. by using only such information (topological or chromatic) as it contains.

There were two main reasons for the choice of an E_B as maximum amplitude of the smallest ordered module:

(1) An E_B is so constituted that each type of information (topology, chromatism on the edges, chromatism on the nodes) can be used at least once for ordering, which optimizes the attainment of order. If we were to limit ourselves to level A, we could only take into account bond and nature chromatism and would lose an important type of selective information: topology.

(2) in the course of physicochemical analyses, one very often notices an attenuation of structural effects beyond the first limited environment (E_B) of an active center (chromophore, reaction center).[25,26]

In an ELCO each site is marked by a topological coordinate or linear order labelling. This results in turn from an extension of the partial order labelling by introducing the position index i or ij of the site. Thus the labels become (A_i) or (B_{ij}) (Fig. 8). Both orders, partial and linear present on an ELCO may then be summed up as follows:

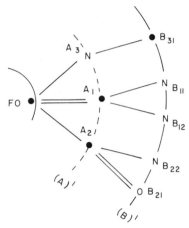

FIG. 8   Linear order and ELCO. *Application of* P_A $:f_{\chi NA} \circ f_{\chi LI} \circ f_{\mathscr{T}} f_{\mathscr{T}}:A_1$ and A_2 (2 neighbours on $(B)^1$) have priority over A^3 (1 neighbour on $(B)^1$). $f_{\chi LI}:A^1$, linked to FO by a double bond, has priority over A^2. *Application of* $P_B:f_{\chi NA} \circ f_{\chi LI}:B_{21}$ has priority over B_{22} (see above). According to P_B, B_{11} and B_{12} are equivalent in the ELCO. They are distinguished from each other by their neighbours in the next ELCO. This is expressed by P_0 rule underlying ELCO propagation.

(a) *concentric partial order on* ELCO

$\sqrt{k}$ and $\sqrt{i,j}$ $A_k < B_{ij}$

A sites precede B sites

(b) *linear order on* ELCO

b1. $A_1 < A_2 < A_3 < \dots A_N$

b2. $B_{11} < B_{12} < \dots B_{1m} < B_{21} < \dots B_{Np}$

The problem, then, given two A sites or two B sites, is to determine which will be generated first. For this purpose, a *complex ordering function* $\mathscr{F}(\mathscr{T},\chi)$ is used which takes into account the topological and chromatic criteria intrinsic to each site. An elementary ordering function corresponds to each of these criteria: $f_{\mathcal{T}}$ (topology), $f_{\chi LI}$ (bond or link chromatism), $f_{\chi N}$ (node chromatism). The kind of order obtained (Major or Finalized) depends on the composition law of the elementary functions which define the ordering function.[7]

2.4.2   Ordering function on ELCO

The major order DARC is obtained by using a modular type ordering function, i.e. one which considers point by point all the information contained in the chromatic graph limited to one E_B. To class two sites (A or B) the composition of the elementary functions is carried out on the level of each type of position. The function $\mathcal{F} = P_B \circ P_A$ results from applying two priority rules P_A and P_B.

Priority P_A (to class A positions) $P_A = f_{\chi N} \circ f_{\chi LI} \circ f_{\mathcal{T}}$ which signifies that if topology ($f_{\mathcal{T}}$) does not suffice to class two sites, link chromatism ($f_{\chi LI}$) is examined and eventually, if need be, node chromatism ($f_{\chi N}$).

Priority P_B (to class B positions) $P_B = f_{\chi N} \circ f_{\chi LI}$ which means that we examine successively link chromatism and eventually node chromatism (topology is not considered since the degree of connectivity of a B position in the ELCO always equals 1).

Those sites not classed by P_A and P_B are equivalent in the ELCO and can be ordered arbitrarily. These ties can be broken during propagation if the ELCO is merely a part of a whole.

2.5   Linear order on environment and ELCO propagation

Since the ELCO constitutes the minimum ordered environment, a linear order is induced on a concentric environment when, in accordance with a specific law, we propagate the ELCO in the same direction as that in which index 1 increases.

   In an ELCO as defined in the last paragraph, some type B sites are origins of bonds issuing from the ELCO. These are regarded as development origins (OD) for an ELCO situated in the next environment E_B.†

   The generation sequence of ELCOs of E_B^p is induced (Fig. 9): (a) by the order in which the E_B^{p-1} ELCOs from which they derive have been generated, (b) (if several E_B^p ELCOs derive from the same E_B^{p-1} ELCO) by the order in which the development origins of this ELCO have been generated.

   In the DARC Major Order, a priority law P_0 does indeed make it possible to class development origins which seem equivalent inside the E_B^{p-1} ELCO ($P_0 = f_{\mathcal{T}}$). In Fig. 8, we break the tie of nodes B of the first ELCO by applying priority law P_0; B_{11} has a substituent in the second ELCO; B_{12} has none.

---

†During propagation the number of type A nodes of the ELCO is one less than the connectivity degree of its origin. For the initial ELCO the number of A nodes coincides with the connectivity degree of the focus.

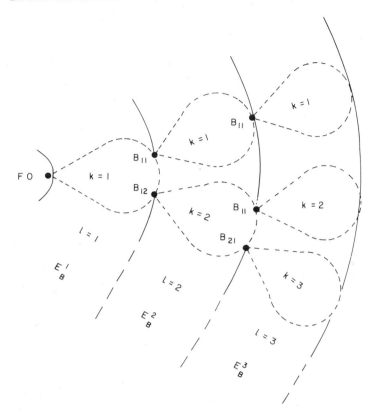

FIG. 9   Propagation of ELCO. In E_B^2 the first ELCO to be generated ($k = 1$) derives from B_{11} of E_B^1:the second ELCO ($k = 2$) derives from B_{12} (proposition (b): $B_{11} < B_{12}$). In E_B^3 the first ELCO derives from the first ELCO of E_B^2 (proposition (a): $B_{11}(k = 1) < B_{11}(k = 2)$). Proposition (b) determines the generation order of the other two ELCOs.

Thus, by propagating an ELCO, linear order is induced on the environment. This linear order is expressed by a label in such a way that each site is spotted by: (1) its A or B type; (2) its topological coordinate A_i or B_{ij} (linear order on ELCO); (3) the rank of the E_B to which it belongs: 1; (4) the ELCO rank in the E_B: k.

To sum up, the linear order labelling (Fig. 10) is defined by

$$(A_i \text{ or } B_{ij})_k^1$$

It should also be noted that while an ELCO is being propagated for the purpose of generating a target compound, a particular compound is associated with each site specified by this linear order label. All these compounds taken together constitute the canonical generation series (Fig. 11).

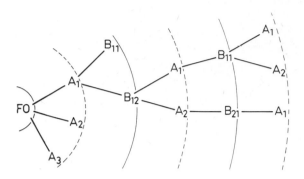

Linear order labelling	Linear order numbering	Neighbours	Topological distance to the focus
FO	1	2 3 4	0
A_1	2	1 5 6	1
A_2	3	1	1
A_3	4	1	1
B_{11}	5	2	2
B_{12}*	6	2 7 8	2
A_1	7	6 9	3
A_2	8	6 10	3
B_{11}*	9	7 11 12	4
B_{21}*	10	8 13	4
A_1	11	9	5
A_2	12	9	5
A_1	13	10	5

FIG. 10   Linear order labelling. Linear order labelling (first column) sum-marizes all of the topological information of the graph and potentially contains the linear order numbering (second column: realizing Morgan's goal) plus other information (ordered connectivity and topological distance).

2.6   From linear order labelling to description

The generation process just described makes an ordered chromatic graph ($G_{\chi 0}$) correspond to any ordinary chromatic graph. If, in addition, one disposes of a set of rules enabling one to unambiguously select the focus (FO),[27] this *ordered* graph will be *unique*. As a result, the linear topological descriptor associated

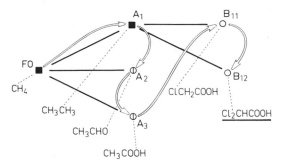

FIG. 11   Canonical generation series of a target compound. During the formal generation of the dichloroacetic acid as target compound, starting from methane as source compound, all the six developed entities are associated to the filled sites of the chromatic graph. As a group, they engender a canonical generation series.

with this graph is unique and non-ambiguous and particularly well adapted for non-redundant mass storage.

This linear descriptor is established modularly ELCO by ELCO and reflects the different levels of order progressively induced on the graph (partial and linear). For a chromatic graph limited to the primary chromatism of the compounds (nature of the links, nature of the atoms), the ELCO descriptive module groups three descriptors: a descriptor of existence (DEX), a descriptor of the nature of the links (DLI) and a descriptor of the nature of the atoms (DNA).

The descriptor of existence reflects the existence of the ELCO linear order label. It is expressed by the following general formula:

$$\text{DEX} : \left(\sum_i a_i \ \sum_i b_{i1} \ \sum_i b_{i2} \ldots \sum_i b_{in} \right)$$

where $a_i = 1$ if a site labelled A_i exists in the ELCO, otherwise $a_i = 0$; $b_{ij} = 1$ if a site labelled B_{ij} exists in the ELCO.

The DLI (respectively DNA) spots by their labels those sites which carry other than simple links (respectively atoms other than carbon) (Fig. 12). At this stage, linear order has been expressed only on the level of the ELCO corresponding to labels A_i or B_{ij}. The ELCO connectivity must still be established. To begin with we express indexes 1 (limited environment number) and k (ELCO number) of labels $(A_i)_k^1$ or $(B_{ij})_k^1$ which translate the global linear order obtained by propagating the ELCO.

For this purpose, one indicates in each ELCO described the number of ELCOs of the following E_B which derives from it or in other words, the number of development origins it contains.

The following diagram indicates clearly that, in this way, indexes k and ι are

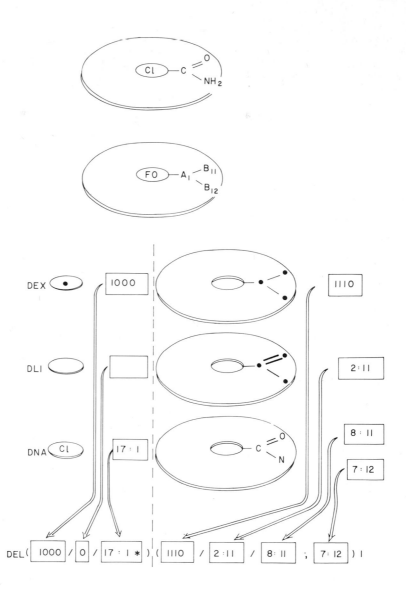

FIG. 12　Description of an aliphatic compound: encoding of the ordered chromatic graph by the DARC/DUPEL method. The FOCUS is a single chlorine atom: $\Sigma a_i = 1 \rightarrow DEX = 1000$; no anterior bond $\rightarrow DLI = 0$; .Cl on A_1 $\rightarrow DNA = 17{:}1$. For the ELCO: $\Sigma a_i = 1$ $\Sigma b_{i1} = 1$ $\Sigma b_{i2} = 1 \rightarrow DEX = 1110$; double bond on $B_{11} \rightarrow DLI = 2{:}11$; 0 on B_{11}, N on $B_{12} \rightarrow DNA = 8{:}11;\ 7{:}12$.

expressed:

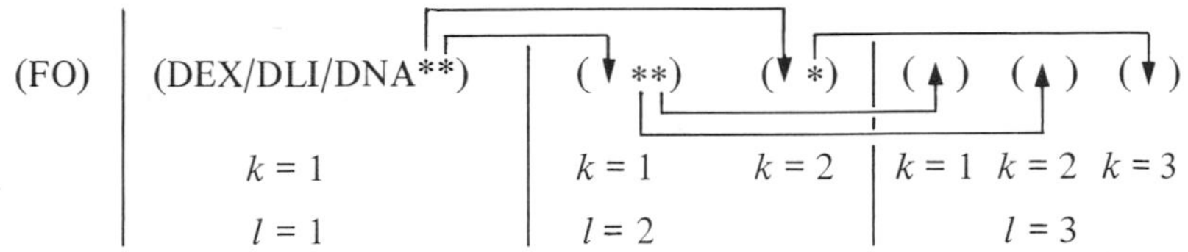

(FO)	(DEX/DLI/DNA**)	(▼**)	(▼*)	(▲)	(▲)	(▼)
	$k = 1$	$k = 1$	$k = 2$	$k = 1$	$k = 2$	$k = 3$
	$l = 1$	$l = 2$		$l = 3$		

Thus, total order is expressed completely. Now in order to obtain a descriptor able to unambiguously reconstruct the chromatic graph, i.e., a truly generative descriptor, one introduces into each descriptive module the ij index of the node of the preceding E_B from which the ELCO under consideration derives; thus, one completes the establishment of the ELCO connectivity. The linear topological descriptor (DEL) (Fig. 13) obtained exactly restitutes the progressive generation diagram of the graph ordered according to the DARC Major Order expressed by the linear order labelling.

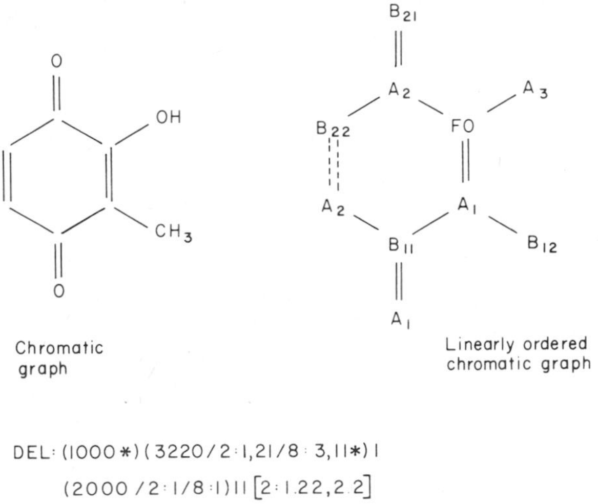

Chromatic graph

Linearly ordered chromatic graph

DEL: (1000 *)(3220/2 : 1,21/8 : 3,11*) I

(2000 /2 : 1/8 :1) II [2 : 1.22,2.2]

FIG. 13   Description of a cyclic compound. A cyclic graph is considered as a spanning tree with extra ring closure links. The spanning tree is first described, and then the ring closure links are grouped within a descriptor of cyclization [DCY] which expresses link multiplicity and location.

This descriptor contains explicitly or implicitly an important amount of information which can easily be derived from it. In particular, we might mention (Fig. 14):

— the adjacency matrix used for theoretical chemical calculations (quantum mechanics),

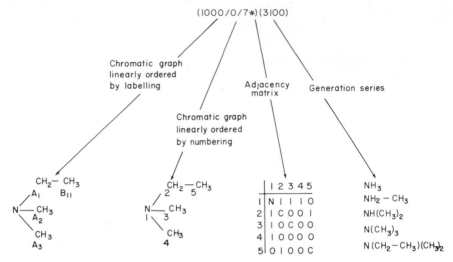

FIG. 14   DEL Information content. The compound $N(CH_2 - CH_3)(CH_3)_2$ has been described by its descriptor DEL. From the DEL: the chromatic graph linearly ordered by labelling or numbering may be unambiguously reconstructed by hand or automatically and displayed on a CRT; the adjacency matrix may be established; the elements of the generation series may be listed in the order of their generation.

— the generation series associated with the target compound which can point out comportment or preparation analogies between compounds,

— the linear order labelling, which among its many other uses, is also helpful for quantitatively and topologically distinguishing between compounds centered on a same focus during correlation research,

— the linear order numbering derived univocally from the linear order labelling and used to optimize fundamental algorithms of graph theory, such as the search for a smallest ring basis.[28]

In the field of chemical machine documentation the topological descriptor DEL is quite convenient for non-redundant storage of large collections of compounds and for retrieval of a given compound in a data bank by simple linear matching.

Furthermore, the following paragraph shows that the much more complex problem of retrieving those compounds only defined by a substructure is solved, easily and economically, by using the ELCO concept and a simple descriptor of the same type as the DEL, generated automatically from the latter.

2.7   ELCO and structure identifying by FREL substructure searching

The problem of localization, within a data bank, of the compound or compounds defined by some of their parts is a valid example of structure identifying and, as such, is related to pattern recognition: *a part must be perceived and recognized in a whole.* On the modelling level, chemistry seems to constitute a privileged field of pattern recognition, because chemical patterns are objectively assimilated to a chromatic graph. However, in practice, treating this problem remains complex, for it is not possible to proceed by static recognition of pre-determined parts recorded in a dictionary. Indeed, drawing up an inventory of all entity parts likely to interest chemists is an infinite combinatorial problem.

The solution we propose is based on a dynamic and flexible fragmentation of the structure. The ELCO concept allows for grasping standard ordered substructure within a structure while preserving the logic involved in grasping the entire structure. The resulting standard fragment called FREL (Fragment Reduced to an Environment Limited to B) is an ELCO rooted on each node with a connectivity degree greater than two. It is, therefore, defined with the same linear order labelling (A_i, B_{ij}) and described by *the same descriptor as if it was an entire structure* (Fig. 15).

When creating a collection of compounds coded by their DEL, the FREL descriptors related to each structure are generated from the DEL and recorded.

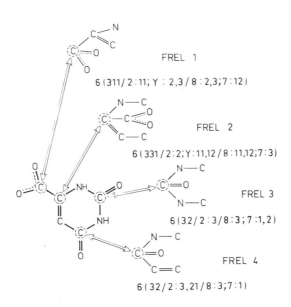

FREL 1

6 (311 / 2 : 11; Y : 2,3 / 8 : 2,3; 7 : 12)

FREL 2

6 (331 / 2 : 2; Y : 11,12 / 8 : 11,12; 7 : 3)

FREL 3

6 (32 / 2 : 3 / 8 : 3; 7 : 1,2)

FREL 4

6 (32 / 2 : 3,21 / 8 : 3; 7 : 1)

FIG. 15   FREL Generation. FRELs (Fragment Reduced to an Environment Limited to B) are ELCO substructures around dense centers (connectivity degree ≥ 3). These four FRELs span the orotic acid with redundancy since they overlap. This overlapping increases the precision of the identification by substructure searching. The simultaneous presence of FREL 1 (included in FREL 2) and FREL 2 indicates that the cycle is not substituted on ortho position of the carboxylic group.

Although FRELs associated with a structure do not describe it exhaustively, since a certain loss of information takes place, particularly for linear chains (no connectivity nodes greater than two), the set of these FRELs nevertheless specifies the topology of the compound and its 'interesting' chromatism with precision.

There are two reasons for this: (1) FRELs are centered on atoms with high connectivity degrees and thus cover the molecule parts richest in chromatism and topology; (2) The loss of information resulting from this dynamic fragmentation is all the weaker according as the FRELs are more numerous and centred on neighbouring atoms. The set of FRELs often spans the molecule and even with some overlapping (the same zone of the molecule is covered by several FRELs). The overlaps illustrate important topological characteristics of the molecule (Fig. 15) with respect to other similar compounds described by the same number of identical FRELs but showing less or no FREL overlapping.

When questioning the collection by a substructure, the set of FRELs of the substructure is generated. The set of the compounds (C) where the sought for substructure appears is contained in the set of compounds $(\tilde{C})$ which include the substructure FRELs. Thanks to the simple form of FREL descriptors, the set $(\tilde{C})$ is established by simple comparison of the substructure FRELs and the registered FRELs (Fig. 16). The topological and chromatic indeterminations of the substructure are easily treated during this comparison and need no cumbersome devices such as the creation of a permuted index.[29] In general, (C) and $(\tilde{C})$ are very close and even coincide as we have shown in an experiment treating some 50,000 compounds of the CAS file.

The use of FRELs for searching by substructure confers a strong heuristic power on them to the extent that they also make it possible to group compounds through fuzzy analogies.

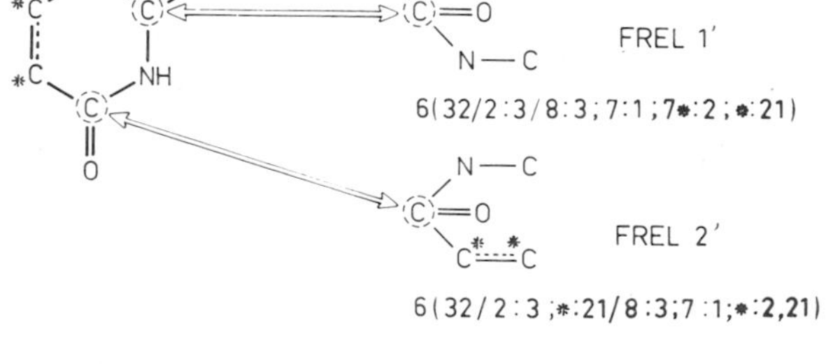

FIG. 16   Substructure searching by FREL. This upper substructure can be matched with orotic acid by computer search through its decomposition into two FRELs 1' and 2' which match precisely with FRELs 3 and 4 of orotic acid (cf. Fig. 15).

3 ORDER AND POPULATION

We have enriched the preorder structure of a chromatic graph by coherently and progressively introducing partial and total order relations on the nodes. The generalization of these ideas, concept levels and methods seem to us indispensable for organizing populations of chemical entities. Indeed the lines of reasoning involved are very often analogical or correlative and bear on the entity situated in a population.

In most problems, the analysis of a set of entities, or population, is carried out either to bring out a general law or to focus on a specific entity. The following examples can be mentioned:

— comportment studies (properties, reactivity, pharmacodynamic activities) with extraction of the local data of a population in view of previsional aptitude correlations;[30-32]
— study of the interconversion of compounds by chemical reactions, by evolution of conformation or configuration (assignment and prediction of reaction mechanisms or metabolic pathways, optimization of synthetic pathways);[33-36]
— the operations of *counting* and *listing* those entities of a population whose structural organization is defined (molecular formula for a family of isomers) and the *access operation* to a target compound (e.g. counting and listing of pathways leading to that target, etc.).[37-39]

All these analyses can be carried out with an arbitrary and episodic ordering for *small populations* of compounds with very pronounced resemblance or grouping relations (isomer families, linear and concentric homology, genetic series, cf. Fig. 17, step 2). For much *larger populations* (tens, hundreds of entities) chance solutions are difficult to find. This complexity is largely due to the absence of any order structure on populations of entities (Fig. 17, step 0).

The fact is that, save for very particular reasons, no equivalent of the chemical entities chromatic graph preorder exists in a magma characterized population. For these populations, the situation is still comparable to when chemical entities had as yet no structural diagram.

To bridge this gap, we have evolved a *general concept of hyperstructure* (which we designate as HS) so as to apply the chromatic graph theory to the study of populations. It is impossible to present all the diverse aspects of hyperstructures rapidly, but we shall here outline some of the general ideas and give some selected examples.

The wide scope opened up by the notion of hyperstructure is suggested by what *reactional hyperstructures* represent. In this type of population, if two compounds X and Y are linked by an edge, the chromatism of this edge symbolizes a definite reaction which governs the passage $X \to Y$. It should be

Population of amines $\mathscr{P}$:

Formula	DEL (FO = 1/1/7:1)	DEL abbreviated
Me₂ NH	(FO*)(2)1	(2)
i–Pr NH₂	(FO*)(111)1	(111)
Et NH₂	(FO*)(11)1	(11)
Me₃ N	(FO*)(3)1	(3)
NH₃	(FO)	(0)
n–Pr NH₂	(FO*)(11*)1(1)11	(11*)(1)
EtMe NH	(FO*)(21)1	(21)
Me NH₂	(FO*)(1)1	(1)

Step 0: NO ORDER
Population $\mathscr{P}$ = magma

Step 1: Weak PARTIAL ORDER
Classification according to molecular weight: distribution among plans of isomers

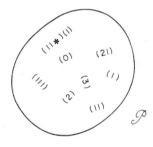

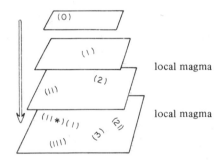

local magma

local magma

Step 2: Stronger LOCAL ORDER − Homology and formal isomerisation

3A. *Linear order on each family of homologues:*

h_1 linear homology
h_c concentric homology

3B. *Preorder on each plan of isomers:*

i_F formal elementary isomerisation (which may be performed by cutting only one N–C or C–C bond)

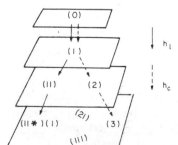

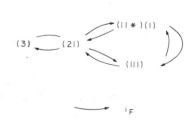

*Step 3: Strong PARTIAL ORDER on
        the whole population 𝒫*
A generalization of 3A

Step 4: TREE ORDER over HS($\mathscr{P}$, s_F)
Two simultaneous partial orderings:
exhaustive and non-redundant
generation of $\mathscr{P}$ and of the vicinity
relations s_F

Formal substitution hyperstructure
HS($\mathscr{P}$, s_F):
    s_F: formal elementary substitution
        (here substitution of H by CH_3)

Anteriology hyperstructure
HS($\mathscr{P}$, s_F, $\mathscr{A}_c$):
    $\mathscr{A}_c$: canonical anteriology relation

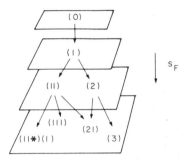
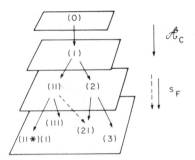

Step 5: LINEAR ORDER over HS($\mathscr{P}$, s_F, $\mathscr{A}_c$)
Three simultaneous orderings: exhaustive, non-redundant
and ordered generation of $\mathscr{P}$ and of the relations s_F

Ordered anteriology hyperstructure: HS($\mathscr{P}$, s_F, $\mathscr{A}_c$, θ_1)
    θ_1: linear and concentric generation order
    (in particular linear order over each plan of isomers)

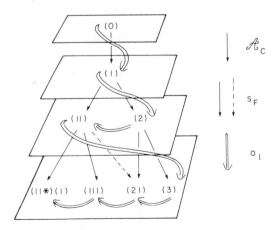

FIG. 17   Definition and nature of a syperstructure HS($\mathscr{P}$, s_F, $\mathscr{A}_c$, θ_1)
Progressive generation of HS starting (step 0) from a no order magma population
($\mathscr{P}$) to a complex ordering allowing labelling and linear ordering (step 5). To
simplify the diagrams, the names of the compounds are abbreviated descriptors
(DEL).

pointed out that for this chromatic graph, some $25,000^{40}$ registered reactions (edges) and a probable four to five million compounds (nodes) exist.†

3.1   Concept of hyperstructure – Definition and impact

Hyperstructure should facilitate the study of the general and local properties of the population (problem of the whole and the part) and should therefore be based on vicinity or locality relations between entities.

Definition: A hyperstructure (HS) is a population (or set) of entities provided with a chromatic graph structure whose nodes are the entities and whose edges express vicinity relations (HS topological proximity) between the connected compounds.

3.1.1   Graph edge and HS – Vicinity relation

The vicinity or topological relation between two compounds, as the elementary edge of an HS graph, may be defined by different criteria. We shall mention two types which lead us to specify certain hyperstructures: (1) *the transformational hyperstructure* (HS_T) whose edges symbolize either a chemical transformation[36] or a physical transformation or evolution (e.g., conformational transformation); (2) *the formal hyperstructure* (HS_F) where the criteria on the edges are structural (e.g., two neighbouring entities have a common substructure SS) (Y is derived from X by replacing a sulphur by an oxygen in its graph).

The importance of the HS_F stems from the need to describe populations of entities so as to tackle 'property – structure' correlations.‡[32]

3.1.2   Generation hyperstructure

When the purpose is to use a graph of an isolated entity, one can describe it by either fragmenting it or, as we have proposed, reconstructing it by a coherent, ordered procedure considered as *the generation or recreation of a target graph*.

In reality a population is a *magma of disjoint objects* on which we impose a graph structure (HS) by 'edge relations', i.e., HS_T or HS_F. If a general law enables us to generate all the edges on a magma, we may extend this generative approach of a graph of an entity to that of a population. This generalization is only possible under certain conditions.

Definition: A hyperstructure is termed generative if provided with a recurrent law of generation and with criteria for stopping this generation.

---

†For entities seven types of bonds are manipulated (edge chromatism) and about 500 atoms, isotopes included (node chromatism).

‡Generally, for lack of a population description (HS), a reference X_0 is fixed and one works on the variation of a substructure in relation to X_0 i.e., inΔ(SS).

HS generation can simultaneously generate both nodes and edges from a source kernel composed of a finite set of entities. Stops are necessary for limiting growth to n possible dimensions. These stops may be usefully represented by target entities or by border entities. The notion of recurrent generation makes it possible to manipulate infinite hyperstructures of which the simplest examples, albeit isolated ones, are linear homology families:

$$CH_3OH \ldots CH_3CH_2CH_2OH \ldots H(CH_2)_nOH \ (1 < n < \infty)$$

3.1.3   Progressive ordering of a hyperstructure

With the concepts thus far introduced, we can induce a chromatic graph structure on a population. This generalization or oneness treatment for entities and populations alike opens the way for populations to an organization where the *access* (A) to nodes by vicinity functions will depend on all order relations, *partial and total*, linked to the preorder of the graph which can all now be induced on the hyperstructure (HS).

The progressive imposition of order or *the synergetic effect of different orders* on the graph and its hierarchized labelling is very complex for the HS, but we shall retain, by extension, that the *ordered generation* (HS topological distance and HS site numbering) provides directly, as a by-product, the listing (L) and counting (C) functions so often sought in other ways.[37, 38]

We shall illustrate these concepts and this methodology of ordered generation by means of formal hyperstructures (HS_F) by using *anteriology hyperstructures* as an example. Moreover, a transformational hyperstructure (HS_T) will be presented by the study of ketone alkylation in an $HS_T \rightarrow HS_F$ relation.

3.2   Formal hyperstructure: anteriology and ordering

Starting with the fairly general concepts introduced during the last century by Laurent and Gerhardt,[31] *formal substitution* corresponds to a *conceptual operation* allowing for the progressive passage from a compound to a direct neighbour without stressing the chemical operation of the passage. For want of mathematical tools, formal substitution is limited, for all practical purposes, to supporting *linear homology* ($NH_3 \rightarrow NH_2Me \rightarrow NH_2Et$) but is not adapted to isomerism.†

Going back to these ideas with the help of the *substitution hyperstructure*, we have found it possible to generalize, and thereby give new dimensions to traditional notions, by extending them to concentric homologues (DARC

---

† Isomerism, as tackled by graph theory, currently leads to important work no longer related merely to isomer counting but rather to listing passages or paths of isomerism. Seen thus, to a family of isomers often corresponds a strongly connected preorder where the edges represent formal or real reversible reactions.[33, 34]

generation law: $NH_3 \rightarrow NH_2 Me \rightarrow NHMe_2 \rightarrow NMe_3, \ldots$) and to generic populations ($MeF \rightarrow MeCl \rightarrow MeCN \ldots$).

3.2.1   Formal substitution and partial ordering

The passages from one family of isomers to another involve very complex combinatorics. We shall confine ourselves to *elementary formal substitution* relations s_F where a methyl group Me is substituted for a hydrogen atom H

$$X \xrightarrow{\;s_F\;} Y \equiv H \xrightarrow{\;s_F\;} Me$$

Definition: A population $\mathscr{P}$ becomes a formal substitution hyperstructure $HS(\mathscr{P}, s_F)$ if two compounds X and Y are linked by an arc oriented from X towards Y each time that Y is derived from X by an elementary substitution $X \xrightarrow{SF} Y$ or again $X (s_F) Y$ (Fig. 17, step 3).

$HS(\mathscr{P}, s_F)$ thus possesses an oriented graph structure to which we can associate partial order defined by: $X < Y$ if a pathway (series of arcs all oriented in the same direction) exists leading from X towards Y (or if $X = Y$ to insure reflexivity). In other words, partial order is defined if Y is derived from X by one or several elementary formal substitutions s_F.

This order relation is such that its trace on a family of linear or concentric homology *extracted* from $\mathscr{P}$ is a *total order* identical to the natural order of homologues (Fig. 17 step 2.A).

We should note that the equivalence classes of $\mathscr{P}$ which cannot be compared by the spatial order relation form isomer families. Figuratively speaking, *families of isomers can be pictured as lodged on superimposed and equidistant horizontal planes.* Substitution relations s_F take place in the form of arcs between these planes. They display a *pyramid* representing the graph ($\mathscr{S}_F$) where the notion of *homology* has a *vertical* tendency (Fig. 17, step 3).

The $HS(\mathscr{P}, s_F)$ are generative and often built up from a source compound. On the levels of isomerism, generation creates too many compounds. One must, therefore, eliminate those which go beyond the bounds defined by structural stopping criteria as well as the redundant compounds generated several times over. The screening takes place at each progression from one level to another for a specific s_F and it should be noted that the choice of development criteria commands progression either towards an infinite HS or a finite one.

To avoid an over-abundant generation of entities, which implies screening at each stage of recurrent generation to eliminate redundancy, we have chosen a more specific generation law, one whose action is more discriminating. This is made possible by taking up again, at this stage, the canonical generation series explained in paragraph 2.5. (Fig. 11).

3.2.2   Canonical anteriology and tree structure

To simplify, we now choose a *source* or generation starting point, identified with *one specific compound* (X_0) (the source compound).

Definition: X is a canonical anteriologue of Y $(X \mathcal{A}_c Y)$ if X belongs to the canonical generation path of Y begun at X_0, thus taken as the source.

In fact, once a source compound X_0 exists, from which the whole hyperstructure HS($\mathcal{P}, s_F$) can be generated, then we can impose on $\mathcal{P}$ the supplementary order relation of *canonical anteriology* defined above.

This order relation is compatible with that defined by s_F since:

$$X \mathcal{A}_c Y \Rightarrow X < Y$$

It provides HS($\mathcal{P}, s_F$) however with a richer structure, since the oriented graph, a but slightly ordered support, receives a tree structure defined by the $\mathcal{A}_c$ relation. As a result, the hyperstructure is more ordered.

Definition: A formal substitution hyperstructure HS($\mathcal{P}, s_F$) becomes a hyper-structure of anteriology HS($\mathcal{P}, s_F, \mathcal{A}_c$) when provided with the canonical anteriology relation (Fig. 17, step 4).

This order introduced by $\mathcal{A}_c$ is a tree order (1 son has only 1 father). All compounds are generated only once, but only those edges corresponding to the s_F linked according to $\mathcal{A}_c$ are generated. We must particularly emphasize the complementary and *synergetic role* played by s_F and $\mathcal{A}_c$ relations in HS($\mathcal{P}, s_F, \mathcal{A}_c$): $\mathcal{A}_c$ controls the generation of all the compounds and of certain edges and avoids redundancy, while s_F insures the exhaustive creation of all the edges.

3.2.3   Ordered hyperstructure generation and linear order

If potentially HS($\mathcal{P}, s_F, \mathcal{A}_c$) is provided with the orders necessary for correlation research on HS, its generative creation and subsequent handling are simplified by access to a complementary linear order. This last stage controls ordered generation.

We have already exposed the principles of imposing this type of order so as to establish an ordered chromatic graph (paragraph 2). Here we generalize them and apply them to the population organized with tree structure HS($\mathcal{P}, s_F, \mathcal{A}_c$) so as to result in an *ordered anteriology hyperstructure* HS($\mathcal{P}, s_F, \mathcal{A}_c, \theta_1$) where the chosen linear order (θ_1) results from a concentric generation (Fig. 17 step 5).†

---

†This type of concentric and modular generation makes possible a homogeneous analysis of graphs of entities and of populations as well as synchronous generation for a target entity both of its linearly ordered graph and of the associated hyperstructure HS composed of those entities traversed and engendered.

Listing (L) and *counting* (C) are the direct by-products of ordered generation. *Access* (A) is then a question of traversing the graph $(\mathscr{P}, s_F)$. The optimization of such traversing is facilitated by the complementary nature of the three different order relations of the $HS(\mathscr{P}, s_F, \mathscr{A}_c, \theta_1)$.

3.2.4 Hyperstructure and conception

For conceptual purposes, the chemist studies a model and a phenomenon interactively with an eye to linking them to each other by a law. It is possible to conceive a methodology which would express the model by a formal hyperstructure and the phenomenon by an experimental hyperstructure. The search for a law is then merely a question of establishing a correspondence between the two, i.e., *a homomorphism of the real on the formal.*

In a reactional hyperstructure, the optimum synthesis pathway for reaching a target compound can be obtained by valuating the edges through thermo-kinetic, accessibility and cost data.

The reactional hyperstructure described here corresponds to a poly-

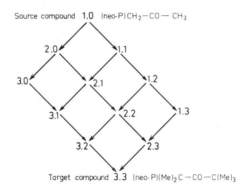

FIG. 18　　Reactional graph of ketone polyalkylation.

This graph is constituted by basic square meshes (b). Each mesh is oriented by an elementary $\alpha - \alpha'$ competitive alkylation (a). Each compound, as node of the

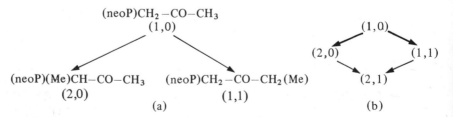

graph and its local mesh, is symbolized by the number of α and α' substituents [e.g. (2,0), i.e., two substitutions in α, none in α'].

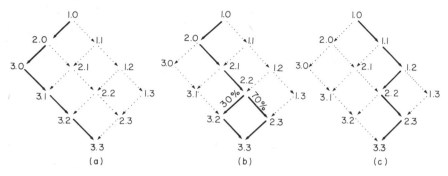

FIG. 19   Experimental reactional graph. The formal hyperstructure graph
($HS_F = HS_T$) from Fig. 18, is shown here by a dotted line. On this graph we
impose: (a) the model pathway controlled by hyperconjugation; (c) the model
pathway controlled by steric hindrance. The real pathway (b) lies between the
hyperconjugation control and the steric control.

substitution with two competitive centres at the origin and at each successive
level save the last. This concerns the synthesis by alkylation of a ketone
$R^1R^2R^3C-CO-CR^4R^5R^6$ through the introduction of R^is on the acetone in
place of hydrogens. All the R^is are different; the alkylation corresponds to an
s_F, and since the order of their introduction can vary, one can identify or
generate 720 potential synthesis pathways (L and C).

The experimental model studied corresponds to syntheses whose source
compound is methylneohexylketone and whose arcs all correspond to a
competitive methylation[36] followed step by step (Fig. 18). Applying experimental results to the reactional hyperstructure† brings out the orientation of
methylation by the ELCO environment of α and α' positions [here $A_i(DD_1)$ or
$A_i(DD_2)$] (Fig. 19,b). The experimental pathway thus obtained results from
hyperconjugation effects (Fig. 19,a) and steric effects (Fig. 19,c) whose powers
of control are illuminated by the imposition of these isolated effects on the
potential HS_T (Fig. 19).

The patterns of a, b, c, on the reactional hyperstructure HS_T are globally
different and diagram (b) points to a more complex interpretation of general
pathway control.

4   CONCLUSION

In this article, we have endeavoured to formalize certain problems of structure
encountered when dealing with entities and with entity populations. The

†For this alkylation reaction, it should be noted that concerning a substitution which can be
carried out, we have $HS_F \equiv HS_T$ since the substitution is of the H $\xrightarrow{SF}$ Me type.

concept of ordered environment ELCO, the use of associated orders and the notion of hyperstructure give added value to the use of the graph theory by grouping information relative to the whole and the parts of a kernel to be described via an ordered chromatic graph.

The ideas proposed by no means negate an important extant body of thought derived by analogy and intuition from experience. A number of solutions linked to present ideas on the chemical function, the chemical entity, and families of compounds are woven into a logical, and we trust, coherent whole. These ideas are consequently revived in the shape of substructure (SS), structure (S) and hyperstructure (HS). *Filiation*, the thread of the chemist's associative thought, so necessary for connecting phenomena and assigning certain structures, is taken up again in the more highly defined '*generation*' concept.

The DARC metalanguage associated with this formalization already makes considerable progress possible in the handling of large documentation files[20] and also allows for 'structure-reactivity' correlations using molecular topology in 'topo-information' expressions.[31,32]

The expected progress along these lines will depend on various future contributions to the work on hyperstructures as well as on '*computer assisted design*' which will facilitate man—machine interaction in the chemist's natural language, that of the structural diagram and, thus, of the ordered chromatic graph.

ACKNOWLEDGEMENTS

The preparation of this article was greatly facilitated by the creative and stimulating discussions I had with D. LAURENT, J. C. BONNET, A. PANAYE and Y. SOBEL. I should like to thank them here most warmly for their efficacious and valuable help.

REFERENCES

1. Berge, C. (1971). "Graphes et Hypergraphes", Dunod.
2. Veal, D. C. (1973). "Retrieval of Information from Literature — Computers in Chemistry", Springer Verlag, New York.
3. Spialter, L. (1963). *J. Am. Chem. Soc.*, **85**, 2012.
4. Balaban, A. T. and Harary, F. (1971). *J. Chem. Doc.*, **11**, 4.
5. Lederberg, J., Sutherland, G. L., Buchanan, B. G., Feigenbaum, E. A., Robertson, A. V., Duffield, A. M. and Djerassi, C. (1969). *J. Am. Chem. Soc.*, **91**, 2973.
6. Grant, D. M. and Paul, E. G. (1963). *J. Am. Chem. Soc.*, **85**, 1701.
7. Dubois, J. E. (1973). DARC System in chemistry: proceedings of the

NATO/CNA advanced study institute on computer representation and manipulation of chemical information, Leewenhorst Congress, (in press).

8. Dubois, J. E., Laurent, D. and Viellard, H. (1966). *C.R. Acad. Sci., Paris,* **263**, serie C, 764.

9. Salton, G. (1968). "Automatic Information Organization and Retrieval", McGraw-Hill, New York.

10. Dubois, J. E. (1972). "Computer representation of numerical and graphic data". Proceedings of the third international conference CODATA, 19 Westendstr. Frankfurt — FRG.

11. Berge, C. (1963). "Théorie des graphes et ses applications", Dunod.

12. Roberts, J. D. and others, (1971). *J. Org. Chem.,* **36**, 2557.

13. Lendeman, C. R. and Adams, J. Q. (1971). *Anal. Chem.,* **43**, 1245.

14. Morgan, H. L. (1965). *J. Chem. Doc.,* **5**, 4.

15. Dubois, J. E., Bonnet, J. C., Laurent, D. and Viellard, H. (1971). *Bulletin de L'IRIA*, n° 6.

16. Dubois, J. E. and Peyraud, J. F. (to be published).

17. Powers, R. V. and Hill, H. N. (1971). *J. Chem. Doc.,* **11**, 1.

18. Sussenguth, E. H. (1965). *J. Chem. Doc.,* **5**, 4.

19. Figueras, J. (1972). *J. Chem. Doc.,* **12**, 4.

20. Dubois, J. E. (1973). *J. Chem. Doc.,* **13**, 8.

21. Cahn R. J., Ingold, C. K. and Prelog, V. (1966). *Angew Chem. Internat. Ed.,* **5**, 4.

22. Dubois, J. E. (1970). Logical pattern description. General progressive valuation and chirality specification within DARC oriented topological system. Proceedings of table ronde Roussel. "Chirality from mathematics to biology". Paris, (in press).

23. Dubois, J. E., Alliot, M. J. and Viellard, H. (1970). *C.R. Acad. Sci., Paris,* **271**, serie C, p. 1412—1415.

24. Dubois, J. E., Alliot, M. J. and Panaye, A. (1971). *C.R. Acad. Sci., Paris,* **273**, serie C, p. 224—227.

25. Maroni, P. and Dubois, J. E. (1956). *C.R. Acad. Sci.,* **243**, 138.

26. Grant, D. M. and Paul, E. G. (1964). *J. Am. Chem. Soc.,* **86**, 2984.

27. Dubois, J. E. and Viellard, H. (1968). *Bull. Soc. Chim. Fr.,* 900—913.

28. Dubois, J. E. and Debouzy, P., (to be published).

29. Milne, M., Lefkowitz, D., Hill, H. and Powers, R. (1972). *J. Chem. Doc.,* **12**, 3.

30. Benson, S. W. and others, (1969). *Chem. Rev.,* **69**, 279.

31. Dubois, J. E., Laurent, D. and Viellard, H. (1967). *C.R. Acad. Sci., Paris,* **264**, 1019.

32. Dubois, J. E., Laurent, D. and Aranda A. (1973). *J. Chim. Phys.,* 1608, 1613.

33. Balaban, A. T. Farcasiu, D. and Banica, R. (1966). *Rev. Roum. Chim.,* **11**, 1205.

34. Gielen, M. (1969). *Meded. Vlaam. Chem. Ver.,* **31**, 185. 201.

35. Hendrickson, J. B. (1971). *J. Am. Chem. Soc.,* **93**, 6847, 6854.

36. Dubois, J. E. and Panaye, A. (1969). *Tetrahedron Letters,* 1501, 3277.

37. Henze, H. R. and Blair, C. M. (1931). *J. Am. Chem. Soc.,* **53**, 3077.

38. Polya G. (1937). *Acta Math.,* **68**, 145.

39. Balaban, A. T. (1966). *Rev. Roum. Chim.,* **11**, 1097.

40. Theilheimer, W. (1972). "Synthetic methods of organic chemistry", Vol. 26, Karger, Basel, New York.

NOTE ADDED IN PROOF

"Main DARC articles appearing since the writing of the present manuscript":

(a) J. E. DUBOIS, "Informatique chimique". Encyclopédia Universalis, Vol. 17 (1974), p. 379.

(b) J. E. DUBOIS, chapter of "Computer Representation and Manipulation of Chemical Information", (1974). Edited by W. T. Wipke, S. Heller, R. Feldmann and E. Hyde. Printed by John Wiley and Sons Inc.

(c) J. E. DUBOIS, "Le système DARC", Courrier du C.N.R.S., n°15, January 1974, p. 19.

(d) J. E. DUBOIS, Main Section Lecture of the 25th IUPAC Congress, Jerusalem, 1975. To be published in a special issue of the Israel Journal of Chemistry. Edited by U. Marcus and I. Eliezer. Foreseen for January 1976.

Author Index

Numbers in brackets are reference numbers and are included to assist in locating references in the text where authors' names are not mentioned. Numbers in italic indicate the pages on which references are listed in full.

A

Acton, N., 86(53), 87(53), *104*
Adams, J. Q., 338(13), *369*
Agosta, W. C., 108(95), 145(95), *173*
Alder, R. W., 264(49), *296*
Alliot, M. J., 344(23, 24), 362(23), *369*
Antoine, G., 80(32), 81(32), *103*
Aranda, A., 359(32), 368(32), *369*
Avram, M., 87(59, 61), 95(79), *104 105*

B

Bailar, J. C., 285(1), *294*
Baird, N. C., 76(24), *103*
Balaban, A. T., 64(3), 66(5,90), 67(5), 69(6), 70(5, 8), 71(8, 9), 73(10, 12), 74(13), 75(13, 15), 76(13, 15, 23, 100, 101), 77(25), 79(28), 80(31), 81(31), 82(34, 37, 38), 83(37, 38, 39), 84(37, 38), 86(5, 37, 38), 87(5, 102), 88(65), 91(65), 96(82), 98(82), 102(34, 103), *103, 104, 105*, 108(1, 2), *170*, 186(2), 208(1), *214*, 269(2, 3, 45), 270(4), 271(4), 272(4), 277(2), 286(3), *294*, 297(9, 10), *298*, 334(4), 359(33, 39), 363(33), *368, 369*

Balandin, A. A., 180(3), 182(3), 185(3), 196(4), *214*
Balescu, R., 224(20), *258*
Banciu, M., 95(81), *105*
Banica, R., 269(2), 277(2), *294*, 359(33), 363(33), *369*
Bellman, R. E., (4), *170*
Benson, S. W., 359(30), *369*
Bent, H. A., 276(6), *294*
Berge, C., 108(5a), *170*, 271(7), 275(7), *294*, 333(1), 338(11), *368, 369*
Berry, R. S., 108(6), *170*, 262(8), 264(8), *294*
Berthier, G., 197(94), *217*
Bethe, H. A., 238(25), *258*
Billes, F., 182(5), 184(5), *214*
Binkley, R. W., 86(43), 87(43), *104*
Bird, C. W., 114(7), *170*
Birkoff, G., 109(8), *170*
Blair, C. M., 3(15, 16, 19, 20, 21, 22, 23), *4*, 21(8), *24*, 27(1), 36(2), *61*, 359(37), 363(37), *369*
Blair, J., 108(9, 10), 111(9, 10), 118(9, 10), 122(9, 10), 126(9), 129(9), 168(10), 169(9, 10), *170*
Bochvar, D. A., 79(27), *103*, 192(9), 195(8), 202(7), 203(6), *214*
Bock, H., 206(70), *216*
Bodanszky, M., 127(11), *170*
Boer, J., de, 257(30), *258*
Bolesov, I. G., 86(48), 87(48), *104*
Boltzmann, L., 224(1), *257*

Subject Index

Synonymous terms are cross-references by: See ..., or are enclosed in brackets; non-synonymous terms are cross-indexed by: See also ...; f denotes: and following pages; GT denotes: graph-theoretical.